Sustainable Development Goals Connectivity Dilemma

Land and Geospatial Information for Urban and Rural Resilience

Sustainable Development Goals Connectivity Dilemma

Land and Geospatial Information for Urban and Rural Resilience

Edited by

Abbas Rajabifard

CRC Press
Taylor & Francis Group
Boca Raton London New York

CRC Press is an imprint of the
Taylor & Francis Group, an **informa** business

CRC Press
Taylor & Francis Group
6000 Broken Sound Parkway NW, Suite 300
Boca Raton, FL 33487-2742

International Standard Book Number-13: 978-0-367-25935-8 (Hardback)

Visit the Taylor & Francis Web site at
http://www.taylorandfrancis.com

and the CRC Press Web site at
http://www.crcpress.com

Contents

Forewords xiii

Preface xvii

Acknowledgements xix

Biographies of Authors xxi

I Setting the Scene **1**

1 Sustainable Development Goals Connectivity Dilemma **3**
Abbas Rajabifard
1.1 Introduction 3
1.2 Addressing SDGs and Land Tenure: The Need for a Roadmap 5
1.3 Book Structure and Overview 6

2 SDGs Roadmap **13**
Greg Scott and Abbas Rajabifard
2.1 Introduction 13
2.2 Sustainable Development 16
2.3 Goals, Targets and Indicators 18
2.4 Geospatial Data and Enabling Technologies 21
2.5 Bridging the Geospatial Digital Divide 24
2.6 A National Strategic Geospatial Information Policy Framework 28
2.7 Vision and Mission 31
2.7.1 Vision 31
2.7.2 Mission 31
2.8 Principles 32
2.9 Strategic Drivers 32
2.10 Goals 34
2.11 Strategic Pathways 34
2.12 Benefits 39
2.13 Implementing the National Strategic Geospatial Information Policy Framework 40
2.14 Conclusions 41
Bibliography 42

3 Marriage of Opposites: Strategies for Public and Private Sectors Working Together in Land Tenure Reform Projects That Support SDGs **45**
Daniel Paez
3.1 Introduction 45
3.2 Background: Land Administration and the Trend of Involving the Private Sector 46
3.3 SDGs and Land Tenure Reform Projects 47
3.4 Land Reform Projects: Achievements and Challenges 49
3.5 Lessons Learnt From Involving the Private Sector in LTS . . 51
3.5.1 Switzerland 51
3.5.2 Canada 53
3.5.3 Australia 54
3.5.4 Philippines 54
3.5.5 India 55
3.6 Strategies to Align Private Participation in Land Tenure Reform Projects With SDGs 56
3.7 Conclusions 59
Bibliography 60

4 Spatially Enabling the SDGs **65**
Maryam Rabiee
4.1 Introduction 65
4.2 Spatially Enabling the SDGs 66
4.2.1 What Does It Mean to Be Spatially Enabled? 66
4.2.2 Sustainable Development Goals in an Interconnected World 67
4.2.3 Integrating Spatial Enablement Into the SDG Framework 68
4.3 Narrowing the Connectivity and Spatial Gap 69
4.4 The Social Impact of Spatially Enabling the SDGs 71
4.5 Land: The Driving Force of Spatial Enablement for the SDGs 73
4.6 Conclusion 74
Bibliography 75

II Enhancing SDGs Connectivity and Disaster Resilience **79**

5 Leveraging National Land and Geospatial Systems for Improved Disaster Resilience **81**
Abbas Rajabifard, Katie Potts, Mika-Petteri Torhonen, Federico Barra, and Ivelisse Justiniano
5.1 Introduction - Supporting SDGs With Land and Geospatial Information 81
5.2 Addressing Global Problems With Land and Geospatial Systems 82

5.3 Global Land and Geospatial Systems 83
5.4 Working Towards the SDGs: Achieving Land Resilience . . . 85
5.5 Global Development Frameworks 87
5.6 A Roadmap for Building Land Resilience 89
5.7 Conclusion 91
Bibliography 92

6 Geospatial Information Technologies in Support of Disaster Risk Reduction, Mitigation and Resilience: Challenges and Recommendations 93
Saied Pirasteh and Masood Varshosaz
6.1 Introduction 93
6.2 Why are technologies alone not enough in disasters loss reduction 95
6.3 Integration of Geospatial Knowledge 96
6.4 Geospatial Rapid Visual Screening for Earthquake Disaster Risk Reduction, Mitigation and Resilience 98
6.5 Human Search and Rescue in Drone Images 99
6.6 An Example of Lack of Laws in Geospatial and Environmental Issues 102
6.7 Conclusions and General Remarks 104
Bibliography 105

7 Application of Unmanned Aircraft Systems for Coastal Mapping and Resiliency 109
Michael J. Starek, Melanie Gingras, and Gary Jeffress
7.1 Introduction 109
7.2 Overview of UAS Technology 110
7.3 Aerial Mapping with UAS 112
7.3.1 Mission Planning 113
7.3.2 Flight Design 113
7.3.3 Image GSD and Overlap 114
7.3.4 Structure-from-Motion Photogrammetry 116
7.4 Regulations 118
7.5 Case Study: Hurricane Harvey Impact Assessment 119
7.6 Conclusion 122
Bibliography 124

III Supporting SDGs: Legal, Policies and Institutional Components and Capacity Building 127

8 Legal and Policy Paths for Effective Sustainable Development 129
Harlan J. Onsrud
8.1 Introduction 129

8.2 Fundamental Economic Policies Germane to Traditional Resources . . . 130
8.3 Role of Legal Controls . . . 131
8.4 Policies and Laws Germane to Digital Economy Resources . 132
8.4.1 Settling Disputes . . . 133
8.4.2 Protecting People Against Excessive or Unfair Private Power . . . 133
8.4.3 Protecting Citizens Against Excessive or Unfair Government Power . . . 134
8.4.4 Ensuring People an Opportunity to Enjoy the Minimum Decencies of Life . . . 135
8.5 Maintaining Order . . . 135
8.6 Open Access to Domestic Government Data . . . 136
8.7 Correcting Unjust Laws and Policies Within Growing Information Economies . . . 137
8.8 Conclusions . . . 139
Bibliography . . . 139

9 Developing a Framework for National Institutional Arrangements in Geospatial Information Management 141
Joep Crompvoets and Serene Ho
9.1 Introduction . . . 141
9.2 Framework Development . . . 144
9.2.1 Introduction . . . 144
9.2.2 Concepts . . . 144
9.2.3 Instruments . . . 145
9.3 Framework Application . . . 151
9.3.1 Introduction . . . 151
9.3.2 Application Approach . . . 151
9.4 Key Examples . . . 153
9.4.1 Introduction . . . 153
9.4.2 Description of Key Practice Examples . . . 153
9.5 Lessons Learnt . . . 157
Bibliography . . . 159

10 Considerations for Institutional Interconnectivity 163
Serene Ho
10.1 Introduction . . . 163
10.2 SDGs as a 'Wicked' Problem . . . 164
10.3 Institutions . . . 166
10.4 Wicked Problems and Institutional Challenges for Coordination and Collaboration in the Public Sector . . . 166
10.5 Challenges of Coordination for Spatial Enablement . . . 168
10.6 Institutional Considerations: Moving Forward . . . 169
Bibliography . . . 171

11 Implementing SDGs in Smart Cities Beyond Digital Tools 175
Zhixuan Yang and Abbas Rajabifard
11.1 Introduction 175
11.2 SDGs and Means of Implementation in Smart Cities 177
11.2.1 Three-Tier SDGs 177
11.2.2 Means of Implementation-Framework 177
11.2.3 Means of Implementation - Data and Indicators 179
11.3 Smart City Context 179
11.3.1 Smart City Concept 179
11.3.2 Argument of Smart City and Sustainability 180
11.3.3 Making Cities Smart and Sustainable 181
11.3.4 Needs of Digital Tools and Living Labs 181
11.4 Key Components Beyond Digital Tools 183
11.4.1 Networked Infrastructure 183
11.4.2 Knowledgeable Community 183
11.4.3 Intelligent Governance 184
11.5 Action Agenda of Smart Cities Towards SDGs Beyond Digital Tools 185
11.5.1 Integration of Innovation Capacity in Smart Cities . . 185
11.5.2 Transformation of Smart Growth in Smart Cities . . . 186
11.5.3 Evolvement of the Socio-Economic Ecosystem in Smart Cities 187
11.6 Discussion and Conclusion 188
Bibliography 190

12 Spatial Enablement to Facilitate the New Urban Agenda Commitments for Sustainable Development 199
Soheil Sabri and Abbas Rajabifard
12.1 Introduction: Background and Driving Forces 199
12.2 Urbanisation; From a Threat to an Opportunity 202
12.3 AFINUA and Its Relation to SDGs and CPI 204
12.4 Spatial Data Infrastructure Advancements and Opportunities 206
12.5 Conclusion and Discussion 208
Bibliography 209

13 The Geospatial Capacity Building Ecosystem - Developing the Brainware for SDI 213
Josef Strobl
13.1 Introduction 213
13.2 Status 214
13.3 Mix of Actors in an Education Ecosystem 214
13.4 Case Study: the Copernicus Master in Digital Earth 216
13.5 Educational Ecosystem Services 216
13.6 Conclusions 217
Bibliography 217

IV Enabling Tools and Technical Components **221**

14 The Role of Geospatial Information Standards for Sustainable Development **223**
Denise McKenzie, Mathias Jonas, Serena Coetzee, Chris Body, Margie Smith, Marcus Blake, Joseph Abhayaratna, Michael Judd, and Marna Roos
14.1 Introduction 224
14.2 Digitization Forces Standardization 225
14.3 The Framework of International Standardization for GI . . . 227
14.3.1 Technical Standards Link Environmental Standards to IT Innovations 229
14.3.2 Standardization Driven by Innovation and Technical Evolution 229
14.3.3 New Information Communities Emerge 230
14.4 Case Studies 230
14.4.1 Australia 231
14.4.2 New Zealand 233
14.4.3 South Africa 234
14.5 Case Studies of Relevant Standards for Specific Goals 237
14.5.1 New Zealand Government Use of WaterML and SOS . 237
14.5.2 Urban Environment - Multiple Urban Implementations Including UK, Singapore, Germany, Finland, Australia, USA, Canada. Key Standards in Use Include CityGML, SensorWebs, SensorThingsAPI and Others 237
14.5.3 Arctic SDP 238
Bibliography 241

15 Urban Analytics Data Infrastructure: Critical SDI for Measuring and Monitoring The National and Local Progress of SDGs **243**
Abbas Rajabifard, Soheil Sabri, Yiqun Chen, Muyiwa Agunbiade, and Mohsen Kalantari
15.1 Introduction 244
15.2 Global Indicator Framework 245
15.3 The Urban Analytics Data Infrastructure 246
15.4 UADI's Contribution to SDGs 250
15.5 Discussion and Conclusion 253
Bibliography 254

16 New Technical Enabling Tools for Data Acquisition and Maintenance of Topographic Data of Urban and High Mountain Areas to Support SDGs **257**
Gottfried Konecny
16.1 Introduction 257
16.2 Global Progress in Mapping From 1900 to 2000 258

16.3 Large Scale Mapping of Urban Areas 260
16.4 Large Scale Mapping in Europe 261
16.5 Future Alternatives by New Technology 262
16.5.1 High Resolution Satellite Imagery 263
16.5.2 Mobile Mapping 263
16.5.3 3D Oblique Imaging via 3D City Models With Automated Object Creation of Buildings 263
16.6 The Use of New Stereo Satellite High Resolution Satellites by China for the Mapping of High Mountain Areas 264
Bibliography 265

17 Night-Light Remote Sensing: Data, Processing and Applications 267

Xi Li, Deren Li, and Huayi Wu

17.1 Introduction 267
17.2 DMSP/OLS Night-light Data 268
17.2.1 VIIRS DNB Night-light Data 269
17.2.2 VIIRS DNB Night-light Data 269
17.2.3 Jilin1-03B Night-light Data 270
17.2.4 Luojia 1-01 Night-light Data 270
17.3 Data Processing 271
17.3.1 DMSP/OLS Intercalibration 271
17.3.2 Improvement of VIIRS DNB Daily Data 272
17.3.3 Saturation Correction 273
17.4 Applications 273
17.4.1 The Applications of Night-light Data 273
17.4.2 Case: The Study of Syrian Crisis 275
Bibliography 277

V SDGs Perspectives: Current Practices and Case Studies 281

18 Why and How Informal Development Should Be Formalized Quickly, Inclusively and Affordably- Experience From UN-ECE Region 283

Chryssy Potsiou

18.1 Introduction 283
18.2 Informality Is Considered a Social, Economic and Environmental Challenge 287
18.3 Fit-for-Purpose Formalization Policies 291
Bibliography 294

19 SDGs and Geospatial Information Perspective From Nigeria-Africa 297

Muyiwa Agunbiade, Olajide Oluwafemi, and Oluyemi Akindeju

19.1 Introduction 297

19.2 Existing Knowledge About Interconnection Within SDGs and Between Geospatial Information 300
19.2.1 The Interconnection Between the SDGs, Geospatial Information, Urban and Rural Resilience 300
19.2.2 Geospatial Information to Support Inclusive Urbanisation, Resilient Development, and the SDGs 301
19.2.3 Approaches to Integrating Geospatial Information and Technologies in the Implementation of the SDGs . . . 302
19.3 Framework and Methods 303
19.4 Findings and Analysis 305
19.4.1 The Interactions Between Various Elements of GIM and the SDGs in African Context 305
19.4.2 Geospatial Information: Strengthening Community, Infrastructure, and Institutional Resilience 305
19.4.3 The Role of Geospatial Data Infrastructures and Services in Achieving the SDGs in African Context . . . 306
19.5 Inferences, Future Direction and Conclusion 309
Bibliography 310

20 Openness and Community Geospatial Science for Monitoring SDGs – An Example From Tanzania 313
Maria Antonia Brovelli, Maria Ilie Codrina, and Serena Coetzee
20.1 Introduction 313
20.2 Open Data and Open Software 314
20.3 Community Geospatial Science 316
20.4 The Use Case and Training Material 318
20.5 Conclusion 322
Bibliography 323

21 Modernizing Land Administration Systems to Support Sustainable Development Goals - Case Study of Victoria, Australia 325
Hamed Olfat and Davood Shojaei
21.1 Land Administration Systems 325
21.2 Selected LAS Works 326
21.3 Land Administration Systems Related Requirements to Support Sustainable Development Goals 327
21.4 Case Study of Victoria, Australia 329
21.5 Conclusion 334
Bibliography 335

Index 337

Forewords

Foreword by Mr. Stefan Schweinfest:

It gives me great pleasure to be able to provide this Foreword as my contribution to this book, which is in itself an immensely important initiative from our UN-GGIM Academic Network, one of our strategic arms of UN-GGIM. This book represents a very tangible and direct means towards bringing together knowledge and experiences from around the globe to build upon and facilitate our strategy, plans and approaches for geospatially enabling the implementation and monitoring of the Sustainable Development Goals (SDGs), and for proposing a potential roadmap to achieving a better world for all and leaving no one behind.

As you will know, the 2030 Agenda for Sustainable Development is intentionally ambitious, providing a transformative and integrated approach to sustainable development, and is anchored by a set of 17 integrated and indivisible SDGs, 169 targets, and a global indicator framework, in order to measure and monitor progress. The 2030 Agenda is a new and universal development agenda for all countries and stakeholders to use as a blueprint for action, and calls for concerted efforts towards building an inclusive, sustainable and resilient future for people, planet and prosperity.

The SDGs are unique in that they universally apply to all countries as we collectively manage and transform the social, economic and environmental dimensions of people and the planet through to at least 2030. The SDGs are a compass for aligning countries' national plans and aspirations with their global commitments. All stakeholders; governments, industry, academia, the private sector and civil society are expected to contribute to the realization of the 2030 Agenda.

In this context, our efforts within the intergovernmental processes of UN-GGIM include raising awareness and paving the path forward for a better future for all. We need strategies that build economic growth and address a range of societal needs, while tackling climate change and environmental protection. From a statistical point of view, governing bodies must develop their data gathering and management operations, populating those metrics with strong, reliable data against which the world can measure its progress with confidence. In a large sense, technical interoperability is not the problem, the institutional aspects of how to get various stakeholders, and the people who have an interest and capability to solve the data puzzle and work together effectively, is the biggest challenge.

The ideas and solutions presented in this book address the SDGs' connectivity dilemma and raise thought-provoking discussions from experts around the world that will help us all work towards achieving the goal of implementing the SDGs. It is exciting to see that by simply bringing people and minds together creates a very positive dynamic, where not only real change can be made, it can be sustained for all. *Stefan Schweinfest*

Director, UN Statistics Division
UN-GGIM Secretariat

Foreword by Dr. Stuart Minchin:

I'm delighted to see this book "Sustainable Development Goals Connectivity Dilemma: Land and Geospatial Information for Urban and Rural Resilience" published as part of the extensive work being undertaken by the UN Global Geospatial Information Management Group of Experts (UN-GGIM) raising awareness and paving the path forward for a better future for all.

This is an exciting time for people in the geospatial industry, government agencies, private firms and global development organisations to work together to achieve the Sustainable Development Goals in a manner than facilitates data sharing and transparency.

We live in a time with unprecedented access to data and technology which gives us all an opportunity to deliver real impacts for all jurisdictions and citizens of the world through providing decision ready solutions that can help to make the world a better place.

This book brings together contributions from around the world and presents an approach towards an SDG Roadmap, with viable solutions for enhancing the connectivity and resilience of SDG efforts. The legal, policy and institutional components are discussed and enabling and technical tools are also presented.

I look forward to seeing how the methods and solutions presented in this book can be put into practice and how we, the geospatial community, can work together to address the global challenges we face.

Stuart Minchin
Chief Environmental Geoscience Division, Geoscience Australia
Australian Principal Delegate to UN-GGIM

Preface

This book is the culmination of the hard work and extensive research undertaken by members of the UN Global Geospatial Information Management (UN-GGIM), its Academic Network and a selected practitioner, to address the challenges our world faces and build a better and more sustainable future for us all.

Both developed and developing countries require an expertise and guidance in geospatial data, methods, frameworks, tools and platforms that can provide reliable, timely and accessible geospatial information in order to progress informed decision-making and ultimately pave the path forward for implementing the Sustainable Development Goals (SDGs).

Achieving the SDGs across different communities and domains faces unique challenges. The solutions in this book not only present an approach towards an SDG roadmap, but also discuss the interconnection between SDGs, geospatial information, the legal, policy and institutional components, technical enabling tools and the way forward to address urban and rural resilience.

This book brings together the expertise of leading geospatial experts, scholars, industry actors, and policy-makers and their perspectives from their respective fields to examine the connection between SDGs, geospatial information, and urban and rural resilience. The themes and objectives of the book are in line with the critical challenges, gaps, and opportunities raised at all UN-GGIM Academic Network forums and events.

The authors in this book, from all around the globe, have worked together in hope of taking steps towards achieving the 2030 Agenda for sustainable development by the United Nations. We hope that together, we can build a resilient future, promote prosperity and make the world a better place, leaving no one behind.

Acknowledgements

This book is the result of a collaborative initiative of the United Nations Global Geospatial Information Management (UN-GGIM) Academic Network, it's members and the wider geospatial community and practitioners. The book has drawn upon the presentations and outcomes of three UN-GGIM Academic Network Forums: “Secure Land Rights and Smart Cities – Making It Work for Sustainable Development” as part of the Seventh Session of the United Nations Committee of Experts on UN-GGIM in 2017, “The SDGs Connectivity Dilemma: Urban Settlements, Resilience, and Sustainability” as part of the Eighth Session of the United Nations Committee of Experts on UN-GGIM in 2018 and “A Sustainable and Resilient World: Capacity Building and Geospatial Research for Implementing the SDG”, as part of the first UN World Geospatial Information Congress UNWGIC in China 2018. The main aim being, to provide interdisciplinary analysis and multi-sectoral expertise on the interconnection between the SDGs, geospatial information, the legal, policies and institutional components, technical enabling tools and the way forward to address urban and rural resilience.

The editor has been privileged to have been involved with the UN-GGIM Academic Network in a leadership role, enjoying the support of the Academic Network members and a vast number of researchers, practitioners and policy makers and geospatial engineers throughout the journey for the preparation of this book.

I would like to acknowledge and thank the contribution of over 40 chapter co-authors from different organisations and countries. I am deeply grateful for the outstanding support and valuable contributions of all participants and speakers of the UN-GGIM Academic Network three International Forums conducted between 2017-2018 as part of UN-GGIM events. In particular, I would like to thank the support of Mr. Stefan Schweinfest, Director of the UN Statistics Division, the UN-GGIM Secretariat Team including in particular, Mr. Greg Scott, Mr. Teo Chee Hai, Ms. Cecille Blake, and Ms. Vilma Frani for their support and facilitation.

I would also like to extend my sincere thanks to Dr Wael Zakout from the World Bank for his contribution to our forums which has led to this publication and his continuous encouragement and support.

I am grateful to the Taylor and Francis Group Press for their support and willingness to publish this work as Open Access, which allows all to use the experiences and research presented in this book to their own best advantage. In particular, I would like to express my sincere thanks to Ms. Irma Britton,

Senior Editor, Environmental Sciences, GIS & Remote Sensing CRC Press - Taylor & Francis Group for her contribution, continuous support and facilitation for publishing this book. I also like to thank Ms. Rebecca Pringle and the rest of the T&F publishing team for their professionalism during the preparation of this book.

Finally, I would like to thank the Department of Infrastructure Engineering and the research team at the Centre for spatial Data Infrastructures and Land Administration (CSDILA) for their support, and in particular special thanks to Dr Ida Jazayeri and Dr Farhad Laylavi from the Centre for their outstanding editorial assistance in preparation of this publication.

I hope this book can contribute to the future of our societies and help in achieving SDGs.

Abbas Rajabifard, Editor
Chair UN-GGIM Academic Network
2019

Biographies of Authors

Abbas Rajabifard is Chair of UN-GGIM Academic Network, and a Professor at the University of Melbourne and Director of the Centre for SDIs and Land Administration (CSDILA). He is an international land and geospatial policy and technological advisor, who has spent his career researching, developing, applying and teaching land administration and SDI to deliver benefits to both governments and wider society. His research is on enabling SDGs using spatial information, 3D urban land administration and cadastre, spatial urban data analytics and 3D platform for smart cities.

Greg Scott is the Inter-Regional Advisor for Global Geospatial Information Management in the United Nations Statistics Division, Department of Economic and Social Affairs. Greg provides high level strategic policy advice and leadership in the coordination and implementation of UN-GGIM initiatives with Member States and related International Organizations involved in national, regional and global geospatial information management.

Daniel Paez is a civil engineer with a PhD in Geomatics from the University of Melbourne. He has experiences working as a public official, university lecturer and private consultant. Currently he is a Senior Industry Advisor for the Centre for SDIs and Land Administration (CSDILA) at the University of Melbourne.

Maryam Rabiee is a researcher working on interdisciplinary approaches to sustainable development. Her work focuses on the Sustainable Development Goals and the social aspects of sustainability in the context of geospatial science, disaster management, and ICTs.

Katie Potts is a post-doctoral research fellow at the Centre for SDIs and Land Administration (CSDILA) at the University of Melbourne and has a background in land administration and disaster risk reduction, focusing on implementing land-related policies and land management strategies to create land security which is vital for the development of wealth and economic health, the privatization of land markets, and sustainable urban planning and land development.

Mika-Petteri Törhönen is a land tenure, policy and administration professional with 25 years of work experience from over 20 countries in Europe, Africa, Asia, Near East and Central America. Mika holds the position of Lead Land Administration Specialist at the World Bank's Global Land and Geospatial Unit. Before joining the Bank in 2011, Mika worked in FAO's Land Tenure and Management Unit in Rome.

Alvaro Federico Barra is a land administration/geospatial specialist at the World Bank in Washington DC. He joined the Bank in 2008 and is currently working on Land Administration/NSDI projects in Sierra Leone, Mozambique, Lebanon and Bosnia, leading a study about improving resilience and resilience impact of national land and geospatial Systems.

Ivelisse Justiniano is an Urban and Disaster Risk Management Specialist at the World Bank. Most of her work focuses on urban resilience, disaster risk assessment and mitigation, risk reduction strategies, land use, and geospatial technologies.

Saeid (Adam) Pirasteh is an Associate Professor at the Faculty of Geosciences and Environmental Engineering (FGEE), Southwest Jiaotong University, China. He is also a Research Scientist Collaborator at the Mobile Sensing and Geodata Analytics Lab, University of Waterloo.

Masood Varshosaz holds a PhD in Photogrammetry from University College London, UK. He is actively involved in various fields including close range photogrammetry, computer vision, panoramic modeling, and laser scanning.

Michael J. Starek is an Associate Professor in Geospatial Systems Engineering at Texas A&M University-Corpus Christi (TAMU-CC) and Director of the Measurement Analytics (MANTIS) Lab with the Conrad Blucher Institute for Surveying and Science. Starek holds a Ph.D. in Civil Engineering from the University of Florida and was formerly a National Research Council Postdoctoral Fellow of the U.S. Army Research Office in affiliation with North Carolina State University. His research focuses on the merging of geomatics, remote sensing, and geospatial computing for precise measurement and analysis of natural and built system dynamics.

Melanie Gingras is a Research Specialist and Manager for MANTIS Lab as well as a Physics Adjunct Professor for Texas A&M University at Corpus Christi (TAMUCC). She earned her bachelor's degree from the University of

Delaware in Geology and her master's degree in Coastal & Marine System Science at TAMUCC where she studied beach morphology using a terrestrial laser scanner (TLS). Her research interests include the union and optimization of TLS and UAS data as it pertains to coastal morphology monitoring with an emphasis on coastal response to hurricanes.

Gary Jeffress is a Research Professor at the Conrad Blucher Institute for Surveying and Science at Texas A&M University-Corpus Christi. He holds a Ph.D. in Surveying Engineering from the University of Maine, Master of Surveying Science (Geodesy) and Bachelor of Surveying degrees from the University of New South Wales, Sydney, Australia. He is a Registered Professional Surveyor in Texas and held surveying licenses in Maine and Australia.

Harlan J. Onsrud is a Professor of Spatial Informatics in the School of Computing and Information Science at the University of Maine. His research and teaching interests focus primarily on the analysis of legal, ethical, and institutional issues affecting the creation and use of digital databases and the assessment of the social impacts of spatial technologies.

Joep Crompvoets holds a chair on Information Management in the Public Sector. He is also secretary-general of the EuroSDR network, which is a not-for-profit European organisation linking national mapping and cadastral agencies with research institutes and universities for the purpose of applied research in spatial data provision, management and delivery.

Serene Ho is the Vice-Chancellor's Research Fellow (Urban Futures) at RMIT University (Australia), a Fellow at the Public Governance Institute at KU Leuven (Belgium), and an Honorary Research Fellow at the University of Melbourne (Australia). Her research examines the implications of disruptive geospatial technologies in terms of public management and social innovation.

Zhixuan (Jenny) Yang, is a lecturer in the School of Investment and Construction Management, Dongbei University of Finance and Economics (DUFE), China. She is currently working in the field of urban governance and real property management, conducting research projects of "3D city governance" and "smart cities in the perspective of SDGs".

Soheil Sabri is a Research fellow in Urban Analytics in the Centre for SDIs and Land Administration (CSDILA) (CSDILA) at the University of Melbourne. His research focuses on enabling spatial information and technological innovation in smart urban planning and design to improve urban quality of life. He investigates approaches for leveraging 3D geospatial data and analytics in planning for future smart cities.

Josef Strobl is head of the Department of Geoinformatics - Z_GIS at the University of Salzburg and as a full member of the Austrian Academy of Sciences (ÖAW) chairs the ÖAW Commission for Geographic Information Science. He is leading international graduate programmes in Applied Geoinformatics as well as the UNIGIS network of distance education programs (www.unigis.net).

Denise McKenzie joined the OGC in 2012 and spends her time managing the Communication and Outreach program globally for the consortium. The program handles the planning and execution of marketing, communication and education to raise awareness and increase implementation of open geospatial and location standards by technology providers and users worldwide. Part of Denise's role is to oversee OGC Alliance Partnerships including representation at the United Nations Global Geographic Information Management (UNGGIM) committee.

Gottfried Konecny is an Emeritus Professor at Leibniz University Hannover. He has a world of experience across continents researching and publishing in the realm of geoinformation. He is one of the pioneers in the field and is considered one of the founding fathers of photogrammetry, remote sensing and topographic mapping.

Xi Li is an associate professor working at Wuhan University. He holds a doctorate degree in Photogrammetry and Remote Sensing from Wuhan University. His research interest includes physical modeling of night-time light as well as night-time light remote sensing applications.

Deren Li is currently working at Wuhan University. He is a member for both the Chinese Academy of Sciences and the Chinese Academy of Engineering. He holds a doctorate degree from the University of Stuttgart in 1985. In 2012, the International Society for Photogrammetry and Remote Sensing awarded him as an Honorary Member.

Huay Wu is Vice Director and full professor with the State Key Laboratory of Information Engineering in Surveying, Mapping and Remote Sensing, Wuhan University. He holds a doctorate in GIS from Wuhan University. His research interest includes Geographic Information System and its applications in smart cities.

Chryssy Potsiou is teaching Cadastre, Land Management and Property Valuation at the School of Rural & Surveying Engineering of NTUA, Greece. She has been: member of the boards of the Hellenic Mapping and Cartographic Organization; Hellenic Association of Rural and Surveying Engineers; the Hellenic Societies for Photogrammetry and for Geographic Information Systems; FIG Commission 3 Chair (2007-2010); FIG Vice President (2011-2014); FIG President (2015-2018); UNECE Working Party on Land Administration (WPLA) bureau member (2001-2018) and Vice Chair (2019-2020). She has published 125 peer reviewed papers.

Mohsen Kalantari is a Senior Lecturer in Geomatics Engineering and the Associate Director for Research at the Centre for SDIs and Land Administration (CSDILA) at the University of Melbourne. He has a proven track record of achievements in several national and international research grants and consultancy projects in the areas of spatial data infrastructures, land administration, and the nexus of geographic information systems (GIS).

Muyiwa Agunbiade is a Senior Lecturer at the Department of Urban and Regional Planning in the Faculty of Environmental Sciences, University of Lagos, Nigeria where he lectures GIS, Spatial data infrastructure and Land Administration courses. After completing undergraduate training in environmental design and management from the Obafemi Awolowo University in Nigeria, he earned master's degrees in both urban planning and Geographic Information Systems at the University of Ibadan, and a doctorate in Spatial Data Infrastructure from the Faculty of Infrastructure Engineering, University of Melbourne, Australia.

Yiqun (Benny) Chen is a Research Fellow in the Centre for SDIs and Land Administration (CSDILA) and the Centre for Disaster Management & Public Safety (CDMPS) at the University of Melbourne. He has a broad research interests including GIS visualisation, spatial analysis, disaster management, geo-intelligence, decision-support system, agent-based modeling and location-based services.

Akindeju Oluyemi, is a Lecturer and a PhD staff candidate in the department of Urban and Regional Planning of the University of Lagos, holds BSc (2005) and MSc (2010) degrees in Urban and Regional Planning from the University of Lagos, with research interest in Urban Morphology, Science of Cities, Complexity Modelling and Environmental Planning.

Oluwafemi Olajide is a lecturer at the Department of Urban and Regional Planning in the Faculty of Environment Sciences, University of Lagos,

Nigeria. He earned BSc and MSc degrees in Urban and Regional Planning from the University of Lagos, Nigeria and PhD in international planning from the School of Architecture, Planning and Landscape, Newcastle University, United Kingdom.

Maria Antonia Brovelli is a Professor of GIS and Digital Mapping at Politecnico di Milano. She is the chair of ISPRS WG IV/4 "Collaborative crowdsourced cloud mapping (C3M)", member of ESA ACEO (Advisory Committee for Earth Observation); chair of the "Capacity Building WG" of the UN Open GIS Initiative, Deputy-Chair of the UN-GGIM Academic Network. One of her main interest is in open source geospatial science and technology, where she is playing a worldwide leading role.

Serena Coetzee is Professor and Head of the Department of Geography, Geoinformatics and Meteorology, and Director of the Centre for Geoinformation Science at the University of Pretoria. Her research focuses on the ever-increasing volumes of geographic information and the challenges of making the information findable, accessible and usable. The research is of an interdisciplinary nature, mainly in geoinformatics and computer science.

Codrina Maria Ilie is a technical geographer, dividing her time between the Technical University of Civil Engineering of Bucharest and, a Bucharest based SME. She is actively working in developing geospatial data services incorporating data from non-traditional sources, such as the open data initiative. Her work at the Technical University of Civil Engineering focuses on geodata standardization and interoperability in groundwater data management.

Hamed Olfat is a Senior Research Fellow in the Centre for SDIs and Land Administration (CSDILA) at the University of Melbourne and is responsible for coordinating the ePlan project in Victoria. His main areas of research include 'electronic land administration', '3D digital cadastre', 'spatial data infrastructure', and 'data discovery'.

Davood Shojaei leads the 3D digital cadastre research at Land Use Victoria in Melbourne. He has extensive experience over the last 15 years in designing and implementing a variety of projects, ranging from GIS, cadastre and BIM, to 3D visualisation. Davood completed his PhD on 3D cadastre at the University of Melbourne in 2014. He is currently working with the ePlan team to develop and implement digital cadastre services in Victoria, Australia.

Part I

Setting the Scene

1

Sustainable Development Goals Connectivity Dilemma

Abbas Rajabifard
Centre for SDIs and Land Administration, The University of Melbourne, Australia

1.1 Introduction

In the context of United Nations Global Geospatial Information Management (UN-GGIM) and the development of Sustainable Development Goals (SDGs), *recalling* Economic and Social Council resolution 2011/24, of 27 July 2011, which established the UN-GGIM to provide a forum for coordination and dialogue among Member States, and to hold regular high-level, multi-stakeholder discussions on global geospatial information, including through the convening of global forums, with a view to promoting a comprehensive dialogue with all relevant actors. Further also *recalling* the establishment of the UN-GGIM Academic Network in July 2016 as a strategic research and training arm for UN-GGIM to assist members, and *recalling* Economic and Social Council resolution 2016/27 entitled 'Strengthening institutional arrangements on geospatial information management' of 27 July 2016, in which the Council acknowledged the considerable achievements of the Committee of Experts including: its contribution to the strengthening of geospatial information management capacities and utilization in developing countries; the efforts to streamline the work of the subsidiary bodies of the Council in the field of geospatial information management; and its role in the implementation of the 2030 Global Agenda for Sustainable Development, the Sendai Framework, and other global development agendas within the purview of the United Nations.

Also *recalling* General Assembly resolution 70/1 entitled 'Transforming our World: The 2030 Agenda for Sustainable Development' of 25 September 2015, which recognizes the need for new data acquisition and integration approaches to improve the availability, quality, timeliness and disaggregation of data, and the use of a wide range of data, including earth observations and geospatial information, to support the implementation of the new development agenda at all levels, while ensuring national ownership in supporting and tracking progress; noting the opening statement of the Secretary-General at this Congress, in which he emphasized that our expertise and guidance in geospatial data, methods, frameworks, tools, and platforms is urgently needed, and that reliable, timely, accessible and disaggregated geospatial information must be brought to bear to measure progress, inform decision-making and ensure effective and inclusive national and sub-national programs that will chart the path towards the 'Geospatial Way to a Better World', to assist in the implementation of the SDGs, and transform our world for the better; and also noting further that United Nations World Geospatial Information Congress (UNWGIC) in November 2018, which has provided a convening, participatory and inclusive environment to enhance the communication, understanding, knowledge and application of geospatial and land information management, to discuss the policy relevance and challenges to advance geospatial science and technology, promote the creation and sharing of more reliable geospatial data, and to enhance value-added applications and services to address local, regional and global challenges; all have highlighted the needs for a roadmap facilitating the achievement of SDGs implementation through the lens of Geospatial information.

With this in mind, this book will provide interdisciplinary analysis and multi-sectoral expertise on the interconnection between the SDGs, geospatial information, the legal, policies and institutional components, technical enabling tools and the way forward to address urban and rural resilience.

Urbanization, natural and human-induced disasters, migration, and technological advancements are among some of the most potent forces that are increasing the connectivity and complexity of the challenges highlighted in the SDGs. Achieving the SDGs across different communities and domains will require the use of geospatial information to overcome challenges such as land rights, food production, disaster risk reduction, safe human settlements, and other social, economic, and environmental issues at local, national, and global levels. Geospatial information and technologies are particularly critical to strengthening urban and rural resilience, where economic, agricultural, and various social sectors intersect.

The SDGs dependency on geospatial information and enabling technologies are mainly due to the primary roles that data and tools for relating people to their location, place and environment, and to measure 'where' progress is, or is not, being made, particularly at sub-national and local levels. However, in the pursuit for sustainable development, many countries continue to face a series of impediments that exacerbate their ability and opportunity to participate fully in the implementation of the United Nations 2030 Agenda, to support national development, economic prosperity, and through that, a global and thriving information economy. These include institutional challenges in data production: having the required human capital and skillsets, effective and sustained access to digital technology, the Internet of Things (IoT), to the provision and exploitation of new data needs, information systems, analytics and associated enabling tools and technologies to support the timely and reliable implementation of the SDGs. Examining the SDGs from a geospatial lens will ensure that the challenges are addressed for all populations in different locations, leaving no one behind.

In addition, identifying the gaps and opportunities in understanding the connectivity between different elements of sustainability and resilience requires input from different disciplines and sectors.

1.2 Addressing SDGs and Land Tenure: The Need for a Roadmap

The achievement of the SDGs for all communities and jurisdictions require a comprehensive roadmap that encompasses all dimensions of data infrastructure, social, economic, environmental and governance ecosystems.

With this in mind, this book provides interdisciplinary analysis and multisectoral expertise on the interconnection between the SDGs, geospatial information, the legal, policies and institutional components, technical enabling tools and the way forward to address urban and rural resilience. In addition, we discuss the security of tenure targets embedded in the SDGs. We stressed the importance of the land administration and surveying profession owning the SDGs selected targets such as Target 1.4 on ensuring that all men and women have equal rights to land and other forms of property by 2030. Very few countries actually know what their baseline on the security of tenure is and even fewer have a program or plan in place for achieving the target by 2030.

Similarly and in a broader context, Goals 1, 2, and 5 of the SDGs have designated targets linked to land tenure rights signify the obligation of incorporating land administration in the process of building sustainable and smart cities for all. Therefore, surveyors and geospatial practitioners should work to change this globally and help countries to adopt national programs on security of tenure while addressing SDGs Goals, Targets and Indicators.

The UN-GGIM 2017-2021 Strategic Framework recognizes the necessity of integrating geospatial information in process of achieving the SDGs and developing future cities. Strengthening local, national, and global cooperation to foster the integration of legal and organizational frameworks including the SDGs, UN-GGIM 2017-2021 Strategic Framework, Sendai Framework, and Habitat III Urban Agenda will positively impact disaster management, development of adequate policies and regulations, climate action, efficient urban planning, and good institutional governance.

The availability of effective and efficient land administration—and its economic, social, and environmental implications—remains a problem worldwide, especially in developing countries where mature land administration systems and formal land registration systems are not available. Therefore, spatial inclusion, secure land rights, and sustainable land use are all major challenges of rapid urbanization that public and private sectors need to address in the development of future smart cities.

Considering this situation, the UN-GGIM Academic Network recognizes the importance of promoting and sharing data acquisition and integration approaches, legal and policy instruments, institutional management models, technical solution and standards, interoperability of systems and data, and access to quality and timely data. As a result, the Academic Network aims to work in the direction resolution 2016/27 of the Economic and Social Council (ECOSOC) by promoting the sharing of geospatial data, enhancing capacity building, and inter-institutional cooperation for sustainable development, disaster risk reduction, and policymaking.

With the above context in mind, the large-scale migration from rural to urban areas, and between countries, affects sustainable development at local, national, and regional levels. In order to identify gaps and opportunities to strengthen urban and rural resilience to global challenges, the book will draw upon the discussions and presentations and outcomes of three UN-GGIM Academic Network Forums: "*Secure Land Rights and Smart Cities – Making it Work for Sustainable Development*" as part of the Seventh Session of the United Nations Committee of Experts on UN-GGIM in New York in 2017, "*The SDGs Connectivity Dilemma: Urban Settlements, Resilience, and Sustainability*" as part of the Eighth Session of the United Nations Committee of Experts on UN-GGIM in New York in 2018, and "*A Sustainable and Resilient World: Capacity Building and Geospatial Research for Implementing the SDG*", as part of the first UN World Geospatial Information Congress-UNWGIC in China 2018. Therefore, in the context of SDGs and urban and rural resilience, the main objective of the book is to bring together the expertise of leading geospatial experts, scholars, industry actors, and policy-makers and their perspectives from their respective fields to examine the connection between the SDGs, geospatial information, and urban and rural resilience.

1.3 Book Structure and Overview

This book is structured in 5 parts, and the themes and objectives of the book are in line with the critical challenges, gaps, and opportunities raised at all UN-GGIM events and UN-GGIM Academic Network forums. Three main themes are the following:

- the role of geospatial information and data infrastructures and services in achieving the SDGs goals;
- the interactions and relations between various elements of the SDGs; and

- the significance of geospatial information in strengthening community, infrastructure, and institutional resilience.

The three main objectives of the book are the following:

- provide interdisciplinary analysis and multi-sectoral expertise on the interconnection between the SDGs, geospatial information, and urban and rural resilience;
- examine how geospatial information will support and inform inclusive and even urbanization, resilient development, and the SDGs; and
- present roadmaps for a more holistic approach to integrating geospatial information and technologies in the implementation of the SDGs.

The rest of this section provides a brief overview of the parts and chapters of this book (Figure 1.1).

Part 1. Setting the Scene

Part 1 provides a context and background to the SDGs connectivity dilemma, with a high level description of what SDGs mean and the impacts of spatial enablement. This part comprises of four chapters, beginning with this Chapter 1, which outlines the context and objectives of the book and discussion about the needs of a roadmap towards achieving SDGs, together with an introduction to the following chapters. Chapter 2 by, Greg Scott and Abbas Rajabifard, put forward the SDGs Roadmap. The chapter addresses the 2030 Agenda for Sustainable Development, anchored by 17 Sustainable Development Goals (SDGs), providing a transformative and integrated approach to sustainable development. With emphasis on measuring and monitoring development progress with reliable policy, science, technology and especially data, the 2030 Agenda presents all countries and the global policy community with a set of significant development challenges that are almost entirely geographic in nature. The chapter also discusses policy challenges, including the strategic leadership, understanding and awareness of national geospatial information policy, frameworks and associated implementation roadmaps. As a tangible means to support the implementation of the SDGs, Greg Scott and I present and discusses the key components of a geospatial roadmap for countries to develop and strengthen their institutional arrangements in national geospatial information management, to bridge the geospatial digital divide, and to measure and monitor development progress. Therefore this chapter will focus on a roadmap towards a sustainable and resilient future for all.

With this, Chapter 3, by Daniel Paez, then addresses the marriage of opposites: strategies for public and private sectors working together in land tenure reform projects that support SDGs. Strategies are presented in this chapter can be used during the design and implementation of future land reform projects both in developed and developing countries as part of empowering SDGs. In order to highlight the role of geospatial information and their ability to enable the environment further, Chapter 4 by Maryam Rabiee addresses the social, economic, and environmental impacts of spatial enablement when spatially enabling the SDGs. Ready and timely access to spatial information is critical to making informed decisions on economic, environmental and social issues. This chapter aims to present the significance of spatially enabling the SDGs and the opportunities it provides for the seventeen goals. The chapter investigates the social impact of spatial enablement for the SDGs and ends with a discussion on land, the driving force of spatial enablement for the SDGs.

FIGURE 1.1
Overview of the Book

Part 2. Enhancing SDGs Connectivity and Resilience

In the context of SDGs connectivity, this part examines how geospatial systems can support disaster resilience, risk reduction and improved mapping for better SDG connectivity and better management. As discussed earlier, geospatial information can help locate the challenges of communities in different regions more visible and integrating geospatial data with other available data can assist with addressing the 'where' component of different social, economic, and environmental challenges. To highlight the significance of geospatial information in strengthening community, infrastructure, and institutional resilience, this part will present ways, in which geospatial information can assist with narrowing the connectivity gap. This part comprises three chapters, beginning with Chapter 5, where Abbas Rajabifard, Mika Petteri Törhönen, Katie Potts, Federico Barra and Ivelisse Justiniano, address the concept of leveraging National Land and Geospatial Systems for improved disaster resilience. In the context of SDGs this chapter is based on a World Bank flagship project let by the authors of the chapter to present a roadmap for exploring the role of land and geospatial information, the function and responsibility of the institutions that govern the data, and the resulting impact that this data has on the overall resilience of society to disasters. Following this in Chapter 6, Saeid Pirasteh and Masood Varshosaz present geospatial information technologies that can support the UN-GHIM in its efforts in disaster risk reduction, mitigation and resilience, particularly those due to earthquakes. The final chapter in Part 2 (Chapter 7), by Michael J. Starek, Melanie Gingras, and Gary Jeffress explores the application of unmanned aircraft systems as as example of tools for coastal mapping and resiliency. The information and applications presented are applicable to a variety of UN SDGs including sustainable land use for "Life on Land" and sustainable agriculture for crop security and "Zero Hunger".

Part 3. Supporting SDGs: Legal, Policies and Institutional Components and Capacity Building

This part discusses the sustainability and resilience challenges that are directly tied to legal, policy, and institutional capacities to make inclusive and effective decisions that positively impact our communities. The Part 3 comprises of five chapters to address legal and institutional gaps and requirements for sustainability and its impact on inclusivity. To begin, Harlan Onsrud puts forward the legal and policy paths for effective sustainable development in Chapter 8. He discusses the requirements and fundamental relationships between legal and policy framework to support SDGs. Next, in Chapter 9, Joep Crompvoets and Serene Ho discuss the issues that have arisen from the UN-GGIM during the sessions over the past 6 years, highlighting the need for developing a framework for national institutional arrangements in geospatial information management. This chapter will provides the rationals and the approach for the development of institutional arrangements in support of geospatial information management as an essential enabler for SDGs. Following this, Chapter 10 by Serene Ho discusses the considerations for institutional interconnectivity. The institutional challenges around the coordination and collaboration in the public sector are reviewed, and a discussion on potential strategies for progression in this space is presented. In Chapter 11, Zhixuan

Yang and Abbas Rajabifard discuss implementing SDGs for Smart Cities as an example to establish the theoretical framework by exploring key components to observe the implementable structure and action of SDGs at the city level. Particularly, the chapter aims to highlight the fundamental foundations beyond digital tools to achieve SDGs. In chapter 12 Soheil Sabri and Abbas Rajabifard explore the UN-Habitat's initiatives in adoption and localising the SDGs through defining the New Urban Agenda (NUA) 2030. They provide a historical account on how the global urbanisation started to be considered as a threat and ended up to be a potential development tool for the future generation established by the NUA. They also explain the Action Framework for Implementation of NUA (AFINUA) developed by the UN-Habitat to localise the implementation, measurement, and monitoring the SDGs and other associated indicators formulated through City Performance Initiative (CPI). This chapter also draws links between spatial enablement concepts and principles and the key elements of AFINUA. And finally, Chapter 13, by Josef Stobl discusses the development of "Brainware" for SDI. This chapter highlights the importance of academic education in geospatial technologies for building and maintaining the brainware components that will make the SDG framework successful across human societies.

Part 4. Enabling Tools and Technical Components

Part 4 addresses technology and tools that can assist with monitoring and measuring progress at different levels of governance and across different regions. There are three chapters in this part, beginning with Chapter 14, contributed by a joint standards team from ISO and OGC, Denise McKenzie et al. addressing the role of geospatial information standards for sustainable development. This is followed by Chapter 15, where Abbas Rajabifard and co-authors describe a new research initiative that sees the development of an SDI to support urban analytics and urban research capabilities focused on Australian cities, called Urban Analytics Data Infrastructure (UADI). The UADI provides opportunity for multi-disciplinary, cross-jurisdiction, national-level analytics, which appeals the requirements of SDGs and potentially can be scaled up to be used by other nations. The chapter explains about the design of UADI architecture, which seeks to provide the urban research community with a digital infrastructure that responds to current challenges related to data access, sharing and driving the SDG indicators. Chapter 16 by Gottfried Konecy discusses technical enabling tools for data acquisition and maintenance of topographic data of urban and high mountain areas. This chapter explores data resolution, global progress in mapping, new mapping alternatives and how these relate to achieving SDGs. The last chapter in this part by Huayi Wu, Xi Li, Deren Li, addresses night-light remote sensing, a tool which can be used towards achieving SDGs. Case studies are presented showing how night-light remote sensing can plat a very important role in the assessment of humanitarian disasters.

Part 5. SDGs Perspectives

This part of the book looks at what the current state of play is for SDGs around the world, and specific case studies and cases are discussed in four chapters. It begins with Chapter 18, where Chryssy Potsiou brings together a wealth of knowledge as a leader of the field, compiling over 20 years of research, in cooperation with the FIG, the World Bank, UNECE and government agencies around the world to address why and how informal development should be formalized quickly, inclusively and affordably. Chapter 19 by Muyiwa Agunbiade, Olajide Oluwafemi, and Oluyemi Akindeju gives a perspective from Nigeria, Africa. This chapter discusses SDGs connectivity by exploring the nature of interlinkages between the SDGs from the lens of geospatial information and geospatial data infrastructure. It also focuses on evolving an integrated framework towards achieving SDGs in developing economies. Chapter 20 by Maria Antonia Brovelli, Maria Ilie Codrina and Serena Coetzee, discusses Openness and Community Geospatial Science for monitoring SDGs, giving an example from Tanzania for SDGs Goal 9. The authors, focus on these two aspects: openness and community geospatial science, presenting concepts and examples of open data and open software with reference to citizen science and volunteered geographic information. The final chapter in this part, by Hamed Olfat and Davood Shojaei examines the Case Study of Victoria, Australia for modernizing land administration systems to support the SDGs.

Sustainable development and resilience are continuous paths towards a better future for all. We must ensure that our contribution to this domain of knowledge has long-term impacts towards leaving no one behind.

2

SDGs Roadmap

Greg Scott
UN-GGIM Section, Environment Statistics and Geospatial Information Branch United Nations Statistics Division, New York, USA

Abbas Rajabifard
Centre for SDIs and Land Administration, the University of Melbourne, Australia

This chapter presents and discusses the major components that will assist our efforts in charting a geospatial roadmap towards the implementation of the SDGs. These being: the goals, targets and global indicators; the role of geospatial data and enabling technologies; bridging the geospatial digital divide; and a national strategic geospatial information policy framework.

2.1 Introduction

In July 2011, recognizing the urgent need to take concrete action to strengthen international cooperation in the area of global geospatial information management, the United Nations Economic and Social Council (ECOSOC) established the United Nations Committee of Experts on Global Geospatial Information Management (UN-GGIM). As the apex intergovernmental mechanism for geography, UN-GGIM makes joint decisions and sets directions with regard to the production and use of geospatial information within national, regional and global policy frameworks; promotes common principles, policies, methods, mechanisms and standards for the interoperability of geospatial data and services; and provides a platform for the development of effective strategies on how to build and strengthen national capacity concerning geospatial information, especially in developing countries. The report of the Secretary-General which led ECOSOC to establish UN-GGIM explicitly mentioned the role of geospatial information in informing sustainable development policies, including their monitoring and implementation [27].

In the past nine years, concerted efforts have been made by UN-GGIM to increase the visibility and awareness of the role of geospatial information, as an essential integrative tool to monitor and measure sustainable development, to policy and decision-makers and the diplomatic community. To this end, in July 2016, following a five-year review of UN-GGIM, ECOSOC adopted resolution 2016/27 on strengthening institutional arrangements on geospatial information management, in which it recognized that UN-GGIM had operated effectively and was well placed to continue to contribute more to the work of the United Nations. The Council decided to strengthen and broaden the mandate of UN-GGIM and invited the Committee to report on 'all matters relating to geography, geospatial information and related topics'. It also stressed the need to strengthen the coordination and coherence of global geospatial information management, in capacity-building and norm-setting, particularly pertaining to the 2030 Agenda, the Sendai Framework, and other global development agendas within the purview of the United Nations [20].

Substantial progress is being made by UN-GGIM in ensuring the intergovernmental coordination and coherence of geospatial information at the technical level, but the challenge of high-level policy awareness remains and continues to resonate. At its fourth session in August 2014, UN-GGIM observed that "the level of understanding and rate of uptake of geospatial information, particularly at the policy and decision-making level, remains less than optimal... many do not understand its value and importance within the context of the sustainable development agenda. The production and use of geospatial information within national, regional and global policy frameworks needs to be mainstreamed in order to enhance the capability for governments, international organizations and researchers to analyze, model, monitor and report on sustainable development, disasters, climate change, and other global concerns" [29].

At its eighth session in August 2018, UN-GGIM noted that "many of the efforts being reported to the Committee at this eighth session, as in past sessions, are aligned to providing the required frameworks, methods, standards and guides to assist strengthening national geospatial information capacity-building in developing countries... as a means to facilitate the strengthening and capacity-building of global geospatial information management in support of the implementation of the 2030 Agenda" [31].

Geospatial information and enabling technologies have emerged as major contributors to economic and digital transformation in many countries, including in the areas of e-government, e-service and e-commerce. The global geospatial industry is witnessing unprecedented growth, driving innovation, knowledge, smart solutions, delivery platforms and a location-based information economy. But with more data and technology available than ever before, many developing countries have yet to have the 'opportunity' to interact with these rapidly emerging capabilities, as the democratization of geospatial information is not being equally shared. Geospatial data, leadership, knowledge and innovation is primarily still limited to the developed countries. While technologies are evolving at a rapid rate, the commensurate capabilities, skills and opportunities in the developing countries are not.

Developing countries are still challenged by issues related to aspects regarding the management of data, and its closely coupled relationship with ICT, the Internet and other technologies. Further, there are institutional challenges related to coordination, leadership managing the value chain, fragmented implementation, diffused policy accountability, and then potentially the lack of skills, tools and mechanisms to properly manage the data supply chain and related technologies. There is still a desperate need for sustained political leadership, resources, commitment, associated frameworks and implementation roadmaps to get access to and exploit the plethora of geospatial data and tools now available.

The global geospatial community still has much work to do in raising awareness of the value and benefits of geospatial information at the policy level, liberating it from the traditional closed information silos at the technical level, integrating across the wider government sector, and establishing new alliances across a much broader and diverse stakeholder community. National policies, geospatial infrastructure, technical capacities and capabilities, need to be developed, better aligned and considerably strengthened so that all countries have the opportunity to develop and contribute to a vibrant national geospatial information ecosystem, and through that, a global and thriving information economy.

This chapter proposes a geospatial roadmap to enable SDGs implementation and, in addition, forms part of that process, addressing the lack of awareness and understanding of geospatial information particularly prevalent at the policy and decision-making levels in developing countries. Framed by the 2030 Agenda for Sustainable Development, and anchored by Figure 2.1, this chapter presents and discusses the major components that will assist our continued efforts in charting a geospatial roadmap towards the implementation of the SDGs. It will first contextualize sustainable development broadly, and its evolution towards the 2030 Agenda, before visiting the goals, targets and global indicator framework in detail. The chapter then describes the role of geospatial data and enabling technologies in contributing to the 2030 Agenda, before discussing the implications of the digital divide that continues to exist today for developing countries, and introduces and describes the 'geospatial digital divide' and the complex challenges that continue to exacerbate the ability for these countries to bridge this divide, to connect to the vast amounts of data and technology, and accelerate human progress. The chapter will conclude with a national strategic geospatial information policy framework as a means to provide the national policy basis and roadmap for countries to develop and strengthen their national and sub-national arrangements in geospatial information management, as they attempt to measure and monitor progress towards the implementation of the SDGs.

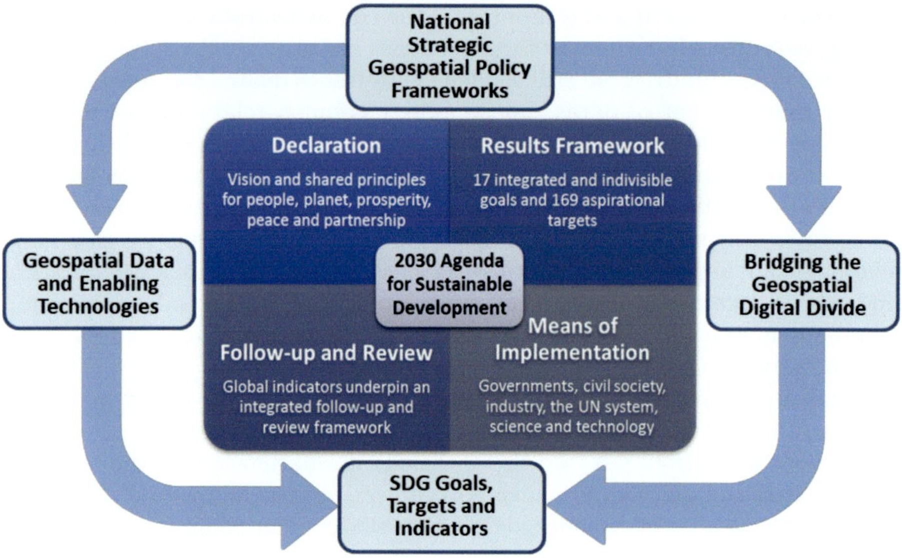

FIGURE 2.1
The major components in charting a geospatial roadmap towards the implementation of the SDGs.

2.2 Sustainable Development

Sustainable development concepts first appeared in the literatures in the early 1960s [24, 6], and quickly advanced to make interconnections between the environment, the economy and social well-being [4]; that the Earth had a finite capacity to sustain human civilization; and that population growth and resource consumption were unsustainable [5, 9, 10]. However, the mainstream theoretical framework and understanding of sustainable development evolved between 1972 and 1992, primarily through a series of international conferences and initiatives led by the United Nations [2].

The United Nations Conference on the Human Environment, convened in June 1972 in Stockholm, Sweden, was the first major international conference to discuss environmental sustainability issues at the global scale. The Conference marked a turning point in the development of international environmental politics, emphasizing that defending and improving the environment must become a goal to be pursued by all countries. Principle 2 of the Declaration specifically alluded to managing the environment for the benefit of present and future generations: "The natural resources of the earth, including the air, water, land, flora and fauna and especially representative samples of natural ecosystems, must be safeguarded for the benefit of present and future generations through careful planning or management, as appropriate" [12].

In December 1983 the United Nations General Assembly (General Assembly) established the World Commission on Environment and Development to formulate long-term environmental strategies for achieving sustainable development. In April 1987 the Commission produced the landmark report "Our Common Future" which advanced the understanding of global interdependence and the relationship between economics and the environment, and introduced and captured what is now the classic definition of sustainable development: "development which meets the needs of the present without compromising the ability of future generations to meet their own needs" [32]. The General Assembly adopted Our Common Future in August 1987, giving sustainable development political salience, and in June 1992 global leaders laid the foundations for its global institutionalization at the 'Earth Summit', the United Nations Conference on Environment and Development in Rio de Janeiro, Brazil. The Earth Summit adopted the Rio Declaration on Environment and Development, and Agenda 21 – a global plan of action for sustainable development [13].

Twenty years on, the United Nations Conference on Sustainable Development, or Rio+20, was convened in June 2012 in Rio de Janeiro. Rio+20 was pivotal in that it sought to initiate the process for a new development agenda for the future to supersede the Millennium Development Goals (MDGs). The focused political outcome document "The Future We Want" [14] contained clear and practical measures for implementing sustainable development, including setting the path to develop a set of Sustainable Development Goals (SDGs) to build upon the MDGs, and to converge with the post-2015 development agenda. The Future We Want captured two critical references to geospatial information within the document's framework for action and follow-up, and through provision of means of implementation. These were specifically in the area of disaster risk reduction: "We further recognize the importance of comprehensive hazard and risk assessments, and knowledge- and information-sharing, including reliable geospatial information" and in the area of means of implementation – technology: "We recognize the importance of space-technology-based data, in situ monitoring and reliable geospatial information for sustainable development policymaking, programming and project operations" [14].

These efforts culminated in September 2015 when the General Assembly adopted "Transforming our World: The 2030 Agenda for Sustainable Development" [19], a universal development agenda for all countries and stakeholders to use as a blueprint for action. The 2030 Agenda is an agreed global and united development policy to guide the way 'all countries' collectively manage and report on the social, economic and environmental dimensions of people, planet and prosperity. With an overarching imperative of 'leaving no one behind', this transformative Agenda requires an integrated and inclusive approach to sustainable development. With considerable emphasis on countries being able to measure and monitor progress with reliable policy, science, technology and especially data, the broad and aspirational nature of the 2030 Agenda has ushered in a new era in thinking about and approaching sustainable development. It has determined a set of 17 SDGs and 169 targets, and defined a process to measure and monitor implementation through a global indicator framework (presently with 232 global indicators) that is highly dependent on diverse, reliable and repeatable data to provide the evidence base for policy, decision-making and reporting on the health and wellbeing of our planet on an ongoing basis. Importantly, the 2030 Agenda intrinsically captures specific and separate global United Nations system outcomes (Figure 2.2) for small island developing States [15], disaster risk reduction [18], financing for development [16], climate change [17], a new urban agenda [20] and oceans [21].

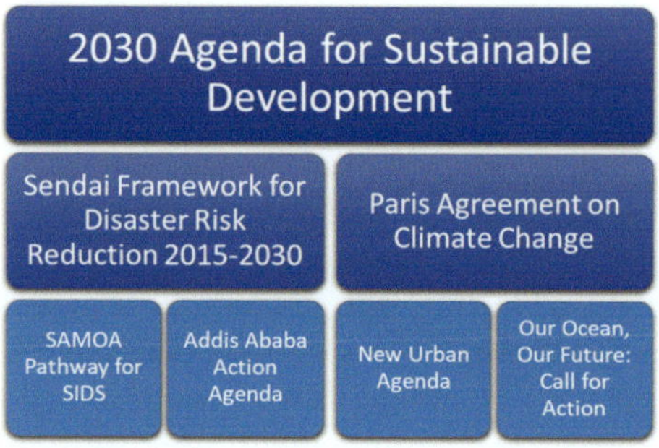

FIGURE 2.2
Overarching broad and universal global development policy agenda accepted by all countries during the 2014 – 2017 period. The 2030 Agenda for Sustainable Development provides the main means and mechanisms for implementation and measuring and monitoring progress through to 2030.

2.3 Goals, Targets and Indicators

The 2030 Agenda presents the global policy community with a set of 17 SDGs – significant development challenges that are almost entirely geographic in nature. It specifically demands the need for new data acquisition and integration approaches, including to exploit the contribution to be made by geospatial information and Earth observations to support the implementation of the SDGs, targets and global indicators. Further, it has articulated the need for counties to increase significantly the availability of high-quality, timely and reliable data disaggregated by income, gender, age, race, ethnicity, migratory status, disability, geographic location and other characteristics relevant in national

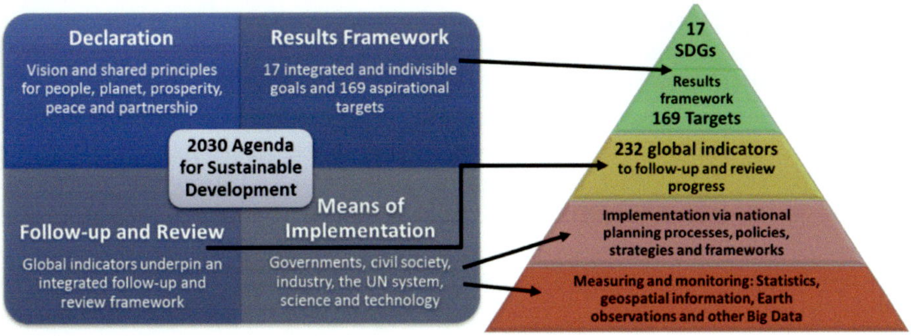

FIGURE 2.3
The 2030 Agenda is an integrated plan of action structured in four main parts: (i) a vision and principles for transforming our world as set out in the Declaration; (ii) a results framework of 17 SDGs and 169 targets; (iii) a means of implementation through governments, society and global partnership; and (iv) a follow-up and review framework of 232 global indicators. Any national SDG implementations will be sub-optimal without strategies and frameworks to integrate geospatial information and other data into the measuring, monitoring and reporting processes.

contexts ([19], Goal 17.18). Meeting these new data requirements is already proving difficult for the most advanced countries, but the 2030 Agenda further demands that by 2020 – in less than 2 years' time – this enhanced data availability is able to support and address the capacities and capabilities of developing countries, particularly African countries, least developed countries, small island developing States, and land-locked developing countries. For these countries, the challenges faced in the collection, processing, production, analysis and dissemination of reliable, timely, accessible and sufficiently disaggregated data for better evidence-based policymaking are significant and not to be underestimated.

As indicated in Figure 2.3, the 17 SDGs of the 2030 Agenda comprise the integrated and indivisible global goals to be achieved by countries, and applicable for both developed and developing countries, balancing the three dimensions of sustainable development. The 169 aspirational targets provide the detailed and actionable objectives for governments to measure progress through to 2030. Each country will set its own national targets, guided by the global level of ambition, and will also decide how these targets should be incorporated into national planning processes, policies and strategies. While the 17 SDGs and 169 targets provide the overall policy and results framework for the 2030 Agenda, in terms of a robust and annual follow-up and review mechanism for its implementation, it is the global indicator framework where the data acquisition, integration and disaggregation is most needed.

The task of determining the global indicator framework was given to the

United Nations Statistical Commission. In 2015 the Commission established the Inter-agency Expert Group on Sustainable Development Goal Indicators (IAEG-SDGs) to develop the global indicator framework as the quantitative means by which national governments can consistently monitor achievement on, and report progress towards, each of the 169 targets. In July 2017 the global indicator framework was adopted by the General Assembly and comprises an initial 232 indicators, which will be reviewed from time to time and especially in 2020 and 2025.

UN-GGIM and the Group on Earth Observations (GEO) worked closely with the statistical community, at a national and global level, to provide inputs into the processes to develop the global indicator framework with the IAEG-SDGs. Through this process, statisticians now better understand that geospatial information and Earth observations are able to provide new and consistent data sources and methodologies to integrate multiple 'location-based' variables to support and inform official statistics and the indicators for the SDGs. These methods are able to fill data gaps and/or improve the temporal and spatial resolutions of data, by bringing together information from various sources, particularly those related to the environment. This information integration is important, as the indicator framework will be the primary conduit to guide and inform Member States, based on individual national circumstances, on how they measure, monitor and report on the SDGs and related targets in the years to come. That said, determining the indicators was just the beginning, as they need to then be appropriately interpreted and implemented via national planning processes and frameworks, and guided by robust metadata and multidimensional data needs (Figure 2.3).

However, it is acknowledged that the constraints faced by many developing countries in producing the data necessary to address the indicator requirements will remain an ongoing capacity issue for some time, if not indefinitely. We still need to democratize the enabling technologies and liberate the associated data in such a way that they are easily reachable and useable by developing countries. Historically, relatively little attention has been paid to the challenges these countries face in effectively collecting and producing data, and in building and strengthening their capacities within the national mapping agencies and statistical offices. With the enabling global mechanism of the 2030 Agenda, the challenge is how to most effectively transfer the available technology, data richness, and connectivity to the technology and data poor.

Entering the fourth year of reporting on the SDGs, countries are realising how difficult it is to translate the shared vision of the 2030 Agenda into national development plans and strategies that ensure no one is left behind. In July 2018, in presenting the annual Sustainable Development Goals Report, the Secretary-General of the United Nations reiterated that without evidence of where we stand now we cannot confidently chart our path forward in realizing the SDGs. This reflects the "challenges faced in the collection, processing, analysis and dissemination of reliable, timely, accessible and sufficiently disaggregated data, calling for better evidence-based policymaking. While today's technology makes it possible to collate the data we need to keep the promise to leave no one behind, we need political leadership, resources and commitment to use the tools now available" [22]. Not only do countries continue to lack important baseline data and enabling technologies to help guide development, national governments remain a considerable distance from fully developing and implementing the required policies and frameworks to ensure that development progress, including appropriate interventions, is effective, measurable, and sustainable.

2.4 Geospatial Data and Enabling Technologies

Disruptive technologies are emerging and affecting our lives in ways that indicate we are at the beginning of a Fourth Industrial Revolution, a new era that builds and extends the impact of digitization in new and unanticipated ways [7]. The concepts and expectations for the rapidly growing global interconnectivity and information societies are being underpinned by both digital disruption and digital transformation – enabling a modern information economy to prosper. Technology is transforming almost every aspect of our lives, and all sectors of industry and the economy at an unprecedented pace and scale, and is similarly having a major impact on the geospatial industry, creating innovative technological enablers and applications, and generating previously unimaginable amounts of location-referenced information. These technologies and processes are not only disruptive, but they are continually evolving, providing new opportunities for innovation and enabling business, industry and governments to be more agile, to adapt and transform their own internal processes, and to scale-up capability more quickly than in the past. In the same vein, increases in the amount and variability of data, combined with rapid advances in digital and communications technologies, have provided the opportunity for geospatial information to be leveraged as a transformative capability for governments to formulate better policy and to respond to national priorities.

The downside of the technology innovation trend is that the lack of robust digital infrastructure, including Internet connectivity and ICT infrastructure, are still major limiting factors to the proliferation of digital, location-enabled services and business models. While many countries have made significant progress in building digital infrastructure with improved coverage and quality in the recent past, most of the developing economies are found to be lagging in this front. The absence of this fundamental prerequisite, known as the 'digital divide' limits the stakeholder's ability to capitalize on many of the basic utilities of the geospatial industry. It is Big Data and the Cloud that are the two dominant technologies driving and accelerating the geospatial industry and capabilities. These are followed by artificial intelligence (AI) and the Internet of things (IoT), but are heavily dependent on wireless and broadband internet applications as the backbone of the digital ecosystem. Its impact on empowering citizens in developing countries requires the necessary infrastructure to bridge the digital divide [11].

The corresponding huge growth in the generation of data has meant that governments now realize the value of digital technologies and data as key strategic assets that lead to valuable and quantifiable results, thus changing the lives of economies and societies around the globe. Yet, reaping the full benefits of the opportunities linked to the value and use of data and technology requires that they are embedded as core components of continuous efforts to modernize and digitalize public sectors, and that they support new ways of working and creating public value [1]. However, all institutions, mechanisms and models are struggling to respond effectively to the pace of change and its distributed nature. New collaborative efforts are emerging across the world – processes that aim to build on both traditional strengths of host institutions but also draw in the expertise of other sectors – whether that be business, governments, civil society or academia [8]. Technological enablers such as the Internet, cloud computing, analytics, Big Data, mobile devices, unmanned aerial systems, and the rapid explosion of location-based services, which bring everyone directly into contact with location information every day, have ensured that people the world over, are beginning to appreciate the need for geospatial information in their consumption of data [28, 30]. As a result, a large proportion of the global community now have an entirely different set of geospatial information uses, needs and expectations than they did even ten years ago, such has been the evolutionary change. In some respects, it also indicates that geospatial information and services are now being driven more and more by users and consumers in response to their contemporary needs as much as responding to technology developments and breakthroughs [25]. Both of these trends are gaining pace as technology creates new experiences and expectations, which in turn creates new opportunities but also seismic shifts in consumer behavior and expectations [3].

Through technology, the potential of geospatial information has rapidly advanced and has now reached a level of maturity that allows this information flow to make a central contribution to the integration of information for many of the current social, economic and environmental challenges facing the world. Scott and Rajibafard (2017) introduced a general integrative sustainable development 'data flow' framework for national information systems to capture the required data elements [25]. It comprises a mix of national data that provides the building blocks and processes for any given country to measure and monitor the SDGs from local real-world conditions to global harmonised reporting through robust and reliable data inputs. Working from the base of Figure 2.4, the building blocks of the data flow framework are as follows:

1. The local to national real-world social, economic and environmental sustainable development challenges, conditions and circumstances that exist

for countries, and that need to be measured and monitored in order to make progress. For the least-developed countries and small island developing States, limited capabilities and resources, including the means to even understand these conditions and their implications, will initially remain a significant capacity challenge;

2. Comprises the many and varied mix of fundamental baseline data resources and inputs, including new data collections that will be required. Many of these may need to be "repurposed" or significantly improved in order to adequately measure and contribute to the determined national indicators;
3. The national information systems that exist within countries will provide the means to ensure access to high quality, timely and reliable data that are structured, organized and managed, ideally in an interoperable and standards-based manner;
4. The national data are then specifically purposed, compiled and disaggregated by a number of SDG metrics; specifically by income, gender, age, race, ethnicity, migratory status, disability, geographic location and other characteristics relevant in national contexts. Such data will provide the SDG metrics for measuring and monitoring progress;
5. The National Statistical Offices will then undertake and finalize the official aggregation and integration into national country indicators, ensuring data integrity and validation;
6. The National Statistical Offices, and in collaboration with specialized United Nations agencies where appropriate, would then provide the final national indicators to the United Nations Statistics Division to be compiled into the global indicator framework with other countries around the world so that the global outputs can be reported. As decided by the Statistical Commission, estimates used for the compilation of global indicators are to be produced in full consultation with National Statistical Offices;
7. Initially as the 169 aspirational global targets; and
8. Finally as the 17 universal SDGs. It is expected that these processes will occur on an annual basis and be formally reported through the annual Sustainable Development Goals Report.

Regardless of logical synergies and linkages over a long period of time the reality is that even today, in a highly data and technology driven global environment, there has been very little connection and fusion between sustainable development and geography, geospatial information and NSDIs at either the political or the technical level. Now that real data is needed to measure and monitor and make evidence-based decision-making, the gaps and the lack of

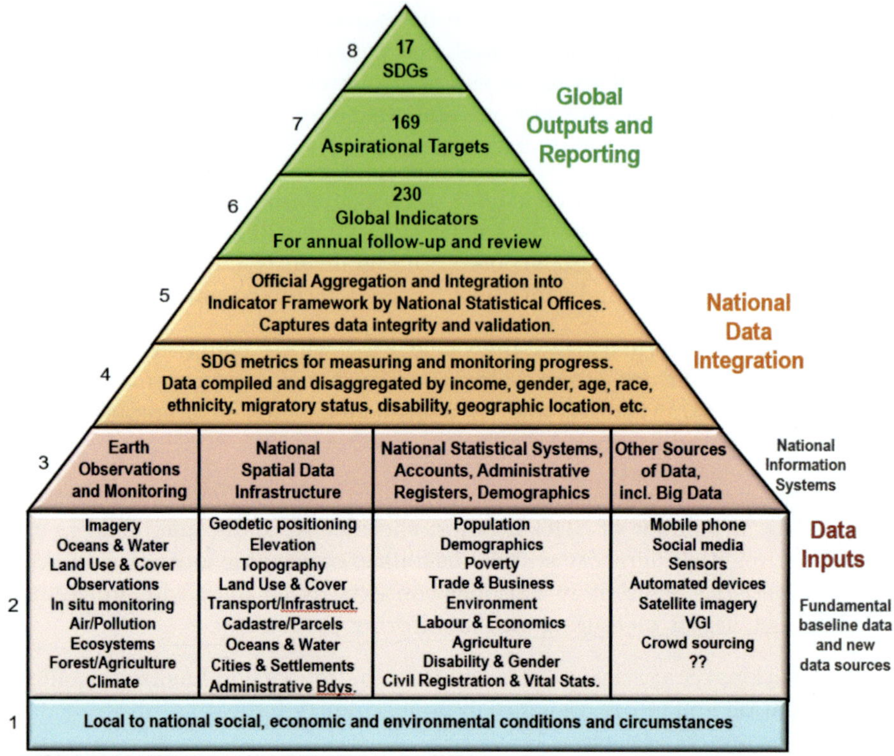

FIGURE 2.4
A general national information systems sustainable development 'data flow' framework that provides the building blocks and processes for any given country to measure and monitor the SDGs from local real-world conditions through to global harmonised reporting [25].

these connections are becoming apparent. Without high-quality data providing the right information on the right things at the right time; designing, monitoring and evaluating effective policies becomes almost impossible.

2.5 Bridging the Geospatial Digital Divide

Although today's technology makes it possible to collate the data we need to keep the promise to leave no one behind, the challenges faced in the collection, processing, analysis and dissemination of reliable, timely, accessible and sufficiently disaggregated data for better evidence-based policymaking are still considerable. An urgent transformational change in both thinking and approach is required – a digital transformation – which is simple in theory, but complex in implementation.

For developed countries, achieving digital transformation is still exacerbated by the lack of awareness and understanding of the role of geospatial information and enabling technologies at the policy and decision-making levels. For developing countries, realising digital transformation remains completely out of reach, as they are yet to attain effective and sustained access to digital technology, the Internet and the corresponding computer literacy and skills that are needed to take part in the information society and to orchestrate transformational change. They are yet to bridge the 'digital divide' before contemplating increased awareness and understanding of geospatial information. For these countries, exploiting the new science, data, technologies and tools to support the implementation of the SDGs compounds the problem. There is a need to extend well beyond the digital divide; to bridging the 'geospatial digital divide', connecting to the vast amounts of data, including geospatial information, and scientific and technological innovation to measure and monitor the 'geographic location' characteristics of the SDGs, targets and global indicators. Although not yet being realised, these real needs bring with them real opportunities for developing countries to raise the awareness and understanding of the role of geospatial information and enabling technologies at national policy and decision-making levels. Achieving sustainable development requires national geospatial policy and digital transformation, but now those developing countries that need it most do not yet know what these transformative technology enablers are able to provide, as they must first bridge the geospatial digital divide.

To illustrate the enormity of the growing data and technology gaps between the developed and developing countries, Scott and Rajabifard (2019 in press) expanded on these concepts further, discussing the digital divide and the fundamental data, skills and technology challenges facing developing countries in a global 'digital transformation' economy, particularly as they pertain to sustainable development [26]. They then introduced the 'geospatial digital divide', an extension of the digital divide, in which the lack of enabling mechanisms, such as ICT and Internet capabilities and access, are compounded and exponentially complicated by a lack of geospatial data and related enabling technology capability and capacity.

In general, the digital divide reflects the gap between those that have access to the newest and most innovative ICTs and those that do not. However, due to the range of criteria which can be used to assess the imbalance and gap, and the lack of detailed data on some aspects of technology usage, the exact nature of the digital divide is both contextual and debatable. Criteria often used to distinguish the gaps between the 'haves' and the 'have nots' of the digital divide tend to focus on access to hardware, access to the Internet, and details relating to both categories. But, as demonstrated in Figure 5, there are multiple layers of additional, and non-linear, criteria to consider. The first is the 'Digital Access Gap', which captures the primary enabling mechanisms for overcoming the digital divide – people's and country's ability to have access to and take advantage of computers, ICTs and the Internet. This digital access requires having both physical access to an Internet-enabled computer, or related device, and the material assets to pay for sustained connection to the Internet, the ongoing costs of certain subscriptions and necessary peripherals for use. Ironically, another key and influencing determinant not often considered is access to reliable electricity, without which there simply is no access to computers, ICTs and the Internet, and enjoyment of the subsequent opportunities and benefits.

The second, the 'Digital Adoption Gap', relates to users possessing the necessary skills to adopt and make use of ICT, computers and the Internet, and to take part in the information society. The digital divide is also a human skills and knowledge transfer divide. In many instances, the lack of such skills is related to digital and Internet literacy and not only exists between countries, but also within countries where segments of society have different knowledge, opportunities and experiences of digital technology and its adoption.

The third is the 'Digital Value Gap' and is governed by the lack of uptake, benefits and realization of the value in creating content and using the available technology, as well as how and for what purposes. This can be viewed as a gradation of inclusion (or exclusion), which can be mapped along the intersections of gender, race, ethnicity, age, education, socioeconomic class (social inclusion), and geography. With respect to geography, these societal elements can be mapped further geographically when aspects such as rural versus urban and industrialized versus developing countries have an effect. Additionally, the broader overarching national social, political, and economic implications of the digital divide, including lack of opportunity to solve societal problems, for countries should be considered in the Digital Value Gap.

In situations where the Internet has recently, and quickly, reached many developing countries, the intensity of use is lower in less technologically advanced developing countries, owing partly to a large within-country digital divide in many of those countries. For example, there are important gaps in access to Internet between men and women, urban and rural areas, and the young and old. One explanation for the between- and within-country divides is that effective use of the Internet is a function of literacy. Hence, closing the digital divide points to the need to focus on basic and secondary education and digital literacy ([23], p. 73).

The geospatial digital divide is an extension of the digital divide, in which the lack of enabling mechanisms, such as ICT and Internet capabilities and access, are compounded and exponentially complicated by a lack of geospatial data and related enabling technology capability and capacity. Building upon the criteria and gaps of the digital divide, there are a number of additional and critical 'gaps' that contribute to the 'geospatial' dimensions of the digital divide as also detailed in Figure 2.5 .

The 'Data Ecosystem Gap' relates to the access to, and exploitation of, data itself – the vast amounts of geospatial data, Earth observations, statistics

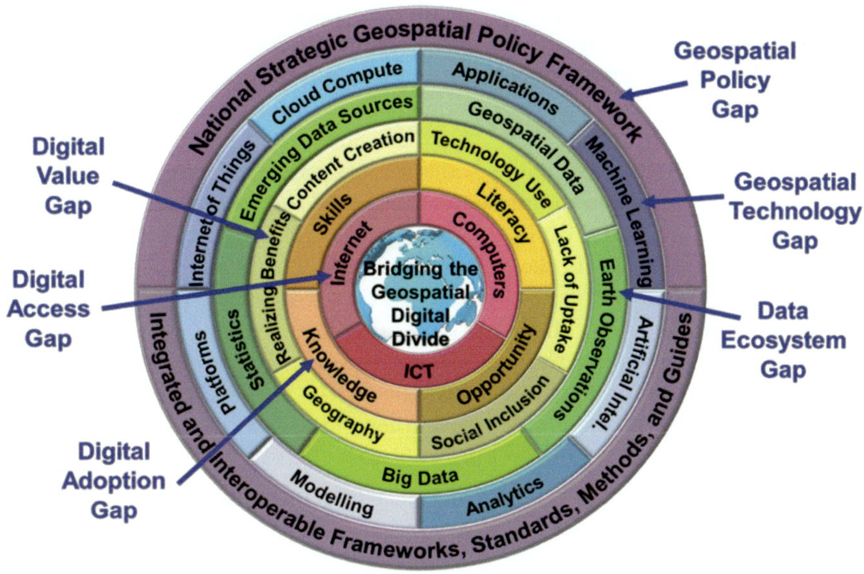

FIGURE 2.5
Building upon the criteria and gaps of the digital divide – the digital access, digital adoption and digital value gaps; there are a number of additional and critical 'gaps' that contribute to the 'geospatial' dimensions of the digital divide – the data ecosystem, geospatial technology, and geospatial policy gaps. Modified from [26].

(socio-economic and environmental), Big Data and the many new and emerging data sources. Much of the traditional 'fundamental data' would ideally reside within national information systems that exist within countries (Figure 2.4), providing the means to ensure access to high quality, timely and reliable data that are structured, organized and managed, ideally in an interoperable and standards-based manner [25]. There are also many local to global data that reside outside these more formal systems – a broader ecosystem that includes additional data contributors from local and regional governments, private companies, academia, civil society, and citizens, including crowd-sourced and volunteer geographic information.

The 'Geospatial Technology Gap' relates to the enabling, and increasingly growing, technology innovations that influence what we can do with the geospatial data ecosystem and how we manipulate its end-to-end use during its life-cycle. These include technology innovations like cloud storage and computing, the Internet of things (IoT), artificial intelligence (AI) and robotics, and machine learning; and delivery platforms, hardware, software, applications and interfaces that enable robust and repeatable modelling and analytics.

Finally, overcoming the many obstacles to close the gaps and bridge the geospatial digital divide requires considerable capacity and capability development, and knowledge transfer. But even more importantly, to be able to deliver and sustain an integrated and interoperable geospatial information environment in developing countries will require the appropriate tools – frameworks, standards, methods and guides – to be developed and institutionalised within countries through an overarching and strategic national framework. Therefore, and providing the means to 'implement' the bridging of both the digital divide and the geospatial digital divide, the 'Geospatial Policy Gap', as the outer and encompassing circle, is the key area in which all of the elements are able to be brought together and consolidated within a National Strategic Geospatial Policy Framework. While the adoption of the 2030 Agenda provides the global policy mandate to exploit the contribution to be made by geospatial information to support the implementation of the SDGs, targets and indicators, does this provide enough for countries to implement the required change – and to bridge the geospatial digital divide?

2.6 A National Strategic Geospatial Information Policy Framework

Recognizing the importance of geospatial information is the first step towards overcoming barriers to implementation and bridging the geospatial digital divide. Scott and Rajabifard (2017) explored the challenges and opportunities to implement and integrate geospatial information into the global development policy agenda in a more holistic and sustainable manner at a national level [25]. The lack of policy and guidance, with commensurate critical gaps and connection points with national geospatial methods and frameworks, was seen as a visible impediment to developing countries and those most affected by the challenges and need to achieve national development. Bridging the geospatial digital divide – enabling people, processes, data and technology to implement national policy – requires a strategic policy realization of the impediments, but in such a way that they are able to be integrated into national strategies and arrangements; so that they can be anchored into national development agendas. A national strategic geospatial policy framework is able to provide this national guidance for developing countries as they attempt to measure and monitor progress towards the implementation of the SDGs.

While each country has primary responsibility for its own economic and social development, the role of national policies and the existence of development strategies cannot be overemphasized. To bridge the geospatial digital divide, countries will need to develop, strengthen and modernize their approaches to geospatial information management, including aspects relating to geospatial information policies, governance, data-driven integration and infrastructure, education, innovation, use and collaboration. To achieve this in line with the

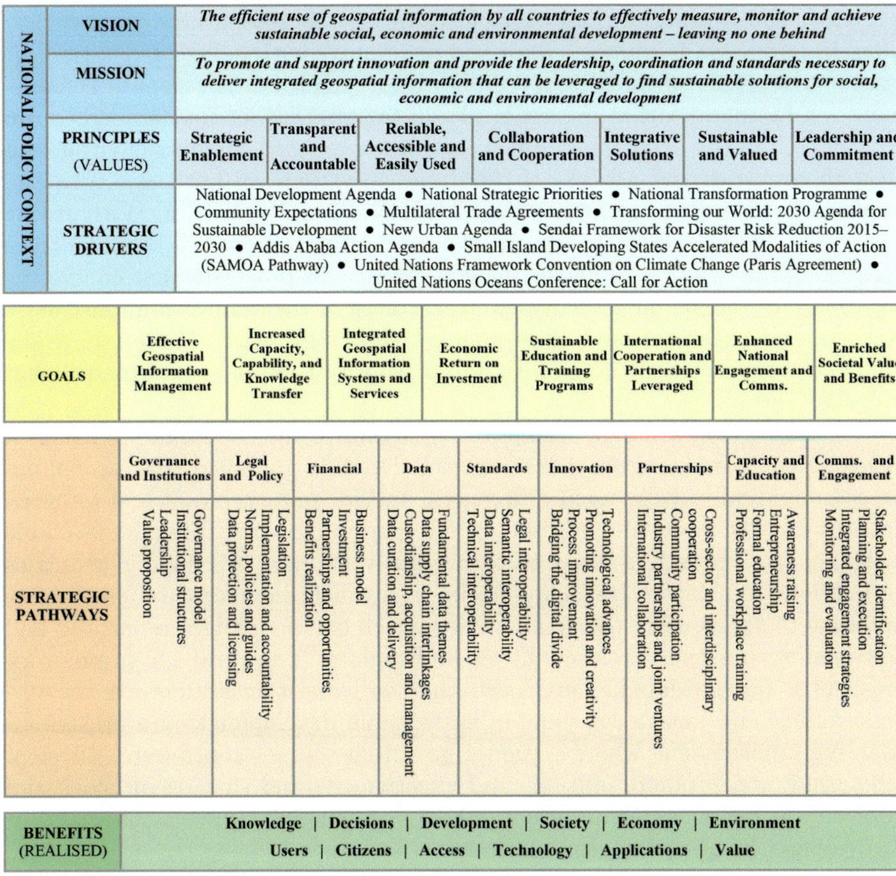

FIGURE 2.6
The components of the National Strategic Geospatial Information Policy Framework

expectations of the 2030 Agenda, countries, through their national governments, need an enabling strategic framework to guide them. Therefore, this section presents a National Strategic Geospatial Information Policy Framework (Strategic Framework), as summarized in Figure 2.6 and developed by Scott and Rajabifard (2019, in press), as a means to bridge the geospatial digital divide, to provide a basis and guide for countries towards developing and strengthening arrangements in national geospatial information management, to meet and overcome complex sustainable development challenges for developing countries, and which builds upon the previous work of Scott and Rajabifard (2017, 2019 in press) [25, 26].

As contextually framed in Figure 2.6, the Strategic Framework, consisting of integrated and interoperable frameworks, standards, methods and guides, addresses the geospatial policy gap in bridging the geospatial digital divide. More importantly, it connects data and technology with policy. This Strategic Framework represents a deliberate policy approach towards geospatial enablement – a Framework that is high level and overarching, is not detailed, is not data or technology driven, but which establishes a broad enabling environment that all countries and aspects of the data ecosystem can both aspire to and implement according to their national situations and priorities. With strong ownership and leadership by government and other national stakeholders, the Strategic Framework establishes the conditions and processes that enable the sustainability of the data ecosystem, regardless of the various administrative settings and organizational environments that may exist. Therefore, a key metric to its success is the quality and strength of national governance that prevails.

The aim of the Strategic Framework is to inform and contribute to national development plans, enable greater national policy guidance that accommodates individual jurisdictional differences, and for more relevant and targeted data to be generated and utilised, while exploiting accessible and available technologies. There is a need to significantly support developing countries and regions to achieve these capabilities in a timely manner through the formulation of appropriate tools and guides which can be piloted, replicated and delivered within and across countries and regions. The Strategic Framework presents a forward-looking approach that will assist in addressing the understanding and knowledge gaps in sustainable development and creates an enabling environment where national governments can coordinate, develop, strengthen and promote efficient and effective use and sharing of geospatial information for policy formulation, decision-making and innovation – and in turn bridge the geospatial digital divide.

The components of the Strategic Framework are designed and structured to be inclusive, universal and transformational. They firstly comprise a high-level national policy context inclusive of a common vision and mission for all national government agencies, underpinning principles or values, and related national to global strategic drivers. The goals and strategic pathways provide the main elements of the Strategic Framework. Achieving the goals will realize the vision, while the strategic pathways outline the actions that need to be implemented to achieve the goals, and the outcomes and benefits necessary to support national development.

The key to structuring the Strategic Framework in such a way is so that each of these elements are able to communicate the intent and purpose of the Framework to stakeholders – these primarily being high level policy and decision makers, institutions and organizations within and across government. It is important that the Strategic Framework can be seen as aligning with government needs, with national development aspirations, and can be seen as an inclusive and engaging mechanism to bring collaboration, coordination and cohesion across a country for the purposes of developing, strengthening and integrating arrangements in national geospatial information management; and as a mechanism that can monitor continuous innovation and improvement. Providing a 'data and information' approach to policy and implementation will bring the analysis and evidence-base to the decision-making process that would benefit all areas of government, and will also provide a consistent monitoring and reporting framework for sustainable development within the national policy context.

2.7 Vision and Mission

The vision and mission statements communicate the overarching aim of the Strategic Framework to stakeholders. The vision statement describes a future state where integrated geospatial information is used to achieve sustainable social, economic and environmental development; while the mission statement is a call to action that will enable governments to achieve the vision.

2.7.1 Vision

The vision is that governments are able to achieve sustainable social, economic and environmental development through the efficient and effective use of national and local geospatial information, systems and capabilities for evidence-based policy and decision-making. The vision statement is a future orientated and aspirational declaration of purpose and being. The vision recognises the responsibility for countries to plan for and provide better outcomes for future generations, and our collective aspiration to leave no one behind. Additionally, it recognizes that any national SDG implementations will be optimized using strategies and frameworks that integrate geospatial information into overall national social, economic and environmental development plans.

2.7.2 Mission

The mission is for countries to promote and support the required innovation, leadership, coordination and standards in order to develop, strengthen, integrate and deliver national geospatial information policy, data, systems, tools, services and capabilities into their national government development policies, strategies and arrangements. The mission is designed to stimulate action towards bridging the geospatial digital divide; to find sustainable solutions for social, economic and environmental development; and to influence inclusive and transformative societal change for all citizens according to national priorities and circumstances.

2.8 Principles

In order to anchor the vision and mission to the national policy needs, and to the more specific goals, objectives, and priority actions, the Strategic Framework requires an enabling and collaborative environment where government organizations can coordinate, cooperate, and thus improve the management and exchange of national geospatial information to support and serve the national interests of all of its citizens. This enables the value of geospatial information to be realized for national and sustainable development. In order to be effective and avoid duplication of technology and resources, the Strategic Framework also needs to be cross-cutting across multiple government agencies, and to leverage existing methods and mechanisms as much as possible, for example NSDI capabilities and methodologies; but must also be able to gather and deliver new data and information capabilities not previously considered.

Therefore, the Strategic Framework identifies seven principles or values. These principles represent the key characteristics and values that are to be used as a guide and reference point when implementing the Framework. How these principles are applied will depend on the implementation approach adopted by each country. The principles are the generic compass for supporting and implementing a policy and data framework, but allow for methods to be tailored to individual country needs and circumstances as required. Adherence to these principles will make complex collaboration possible among multiple agencies, and will deliver consistent geospatial information management, resulting in more open, accountable, responsive, and efficient government. The seven principles that underpin the Strategic Framework are detailed in Table 2.1.

2.9 Strategic Drivers

The strategic drives are aligned to the National Policy Context and will vary from country to country based on national priorities and objectives. They are not exhaustive and are provided as an initial set of global to national strategic objectives. With regard to the global development Agendas, many countries understand that the 2030 Agenda also captures the specific and separate global United Nations system outcomes as illustrated in Figure 2.2. All countries have aligned their national priorities and development outcomes to at least one of these global Agendas. These then cascade down to the national development drivers and strategic priorities that may also include national transformation programmes, multilateral trade agreements, and even community and societal expectations on government.

TABLE 2.1
The seven Principles of the National Strategic Geospatial Information Policy Framework

PRINCIPLE 1 Strategic Enablement	The implementation of the Strategic Framework requires political and financial support, and should therefore align with and support government's strategic direction on issues such as economic growth, social well-being, job creation, natural resource monitoring, and environmental management and preservation.
PRINCIPLE 2 Transparent and Accountable	Government geospatial information is developed and shared according to key accountability and transparency guidelines so that all citizens, government agencies, academia and the private sector have access to this valuable and underpinning national information resource.
PRINCIPLE 3 Reliable, Accessible and Easily Used	Geospatial information is reliable, and made accessible and usable so that it can be leveraged for decision-making, research and development, used to stimulate innovation, and to support the creation of sustainable services and products to advance social, economic and environmental development.
PRINCIPLE 4 Collaboration and Cooperation	Collaboration and cooperation (between government, business, academia, civil society and donors) are factored into the implementation of the Strategic Framework to strengthen information-sharing between providers and users, reduce duplication of effort across the government sector, make for a robust system, as well as providing clarity on roles and responsibilities.
PRINCIPLE 5 Integrative Solution	The implementation of the Strategic Framework is to be integrative in nature – and consider how people, organizations, systems, and legal and policy structures work together to form an effective and holistic system for managing geospatial information and its use.
PRINCIPLE 6 Sustainable and Valued	The implementation of the Strategic Framework will be conducted in such a way that it enhances national efficiency and productivity; is sustainable in the long term; and is deployed in a way that provides improved and valued government services to citizens.
PRINCIPLE 7 Leadership and Commitment	Importantly, the implementation of the Strategic Framework will require strong leadership and commitment, often at the highest level, to enhance the long-term value of investments in geospatial information. This will be achieved through careful analysis, prioritization and sequencing to develop an action plan that carefully applies interventions in the short, medium and long-term, and that can receive high-level endorsement and support by government.

2.10 Goals

To achieve the overarching vision, the Strategic Framework identifies eight goals. The progressive achievement of these goals will move countries towards a future state where they have the capacity and skills to organize, manage, curate and leverage geospatial information to advance government policy and decision-making capabilities, bridge the geospatial digital divide, influence inclusive and transformative societal change, achieve economic prosperity and social development, and ensure effective environmental management. The eight goals of the Strategic Framework are detailed in Table 2.2.

2.11 Strategic Pathways

The Strategic Framework is anchored by nine strategic pathways in three main areas of equally shared influence: these being aspects related to overarching national governance; the underlying data and enabling technology; and the importance of people in the geospatial information life cycle. The objective of these strategic pathways is to provide the 'implementation roadmap' to guide governments towards implementing integrated geospatial information systems in a way that will deliver a vision for sustainable social, economic and environmental development. Although the strategic pathways are presented as separate elements, and recognizing that there are many aspects and dimensions to each individual pathway, it is intended that, when the nine strategic pathways are united as one, the Strategic Framework is connected, integrated and implemented. Figure 2.7 illustrates the nine strategic pathways surrounded by the many benefits that are able to be realized when implemented together.

It is important to note that the strategic pathways are able to readily leverage and build upon existing national NSDI information architectures, capabilities and methodologies. The traditional evolution of NSDIs have been seen as coordinated actions of nations and organizations that promote awareness and implementation of complimentary policies, common standards and effective mechanisms for the development and availability of interoperable digital geographic data and technologies to support decision making for multiple purposes. However, one of the weaknesses is that many NSDIs are still predominately data supply rather than data demand driven and are rarely designed to be strategic frameworks or respond directly to high priority societal policy issues [25].

As shown in Figure 2.7, the strategic pathways represent 3 levels of enabling geospatial maturity. Level 1 maturity (Governance and Institutions, Data, and Partnerships) broadly aligns with the requirements for a typical initial NSDI implementation. The Level 2 maturity (Legal and Policy, Standards,

TABLE 2.2
The eight Goals of the National Strategic Geospatial Information Policy Framework

GOAL 1 Effective Geospatial Information Management	Enabling geospatial information governance, policy and institutional arrangements that ensure effective geospatial information management, accommodate individual organizational requirements and arrangements, and that are aligned to national, regional and global policy frameworks.
GOAL 2 Increased Capacity, Capability and Knowledge Transfer	Mechanisms are established to raise awareness of the value and use of geospatial information, promote capacity and capability, and build an inventive and resourceful mindset across government, industry, academia, and private and community sectors.
GOAL 3 Integrated Geospatial Information Systems and Services	Geospatial information, including community information, is integrated as a national information system and service across the government sector and maximized for evidence-based policy and decision-making.
GOAL 4 Economic Return on Investment	An economic return on investment is realized through best practice management, and the exploitation and innovative use of integrated geospatial information.
GOAL 5 Sustainable Education and Training Programs	Education and training programs are established to grow the number of professionals in the fields of geography, data science and geospatial information technology, and to develop specialist skills related to geospatial financial systems, policy and law, and project management.
GOAL 6 International Cooperation and Partnerships Leveraged	International cooperation and partnerships are leveraged and sustained in a way that fosters the management and exchange of geospatial information in support of national development interests.
GOAL 7 Enhanced National Engagement and Communication	All stakeholder groups, and specifically high-level decision-makers and champions, are fully engaged in the value of integrated geospatial information for decision-making and socio-economic development.
GOAL 8 Enriched Societal Value and Benefits	Social and economic development, and environmental sustainability, is enriched through increased levels of use of integrated geospatial information products and services.

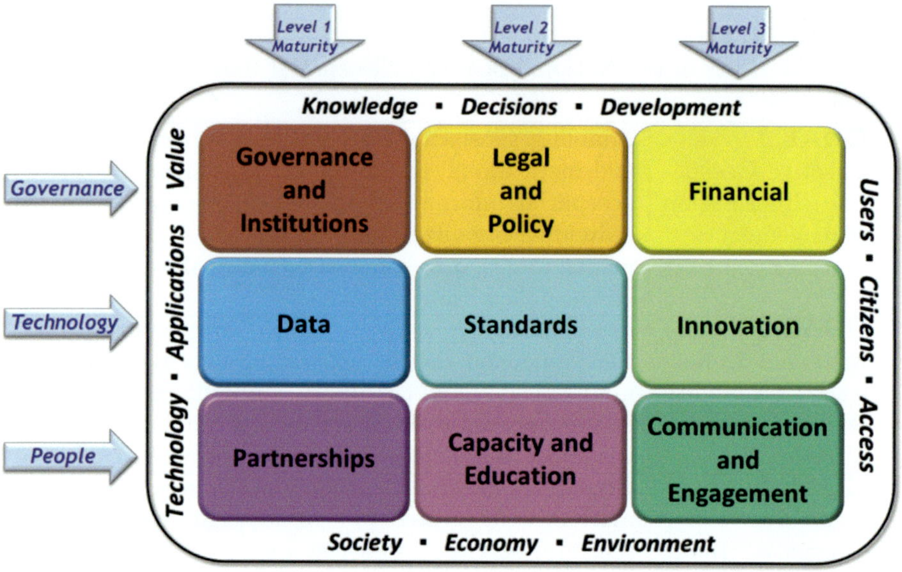

FIGURE 2.7
The nine Strategic Pathways of the National Strategic Geospatial Information Policy Framework

and Capacity and Education) represents what we would typically see as the 'enablers' for an advanced and mature NSDI implementation. Level 3 maturity (Financial, Innovation, and Communication and Engagement) represents the NSDI future state as an integrated 'on-demand' information knowledge infrastructure, in which we can gather and deliver, sustain and communicate new data and integrated information capabilities not previously considered. With key components – such as the governance and institutional arrangements, roles and responsibilities, the very existence of a Strategic Framework, funding models and the legal and regulatory framework – established, it is at this level of maturity that the long-term sustainability and benefits of open data, machine readable data, semantic web technologies and linked data will be realized.

Although not all individually detailed in this chapter, the intent is that each of the nine strategic pathways are able to be explained and elaborated, along with specific detailed elements and objectives, as the particular set of tools that assist in guiding implementation of the Framework and achieving the required results. These objectives are provided at a high level in Figure 10.

TABLE 2.3: The specific objectives of the nine Strategic Pathways of the National Strategic Geospatial Information Policy Framework to assist countries in achieving the required results

STRATEGIC PATHWAY 1 Governance and Institutions	Establishes the leadership, governance models, institutional arrangements and a clear value proposition as a means to strengthen multi-disciplinary and multi-sectoral participation and commitment to achieving the Strategic Framework. The objective is to attain political endorsement, strengthen institutional mandates, and build a cooperative data sharing environment through a shared understanding of the value of the Strategic Framework, and the roles and responsibilities to achieve its vision. Good and consistent governance is critical in countries so that policies and institutional arrangements are able to be insulated and protected from political and administrative change.
STRATEGIC PATHWAY 2 Legal and Policy	Establishes a robust legal and policy framework that is essential to institute appropriate national geospatial information legislation and policy that enables the availability, accessibility, exchange, application and management of geospatial information. The objective is to address current legal and policy issues by improving the laws and policies associated with, and having an impact on, geospatial information management; and by proactively monitoring the legal and policy environment, particularly with regard to designating the official responsibility for the production of data, and with respect to the issues raised by emerging technologies and the evolving innovative and creative use of geospatial information.
STRATEGIC PATHWAY 3 Financial	Establishes the business models, develops financial partnerships, and identifies the investment needs and funding sources for delivering integrated geospatial information management, as well as recognizing the benefits realization milestones that will achieve and maintain momentum. The objective is to achieve an understanding of the implementation costs and ongoing financial commitment necessary to deliver integrated geospatial information management that can be sustained and maintained in the longer term. Investment in all strategic pathways is paramount.

STRATEGIC PATHWAY 4 Data	Establishes a geospatial data framework and custodianship guidelines for best practice collection and management of integrated geospatial information (accurate, logical, consistent, standardized and interoperable) that is appropriate to cross sector and multidisciplinary collaboration. The objective is to enable data custodians to meet their data management, sharing and reuse obligations to government and the user community, through the execution of well-defined data supply chains for organizing, planning, acquiring, integrating, managing, maintaining, curating, publishing and archiving geospatial information.
STRATEGIC PATHWAY 5 Standards	Establishes, and ensures the adoption of, best practice standards and compliance mechanisms that enable legal, data, semantic and technical interoperability, which are fundamental to delivering integrated geospatial information and knowledge creation. The objective is to enable different information systems to communicate and exchange data, enable knowledge discovery and inferencing between systems using unambiguous meaning, and provide users with lawful access to and reuse of geospatial information.
STRATEGIC PATHWAY 6 Innovation	Recognizes that technology and processes are continuously evolving; creating enhanced opportunities for innovation and creativity that enable governments to quickly bridge the digital divide. The objective is to stimulate the use of the latest cost-effective technologies, process improvements and innovations so that governments, no matter what their current situation, may leapfrog to state-of-the-art geospatial information management systems and practices. Acknowledges that government agencies are not normally the first to implement novel and new solutions, and that industry is often leading innovation.

STRATEGIC PATHWAY 7 Partnerships	Establishes effective cross-sector and interdisciplinary cooperation, industry and private sector partnerships, and international cooperation as an important premise to developing a sustainable Strategic Framework. The objective is to create and sustain the value of geospatial information through a culture based on trusted partnerships and strategic alliances that recognize common needs and aspirations, and national priorities.
STRATEGIC PATHWAY 8 Capacity and Education	Establishes enduring capacity-building programs and education systems so that geospatial information management and entrepreneurship can be sustained in the longer term. The objective is to increase the awareness and level of understanding of geospatial information science. This includes developing and strengthening the skills, instincts, abilities, processes and resources that organizations and communities require to utilize geospatial information for decision-making. Recognizes that the human resource asset is the most critical – the people.
STRATEGIC PATHWAY 9 Communication and Engagement	Recognizes that stakeholders (including the general community) are integral to the implementation of integrated geospatial information management systems, and that their buy-in and commitment is critical to success. The objective is to deliver effective and efficient communication and engagement processes to encourage greater input from stakeholders to achieve transparent decision-making processes when implementing the Strategic Framework.

2.12 Benefits

Broad societal benefits that include the citizen, community and country, in the three areas of sustainable development – social, economic and environmental, and leveraging the value of data, technology and innovation to derive outcomes that include decisions, knowledge, development – and ag the end of the day, national prosperity and deliver the vision and mission of the Strategic Framework.

2.13 Implementing the National Strategic Geospatial Information Policy Framework

Noting the 2030 Agenda's promise to leave no one behind, and the commensurate expectations that by 2020 countries will need to have increased significantly the availability of high-quality, timely and reliable disaggregated data, this Strategic Framework has been developed in a rapidly changing environment. Despite the many challenges in developing countries, community expectations are evolving with advancements in technology and the gradual increase in computer literacy. Governments are recognizing that to maintain relevance with the prevailing societal needs and ambitions there is a need to deliver geospatial information in a way that can be visualized and used anywhere, anytime and on any electronic device. Staying abreast of community expectations and having a sense of where the best public value is now and in the future, remains a key responsibility of Government.

This Strategic Framework has responded to these community aspirations and the urgent need for its implementation, and now underpins the United Nations Integrated Geospatial Information Framework (IGIF), adopted by UN-GGIM at its eighth session in August 2018 [31]. Based on the Strategic Framework, the IGIF was developed in 2018 as a collaboration between the United Nations and the World Bank to provide a basis and guide for lower to middle income countries to reference when developing and strengthening their national and sub-national arrangements in geospatial information management and related infrastructures. Prior to its adoption, the IGIF was submitted to all Member States for global consultation, which sought inputs regarding the overall structure and substance, if the approaches and levels of detail were suitable, and if the structure was a reasonable approach to deliver the IGIF to the global community.

The IGIF is seen as being comprehensive and provides a clear vision and mission on how to develop and facilitate the utilization of geospatial information at the national level. For developing countries, it is a valuable tool to be utilized to bridge the geospatial digital divide. Despite its comprehensiveness, the IGIF is still clear enough to be used at the highest level. It was considered valuable to identify the seven underpinning principles, eight goals and nine strategic pathways; several countries even expressed interest to expand these further. Importantly, the IGIF was strongly supported by African countries, whom had the opportunity to provide inputs into an early version in April 2018. This proved a valuable exercise, as the African countries were able to voice their concerns towards ensuring that the IGIF is organized in such a way that it could be readily used as a guidance to establish a geospatial information management system in their countries. They also reiterated the importance of international cooperation, as it is a major goal to have well established international cooperation and partnerships that support national development and capacity-building interests in situations where countries are just beginning to spread the importance of geospatial information across national aspects. It was noted that international cooperation donors require a strong business case and confidence in governance before releasing funds to countries, and that the IGIF provides that confidence.

Member States have emphasized the need for coherent and integrated system-wide strategic planning, implementation and reporting. Policy coherence is crucial for achievement of the SDGs, given the interlinked and inseparable nature of the various dimensions and constituent elements involved – social economic and environmental. At the national level, policy coherence ensures consistency across national policy and programme frameworks, and their alignment in support of national sustainable development efforts.

2.14 Conclusions

Framed by the 2030 Agenda for Sustainable Development, this chapter presented and discussed the major components to assist continued efforts in charting a geospatial roadmap towards the implementation of the SDGs. It first contextualized sustainable development broadly, and its evolution towards the 2030 Agenda, before visiting the goals, targets and global indicator framework in detail. The chapter then described the role of geospatial data and enabling technologies in contributing to the 2030 Agenda, before discussing the implications of the digital divide that continues to exist today for developing countries, and introduces and describes the 'geospatial digital divide' and the complex challenges that continue to exacerbate the ability for these countries to bridge this divide, to connect to the vast amounts of data and technology, and accelerate human progress. The chapter concluded with a national strategic geospatial information policy framework as a means to provide the national policy basis and roadmap for countries to develop and strengthen their national and sub-national arrangements in geospatial information management, as they attempt to measure and monitor progress towards the implementation of the SDGs.

In recognition of its value and urgent need, the Strategic Framework has been used as the basis for the overarching strategic framework for the Integrated Geospatial Information Framework, adopted at the global level by all countries of the United Nations as a means to assist countries in developing and strengthening their national and sub-national arrangements in geospatial information management and related infrastructures – to bridge the geospatial digital divide and to leave no one behind. Many countries are now looking at implementing the Framework with guidance by UN-GGIM and the World Bank, whom are now collectively developing a detailed Implementation Guide and Country-level Action Plans.

Bibliography

[1] *Open Government Data Report, Enhancing Policy Maturity for Sustainable Impact.* OECD Publishing, Paris, 2018.

[2] W.M. Adams. *Green Development: Environment and Sustainability in a Developing World.* Routledge; 3rd Edition, 2009.

[3] Jonathan Auerbach. Why partnership is the business trend to watch. In *World Economic Forum Annual Meeting 2018*, 2018.

[4] Rachel Carson. Silent spring houghton mifflin. *Boston, MA*, 1962.

[5] Paul R Ehrlich. The population bomb: Population control or race to oblivion? Technical Report 363.91, 1968.

[6] Jennifer A Elliott. *An introduction to sustainable development.* Routledge; 3rd Editiion, 2006.

[7] World Economic Forum. What is the fourth industrial revolution?, 2016.

[8] World Economic Forum. Our shared digital future building an inclusive, trustworthy and sustainable digital society. 2018.

[9] Edward Goldsmith and Robert Allen. *A Blueprint for Survival.* Harmondsworth, Penguin, 1972.

[10] Donella H Meadows, Donella H Meadows, Jørgen Randers, and William W Behrens III. The limits to growth: a report to the club of rome (1972). *Google Scholar*, 1972.

[11] Geospatial Media. Geobuiz 2018 report, geospatial industry outlook and readiness index, 2018.

[12] United Nations. *Report of the United Nations conference on the human environment.* United Nations Publication, 1972.

[13] United Nations. United nations conference on environment and development rio de janerio, brazil, 1992.

[14] United Nations. The future we want. *United Nations General Assembly. Resolution 66/288*, 2012.

[15] United Nations. Ds accelerated modalities of action (samoa) pathway. *United Nations General Assembly. Resolution 69/15.*, 2014.

[16] United Nations. Addis ababa action agenda of the third international conference on financing for development (addis ababa action agenda). *United Nations General Assembly. Resolution 69/313*, 2015.

[17] United Nations. Framework convention on climate change: Adoption of the paris agreement. *United Nations. FCCC/CP/2015/L.9/Rev.1*, 2015.

[18] United Nations. Sendai framework for disaster risk reduction 2015–2030. *United Nations General Assembly. Resolution 69/283*, 2015.

[19] United Nations. Transforming our world: the 2030 agenda for sustainable development. *United Nations General Assembly. Resolution 70/1*, 2015.

[20] United Nations. New urban agenda. resolution adopted by the general assembly on 23 december 2016. *United Nations A/RES/71/256*, 2016.

[21] United Nations. Our ocean, our future: call for action. resolution adopted by the general assembly on 6 july 2017. *United Nations A/RES/71/312*, 2017.

[22] United Nations. *The Sustainable Development Goals Report 2018.* United Nations, New York 2018, 2018.

[23] United Nations. *World Economic and Social Survey 2018: Frontier technologies for sustainable development.* United Nations, New York 2018Department of Economic and Social Affairs, 2018.

[24] Walt Whitman Rostow and Walt W Rostow. *The stages of economic growth: A non-communist manifesto.* Cambridge university press, 1960.

[25] Greg Scott and Abbas Rajabifard. Sustainable development and geospatial information: a strategic framework for integrating a global policy agenda into national geospatial capabilities. *Geo-spatial Information Science*, 20(2):59–76, 2017.

[26] Greg Scott and Abbas Rajabifard. Bridging the geospatial digital divide: a means to overcome complex sustainable development challenges for developing countries. 2019.

[27] UN-GGIM. Report of the secretary general on global geospatial information management. *United Nations General Assembly E/2011/89*, 2011.

[28] UN-GGIM. Future trends in geospatial information management: the five to ten year vision. *Discussion document published by Ordnance Survey of the United Kingdom at the request of the Secretariat for the United Nations Committee of Experts on Global Geospatial Information Management. Lead Authors: John Carpenter and Jevon Snell, Ordnance Survey, United Nations, New York. Accessed March*, 15:2015, 2013.

[29] UN-GGIM. Activities related to sustainable development and the post-2015 development agenda. *Committee of Experts on Global Geospatial Information Management. , United Nations, New York*, 2014.

[30] UN-GGIM. Future trends in geospatial information management: the five to ten year vision. Technical report, Discussion document published by Ordnance Survey of the United Kingdom at the request of the Secretariat for the United Nations Committee of Experts on Global Geospatial Information Management. Lead Authors: John Carpenter and Jevon Snell, Ordnance Survey, United Nations, New York. Accessed March, 2015.

[31] UN-GGIM. Strengthening of geospatial information management. *Committee of Experts on Global Geospatial Information Management. , United Nations, New York*, 2018.

[32] SPECIAL WORKING SESSION WCED. World commission on environment and development. *Our common future*, 1987.

3

Marriage of Opposites: Strategies for Public and Private Sectors Working Together in Land Tenure Reform Projects That Support SDGs

Daniel Paez
Centre for SDIs and Land Administration, The University of Melbourne, Australia

In the context of land tenure and SDGs, this chapter proposes strategies for practitioners to involve the private sector in land administration services while at the same time ensuring there is a contribution to the achievement of the SDGs.

3.1 Introduction

SDGs provide a framework for governments, multilateral organisations and donors to drive social investment on sustainable development. Among the 17 SDGs, SDG1(No poverty) SDG2 (Zero hunger) and SDG5 (Gender equality) directly highlight the importance of land tenure systems as a mechanism to achieve sustainable development. These goals focus on encouraging investment in land tenure security (LTS) around socially vulnerable groups (particularly the poor), land productivity and gender imbalances.

For over 20 years governments in developing countries, with the support of organisations such as the World Bank, have invested worldwide on improving the certainty of rights, restrictions and responsibilities by implementing land tenure reform projects. Traditionally efforts have focused primarily on the land right registry and cadastre systems and - more recently - spatial planning and natural resources systems. Most of the investment has focused on large scale systematic registration activities aimed at improving coverage, building information technology (IT) systems, reviewing and reforming institutions, technical capacity building and improving the regulatory framework.

There is an increasing trend across most developed and developing countries to involve the private sector in the delivery of land cadastre and registration services. This trend follows significant results in the infrastructure sector (in areas such as ports, roads and hospitals) where many countries have leveraged new investment by making the private sector and active partner that takes risks, invest resources and achieve social results.

Considering this current landscape for land reform projects, strategies proposed in this chapter are: (i) to target areas where the private sector can do a better job; (ii) find cross-services approaches within the land administration system (iii) build the participation in a trusted private sector; (iv) revenue should be based on achieving tangible social results aligned with SDGs.

Strategies presented in this chapter are proposed to be used during the design and implementation of future land reform projects both in developed and developing countries. Some areas for further research including better mechanisms to strengthening gender equity and synergies between these proposed strategies and others focused on productivity.

3.2 Background: Land Administration and the Trend of Involving the Private Sector

Land administration in this chapter relates to the institutional systems (at a national or sub-national level) that regulate all activities that require the use of land. The core of a land administration system is the cadastre or inventory and description of all parcels in a jurisdiction. Most modern cadastre systems in developed countries have a digital geographic description of the parcels, and they cover the entire area of the country [42].

Depending on the system, the cadastre is contained or be linked to a land registry in where rights, restrictions and regulations are typically stored. In this, tenure rights are fundamental as they underpin the ability of a person or group to conduct economic activities and participate actively in a land market by buying, renting or selling land properties.

In addition to the cadastre and the land registry, land administration systems cover many other areas of a nation or region such as land taxation systems; development and building permits; mining rights; and water concession registries.

Considering the importance of land administration systems for all nations, and particularly a cadastre system with an appropriate land registry, many investment projects have been conducted to improve it. Commonly in the literature, these projects are described as land tenure reform projects as they focus on enhancing government systems related to land rights.

This chapter explores the evolution of these land reform projects and the new trend of involving the private sector as a critical participant in the developing and funding of these projects and more broadly the land administration system. In particular, the focus is on analysing experiences in where public-private partnerships (PPPs) or similar instruments have been used in land reform projects and draws. Lessons are drawn from selected cases studies to develop strategies for ensuring private sector participation in land projects support the implementation of sustainable development goals (SDGs). These strategies, which are the main contribution of this chapter, are applicable in both developed and developing countries and are proposed to be considered during the design of future projects where it is desired to transfer significant responsibilities to the private sector. Ideally, these strategies will support developed and developing countries when reviewing future land reform projects or considering unsolicited proposals submitted by the private sector. Additionally, proposed strategies for private sector involvement in LTS provide assessment parameters for donors and multilateral banks to understand better the contribution of the private sector in a land reform project.

The rest of this chapter is structured as follows: The next section presents an analysis of SDGs and its relation to LTS. Section 3.4 presents the current opportunities and challenges of land reform projects. After this, a critical review of a selected number of case studies where the private sector plays a significant role in the delivery of LTS is presented in Section 3.5 . Section 3.6 gives the main contribution of this chapter which are strategies for PPPs and other mechanisms for involving the private sector in land reform projects in a way that is aligned with SDGs. The chapter finalises with conclusions and general suggestions for further research and development in Section 3.7.

3.3 SDGs and Land Tenure Reform Projects

For over 25 years and since the first Earth Summit, organised by the United Nations (UN) in 1992 in Rio de Janeiro (Brazil), there has been a collective effort to promote a set of goals for all nations, particularly those experiencing poverty and social challenges. In 2016, SDGs replaced the Millennium Development Goals [26].

Although reviews are mixed on the level of achievement obtained by the world in relation to the MDGs [11, 13] there are significant benefits for nations, their development agencies, international donors and multi-lateral banks on having agreed and targeted common goals [4, 27]. Among others, the adoption of SDGs by over 170 nations:

- encourages coordination among different actors involved at the national and international development

TABLE 3.1
SDGs target and indicators directly related to LTS. Data from [19]

Goal	Target	Indicator	Relation to LTS
SDG1: No poverty	1.4	1.4.2	Indicator 1.4.2 directly mentions secure tenure rights in adults and vulnerable groups as a key element to achieve SDG1. This indicator covers both urban and rural land
SDG2: Zero hunger	2.3	n/a	Target 2.3 has a direct mentioned to equal access to land, a concept directly related to LTS. However, no indicator is proposed for measuring this access
SDG5: Gender Equality	5.a	5.a.1	This target and corresponding indicator tracks the participation of women in land ownership, measured based on the LTS
		5.a.2	This indicator assesses the legal framework of countries to determine the level of equality related to land for women

- facilitates the development of indicators and targets
- promotes standards for social responsibility actions by the private sector

SDGs are 17 goals that cover most of all human and natural activities on earth. In previous chapters of this book, many SDGs have been explored, particularly those related to land planning, geographic information systems and land productivity. This chapter will focus on SDGs that are directly related to land administration, mainly those goals where land tenure security improvements play a role in achieving it. Table 3.1 presents a summary of SDGs identified to have a direct relationship with LTS. The author acknowledges that other SDGs have an indirect relationship with LTS. This is the case of SDG11 and SDG15 where the use of land as a critical development resource is promoted. However, and considering the desired focus of this chapter of the practical application of results on land reform projects, only those goals, targets and indicators in where land tenure is directly mentioned and measured has been included in this analysis.

The development of the SDGs and its monitoring and reporting has not been free of controversy. This was the case for land tenure related indicators in where norm contestation, evolution and change occurred during the international consultation process conducted by the UN [36].

Additionally, Agarwal (2018) has questioned the efficacy of SDG5 to empower women as the emphasis of the indicator on LTS needs of women did not consider the other relevant economic factors for making lands productive such as access to credits and irrigation schemes [1]. Therefore, in many cases, improvement of LTS for women (SDG5) and other vulnerable groups (SDG1) would need land reform and also other development projects, such as agricultural productivity, employment and better access to financial instruments.

In some cases, practitioners designing land reform projects in line with SDGs would have to expand beyond the three core LTS goals identified in this section as LTS plays a role in many nexus underpinning achieving the SDGs [41].

In summary, SDGs has provided an overarching objective to LTS to ensure land rights are correctly documented and protected by governments in a universal way. Additionally, and to contribute to SDGs, LTS implemented using land reform projects should address gender land gaps, which is a significant problem today [12]. In practical terms, SDGs dictates that gender equality is achieved in numbers (number of women owning land) and land regulatory frameworks that abolish discrimination.

3.4 Land Reform Projects: Achievements and Challenges

Proper land administration, where the cadastre and land registration information systems underpin significant activities related to land, is vital for sustainable development [42]. Without proper land management, policies and actions needed to address emerging challenges are jeopardized. In a well-developed land administration system, decisions by government bodies related to sustainability, such as the use of natural resources, construction and the implementation of environmental policies, are underpinned by information in national spatial data infrastructure (NSDI) capable of allowing information sharing and cross-referencing using coordinates [34].

The importance of investing in better land information systems in developing countries gained strength in the late 80s and 90s as countries realised proper land administration enables better investment by the public and private sectors while creating a platform for disadvantaged social groups, particularly the poor, to participate actively in market economies [31]. Land reform projects, directly related to improving LTS, have been the prime instrument for advancing land administration systems.

Today, there are significant opportunities for land reform projects including new technologies - such as drones and portable devices - for the collection of geographic information [24, 32]. There is also the digitalisation of data and operations of many government functions including those related to land administration [18]. Automation has also offered as an alternative to improve transaction times while reducing corruption and costs [40].

However, there are challenges today for the government, donors, multilateral organisations and other designing and funding land projects. Primarily, there have been mixed results on the effectiveness of the investment [15] and the borrowing capacity of countries along with their ability to generate public funds has created a strong tendency to involve more actively the private sector as a financial partner of land reform projects. This trend is well expressed by the recent policy "maximising funding for development" from the World Bank [2] and follows the same pattern that has occurred already in other sectors - particularly infrastructure development sector such as port and roads - in where funding projects using public-private partnerships is common [28].

Additionally, there is an increased focus of land reform projects on urban areas due to the urbanisation pattern occurring in most countries in the world [35]. This brings new challenges to practitioners as areas with high population density generates additional social problems [43].

The focus on urbanisation is also an opportunity for future land reform project. Urban land reform projects are likely to be conducted in more economically active areas or where a land market is more feasible compared to rural zones. This allows increased possibilities for land administration system users to pay for the services, a situation that creates possible synergies between urban land reform and private participation in running the system under PPPs.

Multiple studies have reviewed approaches conducted by the World Bank and other multilateral organisations for land reform projects [22, 5, 10, 33]. However, after over 30 years of investment in developing countries land administration systems, it is difficult to identify approaches that are likely to work in all context.

In any case, there are some general views in the international community of crucial principles most land reform projects should follow to be successful and contribute to sustainable development.

Acknowledging this space is always changing; the following list presents some principles today for land reform projects that are considered to be in line with the private participation focus of this chapter:

- Fit-for-purpose: improve land tenure rights by fostering innovation in approaches and new technologies in land administration with the objective of optimising resources by investing in the collection and processing of land information in a way that best fits the specific conditions [9].

- Data and process standardisation: standardisation of data topology and processes within a particular jurisdiction to improve information sharing, optimising resources and reduce errors and duplication [20].

- Community mapping and crowdsourcing methodologies: the use of technologies and methods, particularly portable devices and open public participation, to collect information with the community as an active participating party with roles and responsibilities [6].

- Responsible private sector participation: ensuring the private sector (both involved in the provision of land administration services and in the land market) have partnership participation in land reform projects that find common goals with public policies [22].

In summary, land reform projects are a fundamental investment for governments to achieve sustainable development. New technologies, automation and digitalisation of government services is facilitating the construction of better land administration systems. However, the government has limited resources and borrowing capacities to invest in this project are decreasing. The new trend in the land registration and cadastre sector, as it has happened extensively with infrastructure projects, is to generate new funds for land reform projects using PPPs. Although it is difficult to generalise, there are some high-level principles learnt from over 20 years of land reform projects that are important for considering the participation of the private sector in the implementation. These principles are fit for purpose; data and process standardisation; community mapping and responsible private sector participation.

The next chapter reviews land reform projects that have included a significant component of private sector participation and provides a summary of lessons learnt from these experiences.

3.5 Lessons Learnt From Involving the Private Sector in LTS

The private sector could be included in land reform projects in multiple ways. The World Bank and other international institutions. Grave (2015) explained this participation along with a definition of PPPs [14].

This section presents 5 case studies (Switzerland, Canada and Australia in developed countries and the Philippines and India in developing countries) from where proposed strategies where drawn. The presentation of the case studies focuses more on the lessons learnt applicable to future project design. The author acknowledges that these 5 cases studies do not cover all experiences relevant to LTS and the private sector. Other literature has coated more extensive those international experiences of land reform projects with a significant private sector component [39].

3.5.1 Switzerland

For many years, the Swiss cadastre has been recognised as probably one of the most developed systems in the world. However, only now that there is a trend to involve more the private sector in land administration systems, practitioners have started to recognise that one of the main characteristics of the Swiss cadastre is that for many years it has had significant participation of private surveyors. Switzerland is a small country - both in size and population - compared to international standards. However, it has a complex, decentralised government system with 26 cantons (call states or departments in other jurisdictions) and almost 2600 municipalities in a country with four official languages [38].

This complex environment for the land administration systems required the development of methods and protocols that allowed proper communication between entities at all levels while providing an effective and efficient service to users.

To address this complex environment, the federal government developed a scandalised data model and exchange format for geographic information called Interlis. Initial development of the system started back in 1987, and a consolidated operational version for the cadastre began in operation in 1993 with its adoption by the Federal Government [37].

Today, based on Interlis, the Swiss cadastre has two models of private participation in the creation, running and development of the cadastre system. This participation is based on a strict certification process to surveyors, and the delegation is given to individuals. Based on [3], the models of private involvement in the cadastre system are:

- Delegation: On the German part of the country, an exclusivity arrangement for five years is agreed with a surveyor selected using a tender process. The private surveyor has fixed prices for all official land measurement jobs, and she/he is responsible for hosting and transmitting information.
- Competitive market: On French areas, there is an open market and individuals or companies requiring modifications to the cadastre can contract any surveyor on an open market. However, surveyors generally based their prices on a standard list called HO 33. In these areas the maintain of the cadastre could be done directly by the cantons or delegated to individual surveyors based on a competitive tender

The current scheme of data sharing and participation of the private sector in the Swiss cadastre has contributed significantly to Switzerland economic productivity, stability and well-developed land market [17]. Key lessons, perhaps adjustable to other cases, from the Swiss experiences with the private sector in land administration are:

- *The importance of Interoperability:* Interlis, the legally bounding data topology and format for geographic information, has been the enabling force to allow delegating multiple sections of the cadastre system to the private sector, mainly being the data custodian. Additionally, it enabled participation of the private surveyors at different levels and modified based on the needs of the cantons and municipalities

- *The creation of a trusted private sector:* Federal regulation demand strict education and examination of those intending to become private surveyors in Switzerland. This has created the ability of the government to trust surveyors with complex functions and participate actively in the land administration system

3.5.2 Canada

The literature on the privatisation of the land registration in the provinces of Ontario and Manitoba is extensive [30] as these case has been identified as a critical landmark in the privatisation of land registry services. The case of Ontario is analysed next.

In the 80s, the province of Ontario decided to privatised its land registry services with the objective of providing a better service and reducing running cost as the paper-based system in place could not coup properly with an increased demand in land transactions. The slow progress on government reform, something that has been a constant driver for many land system privatisations, created the appropriate political environment for a change for a PPP of the land registry system.

Based on an open request for proposals, the government offered 50% of the ownership of the land registry for an initial period of 15years. The province signed a contract with the preferred bidder that initially was a lease contract and evolved into a concession. The new entity, called Teranet, had the primary task of delivering significant modernisation of the system including the digitalisation of records and ability to process request online. Later in the process, the public share of the concession was sold, and the operation became 100% own by the private sector.

The primary financial arrangement of the concession was based on a "users pay approach" with a clear return for the private investor and revenue collection for the government. In 2010, after a significant financial and technical success of the first concession, Teranet paid $1 billion for continuing the concession for an additional 50 years.

Key lessons learnt from privatisations of the land registry in Canada, which has been in operation for more than 30 years, are:

- The government managed to generate interest in the private sector in what was an unknown territory by offering to share risk from the beginning of the project. This created the possibilities during the first five years of stabilisation of the project, including reforming the law and adjusting the financial scheme to generate value for money to the government and profit opportunities for the private sector

- An explicit policy of the data ownership allowed for innovation by the private operator while maintaining certainty to the government. In Ontario,

Teranet has been able to develop additional income sources by using the land registry data beyond the traditional scope of LTS.

3.5.3 Australia

Australia is one of the most recent examples of privatisations of land registry. The government of New South Wales (NSW), Victoria and South Australia have completed PPPs of their land registries and are in their initial operation phases. The state of Western Australia is also in the process of completing its concession.

The main characteristic of PPPs in Australia is the focus of the government on selling an asset in the same way back in the 80s and 90s when across the developed world public infrastructure assets (such as ports, public transport and roads) where concession to private operators.

The process has been very similar in all jurisdictions in Australia. An initial scoping study is followed by a legal reform that then enables a tendering process where the asset is sold primarily to the higher bidder.

Even though the process has resulted in financial results that exceeded initial government estimates, the concession in Australia has not been free of controversy. Notably, there have been questioning from professional associations on the financial return focus, and the fact that the resources received have not been invested back on LTS.

In any case, there are significant lessons learnt in Australia that could be used in other jurisdictions. These are:

- The private operator is assuming substantial financial risk as its revenue depends on fees paid by those conducting transactions in the land market. This is expected to create the appropriate incentive to reduce costs and foster innovation, including automation of some transactions which has been proposed for the NSW operator and it is supposed to be included in the other two concessions

- Even though significant risks have been delegated to the private sector, the underpinning principle of a state guarantee for land transactions remains intact. In other words, in Australia today the ultimate responsibility for guaranteeing land transactions still is the government, creating that a vital principle of the torrent system is not being modified.

3.5.4 Philippines

Back in 2007, the land titling computerisation project (LTCP) was implemented in the Philippines [21]. This project was underpinned by two consecutively large-scale systematic registration projects conducted in the country since 2001 [16]. LTCP has as objectives tackle high land registration costs, corruption, slow government reform and low level of taxation. LTCP was built under a PPP model of a build-Own-Operate model. To partly fund this concession, the international finance corporation (IFC) participated in the initial financial arrangement.

Eleazar et al. (2013) evaluated the results of this concession and highlighted some of the problems including high transaction costs that limit the ability of the community, particularly the poor, to access the land titling service [7]. In any case, there significant lessons that could be learnt from what is described as the first land registration PPP in a developing country:

- LTCP run in parallel with the second phase of the land titling project funded by the World Bank. This allowed the concessionaire to explore synergies from the public investment in LTS and, at the same time, generated a sense of sustainability of land reform projects, something fundamental for the country
- The design of the concession highlighted the need to define clear public objectives for a concession beside economic efficiencies and other public sector problems such as corruption. In particular, the Philippines LTCP project is an essential example of challenges to vulnerable groups when market-driven approach as implemented in land administration services

3.5.5 India

The case of Karnataka province in India demonstrates the application of PPP approaches to multiple land administration services at the same time. The project, called Bhoomi, was initially conceived as a pilot project and later evolved as a PPP. In this project three land administration services are a concession to a private operator [25]:

- Digitalisation of records, both new and existing
- Registration of crops so loans and other government services can be provided
- Online transactions

An important objective of the project was to reduce corruption as both LTS and crop registry generated significant economic activity in the region. The digitalisation approach aimed at simplifying the process and securing the data while improving data accessibility and building a robust database that allows cross-referencing with other data sources. Although the results are mixed [25] Bhoomi project is considered a significant experience for the participation of the private sector in LTS and other government functions. Key lessons, relevant to the focus of this chapter, include:

- The success of the Bhoomi project was attributed in many cases to the combination of LTS and crop registry services. These two services needed each other to exist as LTS define the land rights for agricultural activities and the crop registry a market motivation and financial resources for users to request LTS in their land
- There has been a significant in corruption from the project demonstrating that not only financial benefits but the protection of land rights to the most vulnerable, could be achieved using the appropriate mechanism to involve the private sector

3.6 Strategies to Align Private Participation in Land Tenure Reform Projects With SDGs

Although there is significant literature exploring the involvement of the private sector in LTS projects [25], there is limited academic or professional material analysing this private participation and its contribution to SDGs. Meadows, Fairlie et al. (2018) developed an analysis or pre-requisite for the development of PPPs in the land administration sector and Endo, Triveno et al. (2018)analysed the new roles the private sector is playing in land reform projects based on experiences in Latin-American [23, 8].

Based on [29] conceptual design for assessing the viability of PPPs in the land sector, a multi-criteria analysis was conducted to identify areas where strategies were developed. The following table 3.2 shows this multi-criteria analysis. In this table, the first column presents the identified characteristics required for a land PPP to be viable. The second column identifies the relationship with one of the LTS related SDGs discussed previously in this chapter. Finally, a relationship with lessons learnt from previous experience, explore in the previous sections, is included in the last column.

Based on the multi-criteria, gaps on needed strategies were identified, particularly the fact that gender equality has not been correctly addressing as a key condition for PPP viability. Based on these gaps and using the analyses from the previous sections, the following strategies are proposed for designers and implementers of projects where the involvement of the private sector is expected to be significant.

Strategy 1: Involve the private sector in the area where it can best perform

Based on the cases in Switzerland and Australia in where the private participation is segmented to where it best suits the local and national needs, land reform project designers should consider involving the private sector only in segments or areas where it can better help the government in achieving SDGs.

TABLE 3.2
Multi-criteria analysis to identify

Viability condition for a land PPP [29]	**Related SDGs**	**Related case study lesson learnt (previous section)**
State or another guarantee to transactions	SDG2	Australia
Clear definition of properties	SDG1, SDG2	Multiple
Legal delegation to the private sector	SDG2	Australia
The existence of a regulatory authority	SDG2	Australia
Avenues to resolve disputes	SDG1, SDG2	Canada
Process fees and responsibilities for all transactions are well defined	SDG2	India
The regulatory framework considers fit-for-purpose	SDG2	India
Secured and accessible land records and information about the land system	SDG1, SDG2	Switzerland
Mechanisms to handle complains	SDG2	India
Revenue from transactions is clear	SDG2	Philippines, India
There is a strong private sector operating PPPs in other sectors	SDG2	Philippines, Australia, Canada
Risks and reward for the private investor can be clearly defined	SDG1, SDG2	Switzerland
Funds are available to structure a land PPP	SDG1, SDG2	Australia

These areas of segmentation could be either geographical or institutional. In Switzerland, the roles of surveyors vary depending on the canton while in Australia (both in NSW and Victoria) the privatisation occurred on the land registry and not the cadastre.

When doing this segmentation, project designers could better identify if which of the LTS related SDGs could be better addressed. In many cases, there are opportunities to build packages of providing highly profitable areas with those that might not raise revenue but could table better gender equality (SDG5) and title for the poor (SDG1).

Strategy 2: Cross-service approaches

One of the most significant advantages of the Karnataka in India is the delegation of multiple land administration services into one PPP. Designing cross-service approaches in land PPP and other projects could better allow supporting the achievement of SDGs. In particular, joining the land tax collection service with land registry and cadastre services opens the opportunity for the private operator to invest in low-income areas (SDG1) with the prospectus of generating wealth in the long term and improve tax collection. Similarly, like in Karnataka, titling activities could be directly related to other fees such as crop registration, building and development permits and forest licenses, creating additional sources of income to expand LTS.

Strategy 3: Work with the trusted part of the private sector

Success in Switzerland of its delegation of cadastre services to the private sector is underpinned by the trust that exists on the surveyor's registration system. Similarly, privatisations in Australia are considering very successful from a financial point of view because the government has created mechanisms to trust private investment banks and pension funds as owners of land administration services. Both developed countries are currently achieving this trust in the private sector SDG1 and SDG2.

Therefore, as a mechanism to foster achieving SDG1 and SDG2, project designer should consider delegating the land administration services on those segments of the private sector that are well trusted by society. In some cases, it might be the financial sector, the insurance sector, certain professions (lawyers, surveyors) or private associations (such as the chambers of commerce).

Strategy 4: Result-based revenue

A key opportunity politicians have is to define the objective of a project they are willing to support which in most cases should be aligned to the SDGs. Therefore, project designers should ensure that revenue received by the private sector is achieved when these objectives are met. The case in the Philippine demonstrates a situation where this strategy was not used and cause the opposite effect: revenue conditions limit achieving better LTS.

Result-based revenue would be a particularly useful strategy for projects focusing on SDG5 as gender equality is difficult to encourage to a private operator. However, for example, extra revenue could be negotiated with a private operator of a land system to those titling registration that includes women, encouraging this private operator to invest and prioritize equal land ownership in the areas covered by the project.

Other sectors, including water and education, have used this approach (Fritsche, Soeters et al. 2014)and is also referenced in the literature as results-based financing.

3.7 Conclusions

It is not hard to find agreement between researchers and practitioners in need for more investment in land reform project as a mechanism to foster the achievement of SDGs. Although SDGs could be seen as generic and covering all aspects of a nation, there are 3 SDGs (SDG1, SDG2 and SDG5) that directly address the need for better land administration services.

Land reform projects are facing both challenges and opportunities. To address the problem of financial resources and to build from the innovation opportunities, there is a current trend that suggests the involvement of the private sector could potentially foster the development of land reform projects.

Examples around the world where the private sector plays a significant role in running land administration systems have provided valuable lessons. Notably, all future projects need for government to have the right financial, administrative, legal and regulatory framework in place to ensure private participation under a partnership arrangement and not just transfer the problem.

Four strategies are proposed for practitioners and government officials involved in the design and implementation of land reform projects with significant private sector participation:

- Involve the private sector in the segment or geographic areas where it can best the SDGs
- Bundle land administration services with LTS to ensure there are more opportunities to invest in land rights
- Work with the trusted segment of the private sector
- Private sector revenue should be linked to the SDGs that are being addressed

Even though it is challenging to make the sector a decisive contributing factor of the SDGs for the provision of land administration services, there are lessons learnt and strategies that could make this happen. However, the four strategies presented in this chapter address timidly SDG5 (tender equality) as there are limited experiences, particularly in the developing world, where the private sector participation has contributed to women rights. The need to develop strategies to better involve the private sector in gender equality is suggested as an area for further research.

Bibliography

[1] Bina Agarwal. Gender equality, food security and the sustainable development goals. *Current opinion in environmental sustainability*, 34:26–32, 2018.

[2] World Bank. Maximising fiance for development, 2018.

[3] Barbieri. The swiss case: 100 years of ppps., 2016.

[4] Geeta Batra, Juha Ilari Uitto, and Lee Cando-Noordhuizen. From mdgs to sdgs: Evaluating global environmental benefits. *Evaluation Matters*, 1(3):16–23, 2015.

[5] Keith Clifford Bell. Trends in land administration and management with particular reference to world bank support for projects in the east asia region. 2016.

[6] Melissa DeJonckheere, Robin Lindquist-Grantz, Sinem Toraman, Kristen Haddad, and Lisa M Vaughn. Intersection of mixed methods and community-based participatory research: A methodological review. *Journal of Mixed Methods Research*, page 1558689818778469, 2018.

[7] Floradema C Eleazar, Brian Garcia, Ernie Guiang, Annabelle Herrera, Lina D Isorena, Roel Ravanera, and Ernesto Serote. Improving land sector governance in the philippines, 2013.

[8] L. Triveno Endo, V. and E. Pantoja. Contracting out services for land regularization. the new roles of the private sector in land administration, 2018.

[9] S Enemark, KC Bell, Ch Lemmen, and R McLaren. Fit-for-purpose land administration. international federation of surveyors (fig), 2014.

[10] Julio Faúndez. *Good government and law: Legal and institutional reform in developing countries*. Springer, 2016.

[11] Maya Fehling, Brett D Nelson, and Sridhar Venkatapuram. Limitations of the millennium development goals: a literature review. *Global public health*, 8(10):1109–1122, 2013.

[12] Brendan Fisher and Robin Naidoo. The geography of gender inequality. *PloS one*, 11(3):e0145778, 2016.

[13] Sakiko Fukuda-Parr, Alicia Ely Yamin, and Joshua Greenstein. The power of numbers: a critical review of millennium development goal targets for human development and human rights. *Journal of Human Development and Capabilities*, 15(2-3):105–117, 2014.

[14] Carsten Greve. International public-private partnership policies. convergence in themes from asean, the european union, imf, oecd, the un, and the world bank. In *Fourth International PPP symposium, New York, USA, September*, 2015.

[15] Daniel Higgins, Tim Balint, Harold Liversage, and Paul Winters. Investigating the impacts of increased rural land tenure security: A systematic review of the evidence. *Journal of rural studies*, 61:34–62, 2018.

[16] Philip Hirsch. Titling against grabbing? critiques and conundrums around land formalisation in southeast asia. In *International Conference on Global Land Grabbing*, volume 18. Citeseer, 2011.

[17] Lorenz Jenni, LA Guarín, Stefan Ziegler, and BVM Pérez. Development and employment of a ladm implementing toolkit in colombia. In *Proceedings of the 2017 World Bank Conference on Land and Poverty: Responsible Land Governance–Towards an Evidence-Based Approach, The World Bank, Washington, DC, USA*, pages 20–24, 2017.

[18] Georgios N Kouziokas. Geospatial based information system development in public administration for sustainable development and planning in urban environment. *European Journal of Sustainable Development*, 5(4):347–352, 2016.

[19] LandPortal. Land and the sustainable development goals (sdgs), 2019.

[20] Christiaan Lemmen, Peter Van Oosterom, and Rohan Bennett. The land administration domain model. *Land use policy*, 49:535–545, 2015.

[21] Joanna Reyes Mankarios. Land registration and tenure structure in slum areas in the city of manila, philippines. Master's thesis, Norwegian University of Life Sciences, Ås, 2015.

[22] Philip McMichael. The land question in the food sovereignty project. *Globalizations*, 12(4):434–451, 2015.

[23] K. Fairlie P. D Meadows, J. and B. T. An assessment of ppps in land administration: Development of a set of pre-requisites for effective ppp implementation, 2018.

[24] Mihnea Mihailescu, Diana Irimia, Razvan Radulescu, and Razvan Simion. Could we cadaster faster in an integrated it system by using uavs with gis services in a cloud infrastructure. *US-China L. Rev.*, 14:377, 2017.

[25] Prachee Mishra and Roopal Suhag. Land records and titles in india. *PRS Legislative Research*, 2017.

[26] Nam. Transforming our world: The 2030 agenda for sustainable development, 2015.

[27] Robert Newell, Ann Dale, and Mark Roseland. Climate action co-benefits and integrated community planning: Uncovering the synergies and trade-offs. *International Journal of Climate Change: Impacts & Responses*, 10(4), 2018.

[28] Robert Osei-Kyei and Albert PC Chan. Review of studies on the critical success factors for public–private partnership (ppp) projects from 1990 to 2013. *International Journal of Project Management*, 33(6):1335–1346, 2015.

[29] D. Paez and T. Burns. Conceptual design of a private investment scorecard for land administration in developing countries, 2018.

[30] Shiri Pasternak. How capitalism will save colonialism: The privatization of reserve lands in canada. *Antipode*, 47(1):179–196, 2015.

[31] Geoffrey Payne. Urban land tenure policy options: Titles or rights? *Habitat international*, 25(3):415–429, 2001.

[32] Shima Rahmatizadeh, Abbas Rajabifard, Mohsen Kalantari, and Serene Ho. A framework for selecting a fit-for-purpose data collection method in land administration. *Land Use Policy*, 70:162–171, 2018.

[33] Laura Randall. *Reforming Mexico's agrarian reform.* Routledge, 2016.

[34] Greg Scott and Abbas Rajabifard. Sustainable development and geospatial information: a strategic framework for integrating a global policy agenda into national geospatial capabilities. *Geo-spatial Information Science*, 20(2):59–76, 2017.

[35] Mozart Vitor Serra, David E Dowall, Diana Motta, and Michael Donovan. Urban land markets and urban land development: An examination of three brazilian cities: Brasília, curitiba and recife. 2015.

[36] Noha Shawki. Norms and normative change in world politics: an analysis of land rights and the sustainable development goals. *Global Change, Peace & Security*, 28(3):249–269, 2016.

[37] Daniel Steudler. Swiss cadastral core data model—experiences of the last 15 years. *Computers, environment and urban systems*, 30(5):600–613, 2006.

[38] Daniel Steudler. El sistema suizo de administracion de tierras. *Revista de Ingeniería*, (46):16–23, 2018.

[39] Rod Thomas, Lynden Griggs, and Rouhshi Low. Big data and privatisation of registers–recent developments and thoughts from a torrens perspective. 2018.

[40] Rod Thomas, Rouhshi Low, and Lynden Griggs. Designing an automated torrens system—baseline criteria, risks and possible outcomes. *New Zealand Law Review*, 2015(3):425–453, 2015.

[41] Joleen Timko, Philippe Le Billon, Hisham Zerriffi, Jordi Honey-Rosés, Ian de la Roche, Chris Gaston, Terry CH Sunderland, and Rob A Kozak. A policy nexus approach to forests and the sdgs: Tradeoffs and synergies. *Current opinion in environmental sustainability*, 34:7–12, 2018.

[42] Ian Williamson, Stig Enemark, Jude Wallace, and Abbas Rajabifard. *Land administration for sustainable development.* Citeseer, 2010.

[43] Jieming Zhu and Hendricus Andy Simarmata. Formal land rights versus informal land rights: Governance for sustainable urbanization in the jakarta metropolitan region, indonesia. *Land Use Policy*, 43:63–73, 2015.

4

Spatially Enabling the SDGs: The Social, Economic, and Environmental Impacts of Spatial Enablement

Maryam Rabiee
Visitor, Centre for SDIs and Land Administration, The University of Melbourne, Australia

This chapter aims to present the significance of spatially enabling the SDGs and the opportunities it provides for the seventeen goals.

4.1 Introduction

Achieving the Sustainable Development Goals (SDGs) is contingent on a holistic approach that aligns social, economic, and environmental objectives at local, national, and global levels. The significance of the 'where' component of the SDGs is, however, often underemphasized. The Global Goals require a transdisciplinary effort to integrate the geospatial aspect into planning and implementation phases, and UNHCR and ESRI's collaborative initiative, *Rohingya Refugee Emergency at a Glance* is a case in point that utilizes geospatial information and technologies to address various dimensions of sustainable development in parallel with changing circumstances. The interactive platform locates health services, disaster risks, shelter congestion, and water accessibility to sustainably aid Rohingya refugees in Bangladesh [29]. This example demonstrates how integrating spatial enablement in SDGs-related initiatives can create resilient and sustainable circumstances for everyone in every location.

The SDGs have emerged in a world where social, economic, and environmental complexities are intertwined across different geographic locations and at a time when technological advancements encourage the proliferation of real-time and location-based information. Spatially enabled societies, governments, and people can offer a wide range of solutions for complex challenges [23], however how it connects the multidimensional SDGs and its social implications for a comprehensive approach requires further research. The chapter begins by examining the meaning of spatial enablement in the context of the SDGs and is followed by a discussion on how spatial enablement can narrow the SDG connectivity gap. It then investigates the social impact of spatial enablement for the SDGs and ends with a discussion on land, the driving force of spatial enablement for the SDGs.

4.2 Spatially Enabling the SDGs

4.2.1 What Does It Mean to Be Spatially Enabled?

Social, economic, and environmental developments are geospatial processes, because "everything happens somewhere"[1]. Moreover, location is "the fourth element of decision-making" [23], therefore recognizing and utilizing geospatial information to localize solutions is essential to progress. This geospatial need prompted the emergence of 'spatial enablement', a notion to facilitate localization by cohesively engaging technologies, people, and institutions. In broad terms, to be spatially enabled is the ability to use geospatial information and technology to enhance the interactions within that space [8]. In technical terms, spatial enablement refers to geographical identifiers–e.g. geographically tagging records in a database [7]. The term generally entails the use of geospatial information and technologies to ameliorate the social conditions. However, geospatial experts emphasize that spatial enablement is not merely a technical matter, it is an approach that is concerned with the whole of government and society [32]. To this end, spatial information and technologies serve to spatially enable two main entities: society and government.

A spatially enabled government works towards establishing infrastructures that organize and share spatial information required for decision-making and policies, government services, business transactions and community activities [33]. Integrating spatial data infrastructures (SDIs) and better land administration and management (i.e. land governance, property rights, and land policy) allows governments to utilize spatial information to address social, economic, and environmental challenges adequately [30]. In essence, a spatially enabled government is a facilitator for interactions between organizations, technologies, and people by means of a common language using spatial concepts and technologies, and spatial information management processes

[1]This statement is credited to Nancy Tosta in an interview with the Computer World news, GIS: More Than Just A Map. Retrieved from https://www.computerworld.com/article/2582595/gis–more-than-just-a-map.html

[7, 9]. Therefore, spatially enabled governments utilize geospatial knowledge and technologies across different domains and sectors to fulfill societal needs.

Similarly, a society is spatially enabled when spatial information, a common good that prompts creativity, efficiency and product development, is available to governments, citizens and businesses to organize and plan their activities [18, 33]. The International federation of Surveyors (FIG) report, *Spatially Enabled Society* [23], defined the term in the context of availability, accessibility, and usability of spatial information. First, spatial information and services must be available to governments, businesses, and citizens in a free, efficient, and comprehensive manner. Second, tools for spatial information sharing, analysis, and management must be accessible to all sectors of society. Third, spatial information must be used to transparently organize economic, legal, environmental, land, and social activities, and contribute to informed decision-making. The ultimate objective of is to provide value-added services and reinforce sustainable development through six fundamental elements: legal frameworks, sound data integration concepts, positioning infrastructures, SDIs, land ownership information, and data and information.

Spatially enabled societies encourage the collection and processing of spatial information at all levels of society to deliver sustainable development objectives [31]. Land administration systems and SDIs use the collected data to underpin evidence-based decision-making to implement sustainable policies and practices [4, 17]. Therefore, the integration of spatial information and technology in sustainable development can facilitate coherent governance, ensures coverage of more geographic territories, and engaging a larger number of citizens in determining the future of their society. While geospatial information offers of social, economic, and environmental benefits, most countries lack the capacity to manage and share geospatial information and systems to advance sustainable development goals [6]. In spite of limitations in geospatial resources and skills, the SDGs generated a new wave of geospatial awareness at a global level.

4.2.2 Sustainable Development Goals in an Interconnected World

In recent decades, the global community has become interconnected across different geographies, scales, and sectors, driving us into the age of sustainable development [19]. The universality and complexity of our challenges brought together the largest international gathering at the UN Conference on Sustainable Development (Rio+20) in 2012. The Rio+20 outcome document declared that sustainable development goals should be action-oriented, consider different national realities and capacities, and utilize geospatial information for policymaking, programming, and project operations [10]. As a result, the SDGs were put into effect in 2015 to mobilize a transformative course of action towards a sustainable future for people and the planet. The 2030 Agenda requires data-driven action, evidence-based analysis, implementation, monitoring [20] and capacity building that is linked across different sectors [22].

Geospatial analysis, mapping, and modeling, geopolitical policy, and integrative frameworks can fortify networks of capacity building and decision-making practices needed for sustainable development issues [25]. In other words, spatial enablement offers the knowledge and tools needed to spatially connect stakeholders, policy makers, and people to sustainability challenges, resources, and solutions. Thus, integrating geospatial information in the global agenda promotes a holistic approach to measuring and monitoring the SDGs. The collective success of the global community depends on a holistic framework and transdisciplinary action plan that spatially enables the SDGs to deliver social well-being, economic growth, environmental vitality, and good governance.

4.2.3 Integrating Spatial Enablement Into the SDG Framework

The 2030 Agenda is one of the first global frameworks that recognizes the role of geospatial information in sustainable development [21]. The 2016 SDGs Report noted that geospatial information is important for the production of some indicators [12] and falls short of providing any details. The 2017 SDG Report reiterated the same statement without providing further information [13]. Progress was made in the 2018 SDG Report, it called for open and transparent access to integrated geospatial and statistical data, collaboration on the production and dissemination of geospatial and statistical data, and visualizing SDG statistical data within a geospatial context [14]. While the importance of geospatial information is recognized in the framework, there are missing links between spatial enablement and the SDGs.

The Inter-agency and Expert Group on Sustainable Development Goal Indicators (IAEG-SDGs) Working Group on Geospatial Information has examined the SDGs with a "geographic location lens" and proposed that geospatial information has direct contribution to fifteen indicators and supporting contribution to eight indicators [27]. The list of indicators does not, however, encompass all indicators that are geospatially relevant. Target 3.3, for example, intends to end the epidemics of communicable diseases, and geospatial data is necessary for tracking and analyzing disease outbreaks [1]. This indicates that the role of spatial enablement in social, economic, and environmental phenomena has not been comprehensively researched. Therefore, this chapter proposes spatial enablement creates three key opportunities for the SDGs:

Inclusive coverage of SDGs. 'Leaving no one behind' is the key pillar of the SDGs, which requires the integration of where people are and where events occur with demographic data in order to provide geospatial services that are accessible and consistent [24]. Twenty-one of the SDG targets intend to increase access to basic services, land, housing, and other resources, all of which, require spatial data (remote sensing data, satellite data, etc.) to ensure that 'accessibility' is for everyone in all localities.

Participatory implementation of the SDGs. A local-to-global outlook towards the SDGs can incorporate more people in localizing action plans. Spatial enablement and literacy put all members of society in contact with geospatial information and services, consequently engaging more people and data in reaching the SDGs [21]. Citizens are producers and users of spatial data, however, if spatially enabled, they can utilize spatial information to devise solutions and contextualized implementation methods at local, national and global levels.

Comprehensive monitoring of the SDGs. Measuring and monitoring socio-economic SDG indicators, specifically land-related indicators, require spatially enabled datasets (data collected through surveys, crowdsourcing, and censuses) to achieve the SDGs at local and national levels [3]. Statistical and geospatial data are also critical for comprehensive decision-making across different scales of governance [15]. Monitoring the SDGs requires cross-national, cross-sectoral, and cross-scale monitoring, which can be accelerated by incorporating spatial data and technologies.

Spatially enabling the SDGs is the process of uniting the various features of spatial enablement with the SDG framework by employing spatial information, technologies, and services as a common force to confront social, economic, and environmental challenges everywhere. However, the SDGs will not reach everyone if there is a gap in connecting the social, economic, and environmental aspects of challenges, stakeholders and societal actors, resources, and technologies across local, national, and global scales.

4.3 Narrowing the Connectivity and Spatial Gap

There are clear overlaps between spatial enablement and the SDGs: engaging all members of society, addressing the social, economic, and environmental layers of problems, using spatial information to locate events, implement solutions, and monitor progress. However, the SDGs are often examined in silos, overlooking the social, economic, and environmental interlinkages embedded in each SDG [2]. Drawing from the main entities of spatial enablement identified in [7] and the sustainable development data flow outlined in [21], Figure 4.1 illustrates the interlinkages of spatially enabled SDGs. Incorporating these links in research or implementation strategies can narrow the connectivity gap by promoting transdisciplinary planning, cross-sectoral mobilization, and transnational partnerships:

Transdisciplinary planning: the SDGs cover a wide range of issues pertaining to social, economic, environmental, political, technological, and other areas of development. Incorporating the middle layer of Figure 4.1 includ-

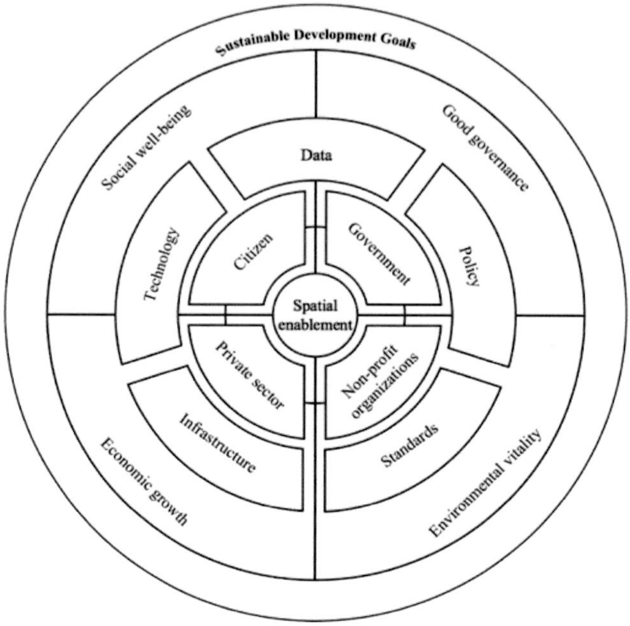

FIGURE 4.1
Integration of spatial enablement in the SDGs

ing data, technologies and systems, infrastructures and platforms, policy, and standards connects people with the tools, methods, and strategies needed to tackle the multidimensional challenges of the SDGs. This connection is created across different scales, sectors, and domains, enabling the continuous and consistent path through transdisciplinary research and action plans.

Cross-sectoral mobilization: the SDGs require all members of society with different socio-political and socio-economic responsibilities to be involved in achieving the goals. Spatially enabling the SDGs engages citizens and breaks the government, private sector, academic, and non-profit silos to facilitate cross-sectoral cooperation to comprehensively address the social, economic, environmental, and governance layers of the SDGs. Therefore, mobilizing sectors and members of society in spatial enablement and sustainable development (inner layer of Figure 4.1) can result in more efficient and effective efforts towards meeting the SDGs.

Transnational partnership and collaboration: poverty, clean energy, water and sanitation, equality and justice, education and other social, economic, and environmental problems in the 2030 Agenda are present, to different degrees, in developed, developing, and underdeveloped nations (outer layer of Figure 4.1). Financial markets rivers, forests, and other natural recourses transcend national borders. Therefore, to fulfill all the SDGs for everyone in all localities requires multinational partnership and collaboration to muster resources, technologies, and people needed to achieve the goals in all nations.

The integration of spatial enablement in the SDGs results in forming connections between the human, technical, and sustainability layers of society order to contextualize all dimensions of a Goal in its local, national, and global circumstances. One of the implications of narrowing the connectivity gap is understanding that the interlinkages within the SDG framework reveal how people and the social aspect are at the core of spatial enablement and sustainable development, and technological aspect is a tool that facilitates progress.

4.4 The Social Impact of Spatially Enabling the SDGs

The SDGs hold different social, economic, and environmental weight and can benefit from spatial enablement in different capacities. In this framing, UNGGIM and the World Bank, 2018 have identified some of the social, economic, and environmental benefits of integrating a geospatial information framework. Building on their analysis, Table 4.1 explains how spatial enablement can socially impact the SDGs and strengthen community resilience.

TABLE 4.1: Social impact of spatial enablement characteristics on the SDGs

SDG	Social impact of spatial enablement
Goal 1	Poverty mapping and engaging citizens in the process can help stakeholders alleviate poverty. Moreover, land tenure security and knowledge of land rights can affect income levels; land is a source of income, food, and shelter.
Goal 2	SDIs can manage and share information about food resources and agriculture to reduce food insecurity among communities and improve agricultural production. Land tenure security improves accessibility to agricultural land as well.
Goal 3	Spatial assessment of disease outbreaks can help governments and international organizations track and prevent the spread of communicable diseases. Spatial platforms can also increase access to medical facilities and supplies.
Goal 4	Integrating spatial literacy and awareness in school curricula can step up citizen involvement in decision-making and implementation phases, in addition to increasing knowledge of land rights, particularly for vulnerable members of society.

Goal 5	Mapping gender inequalities and social phenomena that exacerbate them can support policies that enrich women's welfare. Spatial literacy and knowledge of land rights can empower women and reduce the adverse effects of economic and environmental shocks.
Goal 6	Water management requires geo-referenced information, particularly for location-specific decision-making. Locating transboundary water ecosystems and land management also effects community access to clean water and water resources.
Goal 7	Spatial assessment of resources can identify regions that lack access to clean energy and locate different energy resources. Lack of land tenure can also affect the development of energy infrastructures and access to energy.
Goal 8	Analyzing the spatial distribution of economic activities, gaps, and needs can raise employment rates. Moreover, utilizing geospatial services to enhance access to social services and land information can increase economic growth.
Goal 9	Spatial data can advance soft and hard infrastructures, including SDIs, by connecting transport, energy, financial, health, educational and other services to communities. Secure land tenure will allow the efficient development of physical infrastructures.
Goal 10	Knowledge of the spatial distribution of socio-economic inequalities can contribute to forming laws, policies, and practices that promote equality and equal access to rights, services, and resources.
Goal 11	Authoritative and citizen-generated spatial data can reduce crime rates, identify safe public transport stops, upgrade and transform informal settlements. Securing land tenure can also foster the development of inclusive urban infrastructures and housing.
Goal 12	The spatial tracking of waste management and efficient use of natural resources can reduce the release of contaminants, pave the way for a safer ecosystem and establish sustainable land use patterns for energy and food production.
Goal 13	Disaster risk models, spatial data sharing systems and land tenure can minimize climate impact, accelerated disaster response and early warning systems, managing safe migration, upgrade hazard mapping, and reduce socio-economic vulnerabilities.
Goal 14	Spatial data on fisheries and natural resources can assist with halting illegal marine activities and boost tourism. In addition, information regarding land-based activities can affect marine ecosystems.

Goal 15	Earth observation and geospatial data can map, measure, and monitor land use and natural resource management to improve the conservation of terrestrial ecosystems, which reduces economic loss and secure social and ecological sustainability.
Goal 16	Institutions across different sectors and scales can elevate inclusivity by reforming land administration and land rights, geospatial infrastructures, systems, standards, and policies.
Goal 17	The development of geospatial technologies, availability of spatial data for citizens, industry, and governments, and integration of geospatial frameworks in the SDGs requires cooperation at all levels of government and multi-stakeholder partnerships.

Many of the social elements of the SDG framework in Table 4.1 are connected to land and land administration. Inclusive land management and administration is a requirement for sustainable development [23]. Land-related information, therefore, makes spatial enablement at the government level more inclusive, while empowering communities by creating social equity and economic opportunities [32]. With 68 percent of the world's population living in urban areas by 2050 [14], land information is essential for SDG 11, which tackles both cities and urban communities facing poverty, disaster shocks, housing crisis, and other inequalities. Therefore, land ownership information is fundamental to spatial enablement which can sustainably manage people's relationship to land.

4.5 Land: The Driving Force of Spatial Enablement for the SDGs

Rising urbanization, natural disasters, increasing inequalities, and poverty are some of the pressing global risk of our time [5], which cause severe setbacks for urban communities, land use, and land related activities. Land ownership information, one of the six elements of spatially enabled society, can significantly strengthen the resilience of urban communities by managing and monitoring the multidimension effects of land on urban growth. To foster social cohesion in urban settings, land ownership information significantly contributes to dealing with SDG 1, SDG 5, and SDG 13:

Poverty alleviation (SDG 1): effective documentation of land rights and parcel information can be used to manage land disputes, policy and decision making, and support formal land markets that provide the means to reduce poverty [26]. Community members can get involved in these practices by providing volunteered information where authoritative data is not available [16]. Subsequently, land ownership directly affects income levels and economic growth [11].

Gender equality (SDG 5): ownership and control over land can empower women by reducing dependency on men for financial assistance and increase opportunities for economic and social activities [28]. Registration systems make secure land rights and ownership possible, which in turn facilitates equal access to economic resources, health services, inheritance and other socio-economic advantages. Therefore, equitable access to land, housing, and basic services through land administration functions such as securing and transferring land rights make sustainable urban development more inclusive [26].

Disaster risk reduction (SDG 13): with more frequent and intense disasters, land ownership information is of utmost importance; disasters can destroy land and land records, kill title-holders and erase physical land boundaries. Community participatory methods and satellite imagery can indicate ownership and property location after the occurrence of a disaster [34], which strengthens community resilience to disasters. Therefore, secure land rights can protect community access to shelter, food and other services, when people are most vulnerable after a disaster.

Impediments to even urbanization and secure land rights can have long-term effects that transcend time and location. Land information can strengthen the resilience of our ecosystems by facilitating infrastructure development, cohesive decision-making on best policies, the establishment of just institutions, and disaster prevention strategies that meet community needs. To fulfil these needs by 2030 for everyone, we need to have a comprehensive understanding of the complexity and links between social, economic, environmental, political, spatial, and cultural layers of present and future needs.

4.6 Conclusion

The disturbances inflicted by natural and human-induced events affect the interactions between social, economic, environmental, and governance sectors. Integrating a geospatial framework in sustainable development can reinforce the inclusivity principles of the SDGs by locating, monitoring and overcoming the challenges we face. Further research is needed to identify localized spatial applications and their impact on each of the SDGs. However, realizing the role of spatial enablement in the 2030 Agenda and understanding the impact of spatial enablers such as land administration and management will enable a holistic approach to effective, efficient and innovative solutions for delivering the SDGs.

Bibliography

[1] Dominique Briand, Emmanuel Roux, Jean Christophe Desconnets, Carmen Gervet, and Christovam Barcellos. From global action against malaria to local issues: state of the art and perspectives of web platforms dealing with malaria information. *Malaria journal*, 17(1):122, 2018.

[2] CDMPS. *A blueprint for disaster management RD&D supporting the sustainable development goals.* Centre for Disaster Management and Public Safety, the University of Melbourne, 2018.

[3] Pranab Ranjan Choudhury, Marcello De Maria, and Laura Meggiolaro. Strengthening gis standards to improve monitoring of land indicators for sdgs: Using india as a use case. In *World Bank Land and Poverrt Conference, Washington DC*, 2018.

[4] Stig Enemark. Spatial enablement and the response to climate change and the millennium development goals. In *Proceedings of the 18th UN Regional Cartographic Conference for Asia and the Pacific, Bangkok*, pages 26–30, 2009.

[5] World Economic Forum. The global risks report 2019, 14th edition, 2019.

[6] UN GGIM and World Bank. A strategic guide to develop and strengthen national geospatial information management, 2018.

[7] Peter Holland[1], Abbas Rajabifard, and Ian Williamson. Understanding spatial enablement of government. 2009.

[8] Michel Lussault. *De la lutte des classes à la lutte des places.* Grasset Paris, 2009.

[9] Ian Masser, Abbas Rajabifard, and Ian Williamson. Spatially enabling governments through sdi implementation. *International Journal of Geographical Information Science*, 22(1):5–20, 2008.

[10] United Nations. The future we want. *United Nations General Assembly. Resolution 66/288*, 2012.

[11] United Nations. Global sustainable development report 2016, 2016.

[12] United Nations. *The Sustainable Development Goals Report 2016.* United Nations, New York 2016, 2016.

[13] United Nations. *The Sustainable Development Goals Report 2017.* United Nations, New York 2017, 2017.

[14] United Nations. *The Sustainable Development Goals Report 2018.* United Nations, New York 2018, 2018.

[15] United Nations Expert Group on the Integration of Statistical and Geospatial Information. Global statistical geospatial framework: Linking statistics and place. new york: United nations expert group on the integration of statistical and geospatial information, 2018.

[16] Shima Rahmatizadeh, Mohsen Kalantari, Abbas Rajabifard, Serene Ho, and Ali Daneshpour. How vgi intersects with land administration. In *CEUR Workshop Proceedings*, volume 1570, pages 53–58, 2016.

[17] Abbas Rajabifard. Beyond spatial enablement: engaging government, industry and citizens. 2012.

[18] Abbas Rajabifard, Joep Crompvoets, Mohsen Kalantari, and B Kok. Spatially enabling society. *Research, emerging trends and critical assessment*, 246, 2010.

[19] Jeffrey Sachs. Achieving the sustainable development goals. *Journal of International Business Ethics*, 1, 2015.

[20] Guido Schmidt-Traub, Christian Kroll, Katerina Teksoz, David Durand-Delacre, and Jeffrey D Sachs. National baselines for the sustainable development goals assessed in the sdg index and dashboards. *Nature Geoscience*, 10(8):547, 2017.

[21] Greg Scott and Abbas Rajabifard. Sustainable development and geospatial information: a strategic framework for integrating a global policy agenda into national geospatial capabilities. *Geo-spatial Information Science*, 20(2):59–76, 2017.

[22] Mark Stafford-Smith, David Griggs, Owen Gaffney, Farooq Ullah, Belinda Reyers, Norichika Kanie, Bjorn Stigson, Paul Shrivastava, Melissa Leach, and Deborah O'Connell. Integration: the key to implementing the sustainable development goals. *Sustainability Science*, 12(6):911–919, 2017.

[23] DANIEL STEUDLER and ABBAS RAJABIFARD. The role of land administration, land management and land governance in spatially enabled societies. 2012.

[24] Andrew J Tatem. Worldpop, open data for spatial demography. *Scientific data*, 4, 2017.

[25] UN-GGIM. Monitoring sustainable development: Contribution of geospatial information to the rio+20 processes, 2012.

[26] UN-GGIM. The application of geospatial information – land administration and management, 2015.

[27] UN-GGIM. Results of analysis of the global indicator framework with a "geographic-location" lens, 2017.

[28] UNDP. Human development report, united nations development programme, 2016.

[29] UNHCR. Rohingya refugee emergency at a glance, 2018.

[30] Jude Wallace, IP Williamson, Abbas Rajabifard, and Rohan Bennett. Spatial information opportunities for government. *Journal of spatial science*, 51(1):79–99, 2006.

[31] I. Williamson, A. Rajabifard, and P. Holland. Spatially enabled society. 2010.

[32] Ian Williamson, Stig Enemark, Jude Wallace, and Abbas Rajabifard. *Land administration for sustainable development.* Citeseer, 2010.

[33] IP Williamson, Jude Wallace, and Abbas Rajabifard. Spatially enabling governments: A new vision for spatial information. United Nations Regional Cartographic Conference for Asia and the Pacific, 2006.

[34] Jaap Zevenbergen, Norman Kerle, and Arbind Tuladhar. Spatial information for addressing and assessing land issues in disaster risk management. *Land tenure journal*, (1), 2015.

Part II

Enhancing SDGs Connectivity and Disaster Resilience

5

Leveraging National Land and Geospatial Systems for Improved Disaster Resilience

Abbas Rajabifard
Centre for SDIs and Land Administration, The University of Melbourne, Australia

Katie Potts
Centre for SDIs and Land Administration, The University of Melbourne, Australia

Mika-Petteri Torhonen
Social, Urban, Rural, and Resilience Global Practice, The World Bank, USA

Federico Barra
Social, Urban, Rural, and Resilience Global Practice, The World Bank, USA

Ivelisse Justiniano
Social, Urban, Rural, and Resilience Global Practice, The World Bank, USA

This chapter presents a roadmap for exploring the role of land and geospatial information, the function and responsibility of the institutions that govern the data, and the resulting impact that this data has on the overall resilience of society to disasters.

5.1 Introduction - Supporting SDGs With Land and Geospatial Information

National land administration systems and geospatial data infrastructure are fundamental for disaster risk management. They play a key role in facilitating tenure, land use, land valuation and zoning information, for planning, monitoring and implementing responses before, during and after disasters. The input of this information enhances resilience capabilities and enables stakeholders to carry out actions required for disaster mitigation and preparedness. With disaster events around the world increasing in frequency and severity, better access to this information is critical to disaster risk management activities.

Achieving the SDGs is parallel with establishing safe and resilient communities that have effective disaster risk management practices in place. Every SDG is related in some way to disaster risk management and requires disaster resilience to some degree. Land and geospatial information is critical to the successful implementation of the SDGs through the provision of reliable land data that provides land tenure security for owners and individuals with interests in land, and for land value, use and development dimensions at the local scale that can guide resilient actions [7]. By addressing the maturity and completeness of land and geospatial systems, and the level of integration into disaster risk management activities, the progress towards establishing a foundation for best practice land management can be understood, and areas for focus can be identified.

This chapter is a resource for enhancing resilience – and in particular, resilience to disaster events in a specific country context by improving the impact of existing land and geospatial systems. It explores the role that land and geospatial systems play within a country, and highlights ways that disaster resilience can be significantly improved for stakeholders, particularly at the community level, through use of existing land and geospatial information and resources. Resilience is the ability of a system, community or society exposed to hazards to resist, absorb, accommodate to, and recover from the effects of a hazard in a timely and efficient manner [9]. In the context of land, land resilience translates to the resilience of land and property and the people to land relationships that exist to recover to the extent that land tenure, value, use and development activities can effectively take place.

5.2 Addressing Global Problems With Land and Geospatial Systems

National land administration systems and geospatial data infrastructure act as the country's source for recording people to land relationships. The stability of these systems and the security of tenure they provide enables those who have legal rights in land to be confident that those rights can be assured even in the event of a disaster. This security supports wider resilience by providing confidence that if land is physically left – as required by disaster events that demand evacuation, that owners are protected against land grabbing and other activities that would otherwise threaten their right to land.

In addition to providing security to the community, the land records themselves need to be made resilient from any physical impacts that could destroy them. It is critical that best practices for data management are adopted to ensure information is digitally recorded and backed up so wider resilience of the community can be supported.

When disasters displace people, land records and geospatial data are key to protecting property rights and building resilience. In 2017, droughts, floods, hurricanes, and other disasters displaced over 18 million people [2]. When people are forced to leave their homes behind, land records offer critical protection of their property rights. This is crucial, since land and homes are usually the main assets that people have. Land and geospatial information are key to ensuring that land records are comprehensive and secure. It informs the what, who, where, how much, and other key attributes of a property. Without this information, it is almost impossible for countries to develop proper disaster response or preparedness plans.

In a disaster situation, comprehensive land and geospatial information and systems can secure the recovery of economic activities by providing accessible and instantaneous data about a disasters impact, value of losses, beneficiaries, as well as the levels of appropriate compensation and investment required to restore activities. In fact, land and geospatial information play an important role in all phases of disaster risk management, namely: disaster prediction (simulation and visualization), prevention, preparedness and mitigation, emergency response, evacuation planning, search and rescue, shelter operations, and post-disaster restoration and monitoring. Moreover, robust land and geospatial systems can help increase resilience by providing detailed and comprehensive information about the earth's surface. This information demonstrates physical hazards with detailed geographical impact areas, as well as tenure and use, and property assets and their values, to guide development of more effective policies, land use planning, and investments.

5.3 Global Land and Geospatial Systems

Global land and geospatial systems are important national resources. They contribute to stability and economic growth by providing security and surety around people's greatest assets – land and property. A good land system is made up of an effective land administration system and supported by geospatial information and systems.

When building resilience to disasters, establishing a mature land administration system prior to a disaster event is essential to ensuring a fast and effective recovery. Disasters can result in the loss of official records concerning land ownership, which is why land administration systems are essential to early recovery. They can support tenure security, settlement planning and the transition to sustainable development. Improvements in tenure security and land use practices can foster resilience to disasters through increased food security and environmental sustainability. Conversely, the mismanagement of these issues can increase vulnerability to disaster through unsustainable land use and insecurity of tenure [8]. Additionally, secure property rights include the ability for betterment of societal infrastructure, such as road paving, street light installation, and the development of sewerage systems, all of which are made possible through land tax revenue [6]. There also needs to be more institutional collaboration, interoperability and integration at the national level. This needs to happen across the various national data information systems and platforms that exist to support the development and maintenance of geospatial information for improving societal infrastructure [3].

In striving to achieve a mature land administration system that supports resilience to disasters, a number of issues can arise. For example, an inefficient and ineffective land registration process, which can be compounded by an inoperative land information system, an incomplete and/or outdated cadastre, a lack of trained surveyors to conduct high-quality land surveying, and absence of geospatial data sharing protocols. Situations like this contribute to difficulties in tax collection, distort land markets, result in poor urban planning, and also undermine the associated disaster risk management activities.

In cases like this, and in order to improve resilience to disasters, countries should aim for:

- A complete cadastre
- Establishment of effective land and property rights
- Establishment of appropriate land policy
- Restoration of land records
- Development of a legitimate legal framework and adjudication
- Protection of women's land rights

Working towards measures like these can result in positive outcomes, such as:

- Addressing current land and property disputes, evictions and discrimination
- Developing proposed institutional and normative frameworks, including housing, land and property Directorates
- Allocation of land use for temporary purposes (such as shelter)
- Identifying and securing the land records
- Servicing and management of the emergency
- Supplying information to those who have lost their property rights
- Assessing the state of the land records, institutions and problems.

- Reduction of land disputes

National land and geospatial information can help build disaster resilience. However, we need to better understand the role of such information at the local level, the responsibilities of the institutions that govern land data, and the impact of land and geospatial data on the overall resilience of society.

5.4 Working Towards the SDGs: Achieving Land Resilience

Land is the single greatest resource in most countries, and access to land, security of tenure and land management have significant implications when considering the challenges faced by humanity today. As a finite resource, land – and property, are the main assets of people, and therefore the impact of disruptive events such as disasters, have significant effects on the livelihood of citizens worldwide.

These major disaster events cause large numbers of people to be displaced. Between 2016 and 2017, over 18.8 million people were displaced as a result of disaster events. When disasters displace people, land records and geospatial data are key to protecting their property rights and building resilience.

Disasters are events increasing in frequency and severity and providing countries, worldwide, with an increased impetus to address these events. They are not only having devastating impacts on the world's economies, but, most importantly, on the main assets of their citizens: land and property. In addition to the initial impacts of a disaster, the ongoing and secondary impacts, which can cause major disturbances, need to be considered too. For example, it is often necessary for homes and fields to be abandoned during disasters, however, returning may be restricted due to insecure tenure and the inability to prove prior occupation. Once access to land (a core social safety net) is lost, resuming livelihoods becomes challenging or even impossible, which consequently, increases vulnerability. Families face the prospect of duress selling of assets at reduced prices and moving to informal urban settlements.

There is wide recognition that national land administration systems and spatial data infrastructures are fundamental for disaster risk management. They play a key role in facilitating pre and post disaster tenure, land use, land valuation and zoning information within a unified geospatial platform for planning, monitoring and implementing responses. The input of this information enhances resilience capabilities and enables stakeholders to carry out required mitigation and preparedness actions.

Land and geospatial information can also assist with disaster reduction, risk reduction, preparedness, mitigation and emergency response. It can also expedite recovery operations by providing data on the impact, value of losses,

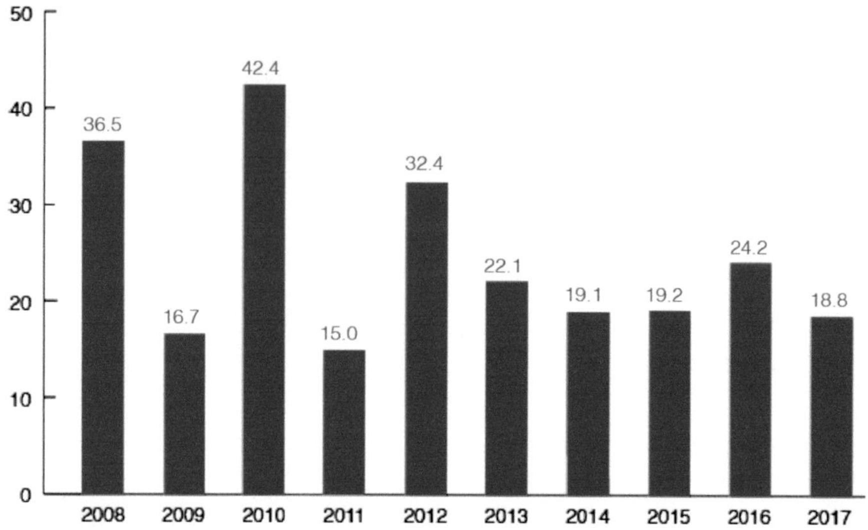

Source: Extracted from Internal Displacements Monitoring Centre and Norwegian Refugee Council (2018)

FIGURE 5.1
New human displacements due to disasters (2008-2017) [2]

and the investment needs for recovery and reconstruction. Better access to information, along with more secure tenure, yields land use and management decisions that take resilience into account and reduce vulnerability, which can result in improved land resilience and overall resilience to disasters.

Sharing this information with disaster risk management agencies and enabling them to harness this valuable data in their planning and operations enhances the overall process and supports government-wide agendas. However, in many contexts, there is a disconnect between a number of these key elements. In order to achieve land resilience, available land and geospatial resources need to be applied and continually improved upon to meet the needs of the community through the application towards disaster risk management activities.

In order to get to this point and achieve land resilience, three critical elements need to be founded within a country: a mature land administration system, comprehensive geospatial data and systems, and established relationships for sharing with disaster risk management agencies.

In addition to having these elements in existence within a country context, the land administration system, geospatial data and systems need to also be physically resilient to disaster. In contexts where land administration is primarily paper based on not digitally recorded a large vulnerability is present.

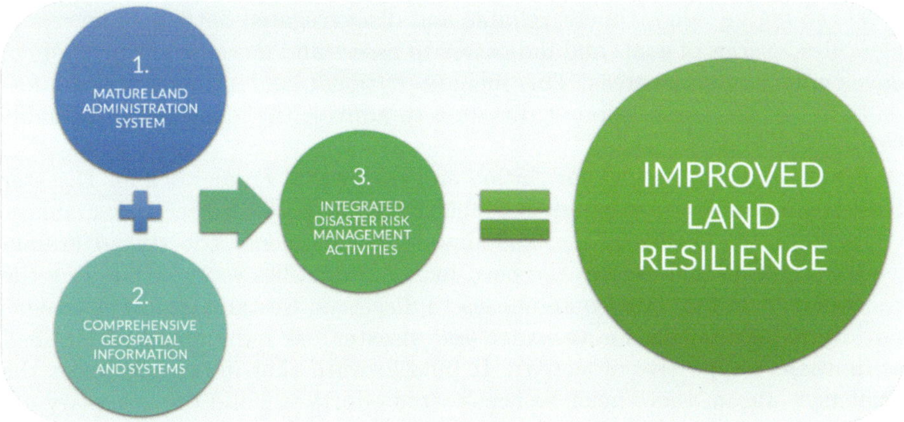

FIGURE 5.2
Key elements for improved land resilience

Equally, geospatial data and systems that are not adequately maintained or backed up are not resilient to an event that may impact its physical location. Securing both the information itself to make it resilient and leveraging the information to support resilience activities are key priorities.

5.5 Global Development Frameworks

Several key initiatives aiming to build resilience to disasters have emerged around the world in recent years. Many of these initiatives tackle a broad range of issues at a number of levels ranging from global or national down to local and community levels. In particular, the 2030 Agenda for Sustainable Development, the Hyogo Framework for Action, and the Sendai Framework for Disaster Risk Reduction aim to substantially reduce the risk of disaster and losses through the implementation of strategic goals and integrated and inclusive measures that prevent and reduce hazard exposure and vulnerability to disaster, increase preparedness for response and recovery, and thus strengthen resilience. They also outline key points that relate to improving resilience to disasters, as well as highlighting the positive effects that national land and geospatial systems can have. In addition, the Integrated Geospatial Information Framework [3] builds on many of these ideas with a focus on geospatial information and how it can be improved to support global development.

The 2030 Agenda for Sustainable Development outlines a need for new data acquisition and integration approaches, including supporting developing countries to strengthen the capacity of their national data systems to ensure access to high quality, timely, reliable and disaggregated data [5]. The report identifies a series of goals and indicators to assess and measure the progress of development in these areas. This includes national land and geospatial information, and the application of this data to address the identified sustainable development goals (SDGs).

The Hyogo Framework for Action and the Sendai Framework for Disaster Risk Reduction both respond to global issues around disaster risk management, improved resilience and sustainable development. The Hyogo Framework for Action underscores the need for, and identifies ways of, building the resilience of nations and communities to disasters. Sustainable development, poverty reduction, good governance and disaster risk reduction are identified as mutually supportive objectives. It puts forward that in order to meet the challenges ahead, there must be accelerated efforts to build the necessary capacities at the community and national levels to manage and reduce risk [9]. Further to this, within the Hyogo Framework for Action, land issues were established as one of the key priorities for the period of 2005-2015, and have been gaining momentum within the disaster risk management community in recent years.

The Sendai Framework for Disaster Risk Reduction follows on from the Hyogo Framework for Action. It aims to substantially reduce the risk of disaster and losses in lives, livelihoods and health and in the economic, physical, social, cultural and environmental assets of people, businesses, communities and countries. It works to achieve this through the implementation of integrated and inclusive measures that prevent and reduce hazard exposure and vulnerability to disaster, increase preparedness for response and recovery, and thus strengthen resilience [4]. The Framework has identified seven global targets which address areas, including: global disaster mortality; number of people affected by disaster; direct disaster economic loss; damage to critical infrastructure and disruption to basic services; creation and implementation of disaster risk reduction strategies; international cooperation; and availability and access to multi-hazard early warning systems and disaster risk information and assessments. Learning from the experience of implementing the Hyogo Framework, the Sendai Framework has identified four areas requiring further focused action within and across sectors by States at local, national, regional and global levels. These four areas are:

1. Understanding disaster risk: comprehending all the dimensions of vulnerability, capacity, exposure of persons and assets, hazard characteristics and the environment so that the knowledge can be used to inform risk assessment, prevention, mitigation, preparedness and response.

2. Strengthening disaster risk governance to manage disaster risk: Fostering collaboration and partnership at national, regional and global levels.

3. Investing in disaster risk reduction for resilience: Essential investments to enhance people, communities and the environment.

4. Enhancing disaster preparedness for effective response and to 'Build Back Better' in recovery, rehabilitation and reconstruction: taking the opportunity to strengthen and enhance all phases of disaster risk management

The overall focus is to prevent new disasters and reduce the existing disaster risk through the application of prevention and reduction measures to economic, structural, legal, social, health, cultural, educational, environmental, technological, political and institutional areas. To achieve this effectively, enhanced implementation capacity, and strong country commitment that is facilitated through political leadership is required.

In addition to the above frameworks, the Integrated Geospatial Information Framework released in 2018 by the United Nations and the World Bank complements the above agendas, which unequivocally call for globally coordinated actions in new data acquisition and integration approaches [3]. The vision and mission of the IGIF is to address the SDGs and note that strategies and frameworks around the use and management of geospatial information will be required to realize some of these goals, and within the context of disaster risk management, geospatial information will play an important role in developing policies, strategies and legislative arrangements to future challenges faced.

The purpose of the IGIF is to guide the development and strengthening of geospatial information, as well as the management of relevant infrastructures in developing and developed contexts. Through the nine strategic pathways outlined, we can glean a way to deliver sustainable social, economic, and environmental development through the implementation of integrated geospatial information systems. The strategic pathways are underpinned by a series of principles for geospatial information that represent key characteristics and values: Strategic enablement; Transparent and accountable; Reliable, accessible and easily used; Collaboration and cooperation; Integrative solution; Sustainable and valued; Leadership and commitment. These principles promote consistent geospatial information management, resulting in more open, accountable responsive and efficient governing [3].

5.6 A Roadmap for Building Land Resilience

A Roadmap for improving land resilience within a country context has been developed. The roadmap utilizes a number of tools developed from the project *Improving Resilience and Resilience Impact of National Land and Geospatial Systems* [1] to assess the maturity of land and geospatial systems within a country context. The Roadmap is shown below, and is achieved through the implementation of the land resilience tools:

- The Contextual Analysis Questionnaire is an operational tool used to assess the current status of land and geospatial systems within a country.
- The Pre and Post Disaster Recommendations for Land Resilience identifies key resilience indicators for land and geospatial organizations.
- The Land Resilience Maturity Index Assessment is a technical tool for quantifying the maturity of a country's land and geospatial systems in relation to land resilience.
- The Country Action Plan Template brings together the outputs of the three tools to delineate the dimensions to focus on for improving and enhancing the overall land resilience of a country.

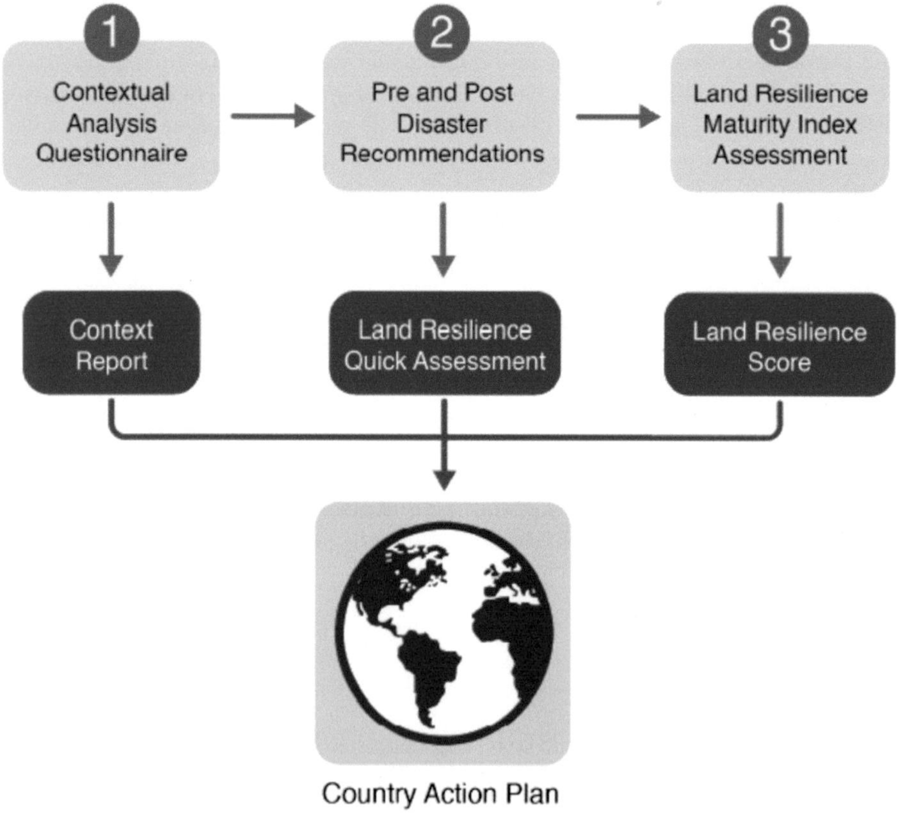

FIGURE 5.3
The Land Resilience Roadmap

The Roadmap facilitates the understanding of:

- how resilient land and geospatial systems are to disasters events

- to what extent land and geospatial systems are able to contribute support to external applications such as disaster risk management activities
- areas that land and geospatial systems could improve or enhance to support disaster risk management functions

By following this Roadmap, the current level of maturity of the land administration systems, the comprehensiveness of geospatial systems, and the level of integration of these systems with disaster risk management activities within a specific country context can be determined which enables areas requiring attention to be identified and addressed, which is turns supports improvements to current practices and overall improved land resilience.

5.7 Conclusion

Improving land resilience starts with a desire from the community for a better approach to managing land. In the context of disasters, often specific events highlight the need for a change or outline situations where significant problems arise. This chapter highlights the importance of land and geospatial information in achieving land resilience and presents a way to understand the current arrangements of land and geospatial information to improve current practices.

Action is required though. Individuals and organizations much work to overcome sharing, integration and interoperability challenges, make better decisions, promote transparency and act cohesively to improve land resilience. In particular, the physical land and geospatial information itself. Though this information and systems may be vulnerable to a wide range of hazards and disaster events, there are many ways to enhance their resilience and significantly reduce any potential loss. Best practice guidelines along with suggestions offered in this chapter are key to securing the valuable land and geospatial information that upholds many other land resilience practices.

The rich land and geospatial resources give us better tools to anticipate, plan and respond to disaster related problems. But decision makers and stakeholders need to work hard to help to ensure that all of this information leads to effective action. This involves developing better integration strategies across organizations at all jurisdictional levels to ensure that information can be utilized for activities where it can benefit the larger community. It also means developing ways of encouraging participation in establishment and improvement in wider resilience activities. Finally, it means better tracking of outcomes to keep organizations and stakeholders accountable for their promised actions and so improvements can be observed and celebrated.

Bibliography

[1] World Bank. Improving resilience and resilience impact of national land and geospatial systems, 2019.

[2] Internal Displacement Monitoring Centre. Global report on internal displacement, 2018.

[3] UN GGIM and World Bank. A strategic guide to develop and strengthen national geospatial information management, 2018.

[4] Francis Ghesquiere, Jan Kellett, Jack Campbell, Shyam Kc, and Robert Reid. The sendai report: managing disaster risks for a resilient future. Technical report, The World Bank, 2012.

[5] United Nations. Transforming our world: the 2030 agenda for sustainable development. *United Nations General Assembly. Resolution 70/1*, 2015.

[6] David Palmer. Security, risk and registration. *Land use policy*, 15(1):83–94, 1998.

[7] UN-GGIM. The application of geospatial information – land administration and management, 2015.

[8] United Nations Human Settlements Programme (UN-HABITAT). Land and natural disasters: Guidance for practitioners, 2010.

[9] U UNISDR. Hyogo framework for action 2005–2015: Building the resilience of nations and communities to disasters. In *Extract from the final report of the World Conference on Disaster Reduction (A/CONF. 206/6)*, volume 380. The United Nations International Strategy for Disaster Reduction Geneva, 2005.

6

Geospatial Information Technologies in Support of Disaster Risk Reduction, Mitigation and Resilience: Challenges and Recommendations

Saied Pirasteh

Southwest Jiaotong University, China

Masood Varshosaz

KN.Toosi University of Technology, Iran

This chapter presents Geospatial Information Technologies in the context of sustainable development goals (SDGs), in particular, SDG 11 and SDG13. It contributes to the United Nations Committee of Experts on Global Geospatial Information Management (UN-GGIM) in its efforts in disaster loss reduction, mitigation and resilience. Particularly, the chapter discusses what these technologies can bring to support the implementation of Sustainable Development Goals (SDGs).

6.1 Introduction

SDGs are conceived for a wide range of issues in local, national, regional, and global contexts. These include the major development issues related to poverty, hunger, health, natural hazards, agriculture, education, and gender inequality. In addition, SDGs cover specific topics such as energy, infrastructure, economic growth and employment, inequality, cities, sustainable consumption and production, climate change, forests, oceans, and peace and security (Figure 6.1).

The objective of this chapter is to (1) participate in the global sharing of experiences on utilizing geospatial information technologies to address disasters

FIGURE 6.1
The global goals for sustainable development

resilience and challenging issues of determining the vulnerability of buildings; and (2) demonstrate examples of the support-integrated geospatial information technologies including earthquake and structural engineering disciplines followed by utilizing drone images for rescue.

This chapter briefs an example on the need of environmental and geospatial information laws and regulations. Finally, recommendations are proposed that might be helpful to other countries having similar issues.

Today's challenges push everyone to ponder how we, as people living on the Earth, should play a role in SDGs to transform research into real life practices for improved condition and a better way of living together in a better world. Geospatial information technology is a collection of information communication tools [34], to store, capture, manipulate, transform, analyze, and generate information related to our planet. This technology is used for global management, and there is no doubt that, in this era of multilateralism with integrated technologies, there is a crucial need today to work together and face global challenges within a defined proposed ecosystem in the SDGs global context.

Based on a UN report, there is an increase in naturally occurring disasters [38]. Currently, earthquakes can occur almost anywhere in the world including Australia, Canada, Chile, China, England, India, Iran, Mexico, New Zealand, Pakistan, the US and many other countries in seismic zones such as Italy and Nepal. While very large earthquakes may still not occur frequently, the frequency of minor or medium-sized earthquakes is increasing. Indeed, some countries are exposed to many minor and major earthquakes annually. It needs significant consideration to formulate the strengthening of buildings against future earthquakes. At present, there are numerous weak structures in many countries particularly in developing countries that are not able to withstand an earthquake. Public and private developers intend to use the scientific methods to prioritize and optimally allocate budget in order to reinforce the structures, because of limited financial resources, time and availability of an appropriate model. It seems logical to renew our structures with mitigation resilience utilizing the integration of geospatial information technology and engineering. Nowadays, there is sufficient data on the effects caused by earthquakes. And recently, there is greater research on the challenges at various scales by using

Geographical Information System (GIS) for emergency response to disasters [7, 25, 10, 11, 40, 33, 32, 31, 37, 4, 13, 36, 3, 36, 20, 16, 21, 24, 39, 29]. For example, Liu (2018) in [30] studied the seismic identification and reinforcement design of building structures of China after the earthquake in Wenchuan in 2008. She suggested some reasonable improvement methods for the future development of buildings. Also, [35] evaluated a number of buildings in Esfahan city and classified the buildings from low-to-high based on their vulnerability to seismic activity and collapse due to earthquakes. The result was that, at present, numerous structures are still not able to withstand an earthquake.

Nevertheless, the integration of geospatial information technologies with engineering parameters to develop a platform has been an interest of researchers for mitigation resilience, quality efficiency, saving costs, and enriching the quality of the hazard mitigation for loss reductions. Most of the current systems are local and have various limitations mainly because of adaptation of the local instruction of building code or any other infrastructure instruction codes [22, 23, 18]. For example, maps cannot be retrieved to represent buildings geospatially correlated with the evaluation parameters. Tools are not in place to screen buildings for potential seismic hazards with the implementation of 3D reconstruction models. The local system is also often based on a particular region and the scalability remains a challenge. In addition, recent advances in computer vision, artificial intelligence, machine learning and robotics combined with geospatial information using digital cameras are suitable tools for many applications such as rescue and emergency management. The use of cheap platforms with low-end imaging cameras makes this technology available to the public. The amount of data in the images is high but this can be provided online during the search and rescue operation particularly when using drones.

In this chapter, examples are presented showing how geospatial visual screening works for pre-earthquake and post-earthquake preparation. There is also a brief section on how drone images help in human search and rescue.

6.2 Why are technologies alone not enough in disasters loss reduction

To protect the environment through technology, there must be global cooperation. Although science and technology have progressed, there has been little changes to the sustainability of the environment and the preservation of our planet. Though technology has been advancing rapidly, many events such as the catastrophic flood of the Dez river in Iran, and the deforestation of mangroves and salinity of water in the Persian Gulf in Iran (Figure 6.2) happen due to the ineffective national and regional policies including a mismanagement of decision makers in the pursuit of nature and the lack of use of appropriate technology. In conclusion, it seems that at a global level, earth and the environment are at higher risk than before, and the number of disasters are rising all over the world: Natural disasters like earthquakes, floods and storms as well as human related emergencies such as wars and deforestation are ever-increasing.

(a) (b)

FIGURE 6.2
(a) Recent flood in Iran, (b) plantation of mangroves for sustainable environment and desalinization in Booshehr, Persian Gulf, Iran

6.3 Integration of Geospatial Knowledge

With rapidly growing technologies, critical thinking and problem-solving, including a momentum to sustain the continuing development and enhancing skills in practice, are not only supporting the representation of the real-world but also present as a challenge.

These days the requirements of a desired software and hardware product emphasizes on using artificial intelligence and deep machine learning; however, they depend on the applications and needs of the people for creating a sustainable environment and the economic growth in various sectors. It is required to work with big geo-data and geospatial technologies in conjunction with innovative soft computing solutions to develop products in many fields such as engineering, environmental issues, hazards, mapping, and construction support, augmented reality, and real-time asset management to support the business processes and to develop various types of mobile applications and software.

To utilize and implement the integration of geospatial data and techniques with other technologies requires a conceptual framework and also building platforms to create a full global geospatial ecosystem. Building platforms and services such as GeoEngine (a geospatial rapid visual screening of buildings for developing ecosystems) is allowing digital transformation and enabling countries to support the implementation of SDGs. Different supporting integrated platforms are required to fulfill the needs of stakeholders, communities, citizens and the public at large. In this multilateralism era, the needs for geospatial information in required in daily life.

However, the author's experience in research and development indicated that the software and hardware engineering processes consist of many activities and require application based knowledge. They are requirements and analysis, specifications, software architecture, implementation, testing, documentation, training and support, and maintenance. Therefore, the integrated technologies, including machine learning, computer vision, and artificial intelligence are the best choices towards creating platforms and working to develop smart systems to connect all disciplines. For example, we can connect earth observation systems and geospatial information technologies for full infrastructure projects, and asset management systems implementation. This builds an intelligent world for the future generations in a GeoEngine platforms' ecosystem. Figure 6.3 shows an example of how geospatial information helps rapid visual screening processes and determines the vulnerability of buildings and estimation of risk. Another example is the platform on representing maps and disaster responses. The snapshot of architecture design of the model is illustrated in Figure 6.4.

FIGURE 6.3
Screening a building (Geospatial Rapid Visual Screening)

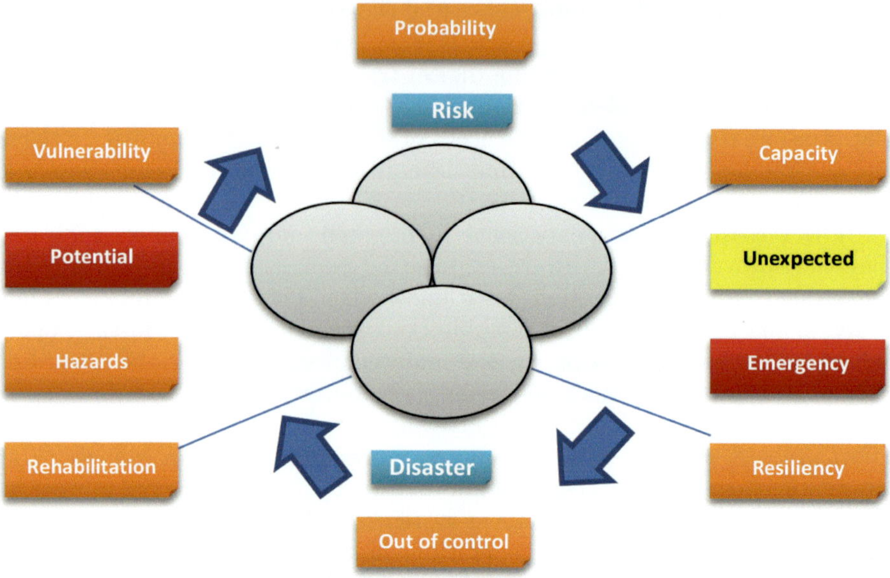

FIGURE 6.4
Architecture design of the disaster responses of the platform [1]

6.4 Geospatial Rapid Visual Screening for Earthquake Disaster Risk Reduction, Mitigation and Resilience

The modern space-based remote sensing integration with geospatial information technology has opened up efficient means for disasters risk reduction and resilience. The maximum economic casualties caused by natural disasters such as earthquakes, flood, tsunami, and landslide, between 1900-2018, occurred in Asia and the Pacific (https://reliefweb.int/map/world/major-natural-hazards-asia-and-pacific-0). Nevertheless, this section explains how geospatial information supports building inspection and safety for earthquake disaster risk reduction, mitigation plan and resilience.

In the context of SDGs, geospatial information rapid visual screening techniques can be interpreted in SDGs 11 and 13 in the aspects of disaster resilience, creating a safe and livable city and homes with the focus on housing condition. Geospatial rapid visual screening techniques allow screeners to determine the vulnerability of buildings and estimates risk with the potential to generate a 3D model for building and disaster applications beyond which it might be useful in building information management system (BIMS).

However, this technology aims to integrate geospatial technologies with engineering for risk assessment and 3D modelling utilizing computer vision techniques for disaster risk reduction, mitigation and resilience.It seems that implementing geospatial rapid visual screening techniques can strengthen the buildings, increase mitigation of hardware and software infrastructures for disaster monitoring and warning. It also supports a disaster prevention process within the context of SDGs 2030.

For exmaple, we attempted comprehensive sophisticated computing process to generate a damage index (DI) and building score utilizing the geospatial rapid visual screening technique with the Damage Index of Building (DIoB) algorithm and the Federal Emergency Management Agency (FEMA) approach [36, 35]. The damage index model (DIM) incorporation with geospatial information can be interactively utilized into the GIS to compute relevant engineering parameters for analyzing data and a better quality management on the web andcloud. Its resilience ability allows an easy accessibility of the geospatial data to evaluate buildings for earthquake mitigation and preparedness. This geospatial rapid visual screening techniques is a new combination of a geospatial integrated system and engineering disciplines for rapid evaluation of buildings, and for city planning by further combining GIS with building structural information, civil engineering and industrial engineering as required by the FEMA. All this in addition to the general information of the building, visualizing the 3D model of building, seismic data, soil data, land use data, structure, parcel, material type, foundation, ceiling, wall, floor, the interior and the exterior.

TABLE 6.1
Vulnerability scores on residential building

Score	Level of vulnerability
1–10	Demolishment
10–40	High vulnerability
40–70	Medium vulnerability
70–90	Low vulnerability
90–100	Invulnerability

6.5 Human Search and Rescue in Drone Images

The utilization of drones has become one of the recent good examples of such technologies empowered by machine learning and robotics.

Events, such as earthquakes, fires, floods, avalanches, and landslides occur all over the world. According to the Red Crescent instruction, the first and most important part of a search and rescue operation is to determine the location of injured people [14]. This stage can be very time-consuming, and

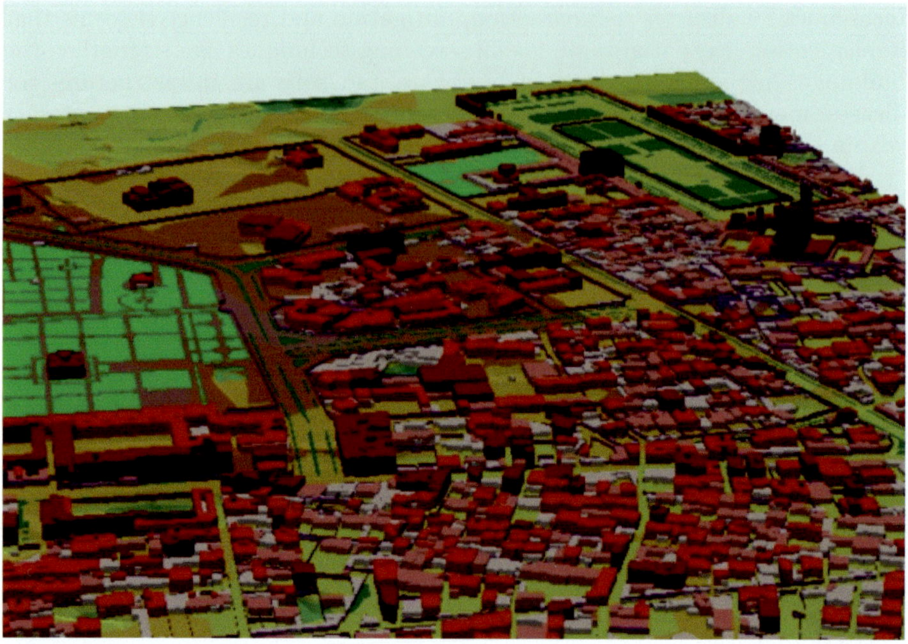

FIGURE 6.5
Most vulnerable buildings shown in dark red color

sometimes only a few extra minutes could be all that's needed to save the life of a person. In the event of avalanches, the use of trained rescue animals such as dogs is a common method, which, unfortunately, may not be enough to have a fast and accurate search operation [8].

Nowadays, with the advancement of technology, the application of intelligent relief equipment has become more possible than ever before. Therefore, the development of spatially enhanced tools that can automatically locate people in need of help is a key element for post-accident search and rescue tasks. In this context, drones are becoming very popular as they are capable of carrying out rescue missions automatically (Figure 6.7). In addition, they can perform tasks that may not be possible for a human operator to do such as collecting the required information without having to interfere with the working environment.

One of the major accessories used by such unmanned aerial vehicles UAVs are cameras. To date, a lot of research has been carried out on finding humans in images [2, 9, 28, 39, 29]. (2013)[12] reviewed the performance of the histogram of oriented gradients (HOG) descriptor in drone images and improved its performance for human detection. They considered the different viewing angles a person can be seen in a drone image. Following this, the Sector-ring Histogram of Oriented Gradients (SRHOG) feature [29] was developed. The

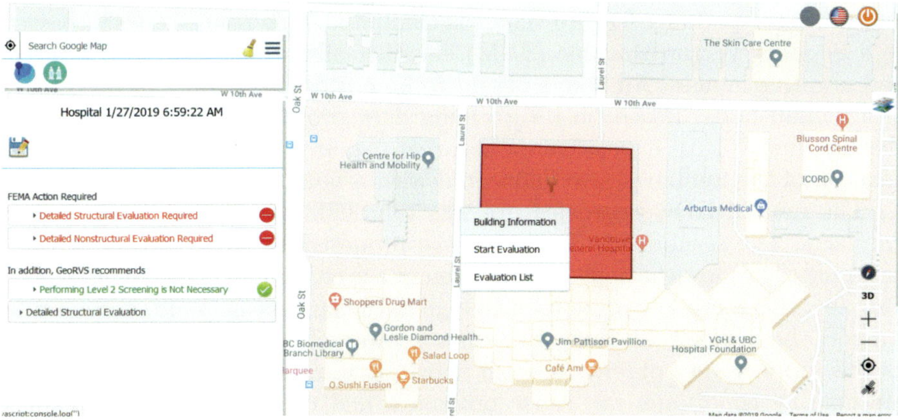

FIGURE 6.6
Recommendations of the building based on scores

FIGURE 6.7
Sample images showing how humans may appear on drone images

SRHOG focus is to change the way the gradient is calculated making it more resistant to rotation distortions. It has also a different way of forming image blocks and interpolating their values. As a result, it is able to create a rotation-invariant feature. The robustness of this feature descriptor makes it a suitable to be used in drone images where humans are imaged differently with respect to those taken on the ground.

Later in 2018, the SRHOG was improved by adding two additional filters to the gradient calculation process [19]. One of the filters was used for calculating the gradient in the radial direction, whereas the other was applied along the tangential direction. This feature was also used to develop an algorithm for human detection within drone images. This algorithm takes an input (test) image and outputs the position of the injured person, if any. For this, first a pyramid of images is created, each of which has a different resolution. At each resolution, a search window is moved along various parts of the test image and the corresponding feature vector is computed. The result is, then, passed to a Support Vector Machine (SVM) classifier already trained with positive and negative images. An image is said to be positive if it contains all or part of a human body. Based on the training data provided, the classifier labels the test image with a positive or negative tag. If the image is positive, the location of the human is also delineated with a bounding box.

The new SRHOG-based feature was thoroughly evaluated using several data sets. The results suggested that the proposed method has a very good accuracy (90%) in situations where the human is either lying down or standing. Perhaps, this method can be used in a global scale for emergency response to support the SDGs 2030. However, it has a lower efficiency for sitting persons, as they appear very differently from above. It was observed that inappropriate lighting conditions can cause problems, and could be resolved to some extent using a preprocessing stage. In this context, an examination of the preprocessing techniques is required in future studies.

6.6 An Example of Lack of Laws in Geospatial and Environmental Issues

Governments need a set of binding legislation to regulate the relationships between their members and the environment, sometimes utilizing geospatial information technology and increasing spatial enablement [17]. This section briefly discusses an example of a lack of rules or inadequacies in the environmental and geospatial information field relating to agricultural-environmental issues and sustainability [15]. The detailed study by Ansari and Namadrian (2016)[5]; Arshadi and Pirasteh (2019)[6]; and (KWPA 2019)[27] showed that some environmental and geospatial regulations need to be revised or redefined more explicitly to reinforce their effects.

Today, the growth in population, irrational utilization of natural resources, biodiversity reduction, pollution and other causes of environmental destruction have impacted the world adversely, and for this reason geospatial information for achieving the 2030 SDGs has to play a significant role. The present quality of natural human life is a result of an unbalance and destruction of environment. It has led governments, organizations, and international communities to develop technologies such as geospatial information technology and to use them for disasters risk reduction and resilience. These bodies formulate and execute regulations to mitigate the environmental pollution and damages.

However, a case study attempted by Arshadi and Pirasteh (2019)[6]; and KWPA (2019)[27] have presented that the following are challenges and to be considered when the policy-makers decide to enrich the laws [26].

1. The authors determined that the traditional rules of environmental assess-

ments and data sharing, including geospatial information, are powerless. The provision of specific legislation in this regard is necessary.

2. The change in the legal sense "beneficiaries" and strengthening the role of non-governmental organizations are one of the most important strategies to support geospatial data sharing and protect the environment in the prosecution of cases.

3. There are no concrete and appropriate geospatial information regulations of data sharing. For example, during testing the geospatial rapid visual screening algorithms, we have not found adequate geospatial data of land use, seismic, soil and other relevant data for the city and building. Also, in one of the projects completed, there were severe challenges with collecting standard data for agricultural cadastral mapping and crop identification of Dezful, Iran (Figure 6.8). These included challenges such as flight permission and security when using drones for scanning the agricultural lands. Furthermore, during the agricultural cadastral mapping and crop identification of the Dezful project, it was concluded that there is no difference between drinking water and irrigation water, when distributing water. However, since the effects of contamination of drinking water are assessed much more robustly than contaminating agricultural water, the current existing law should have a clear definition of the law for all types of water.

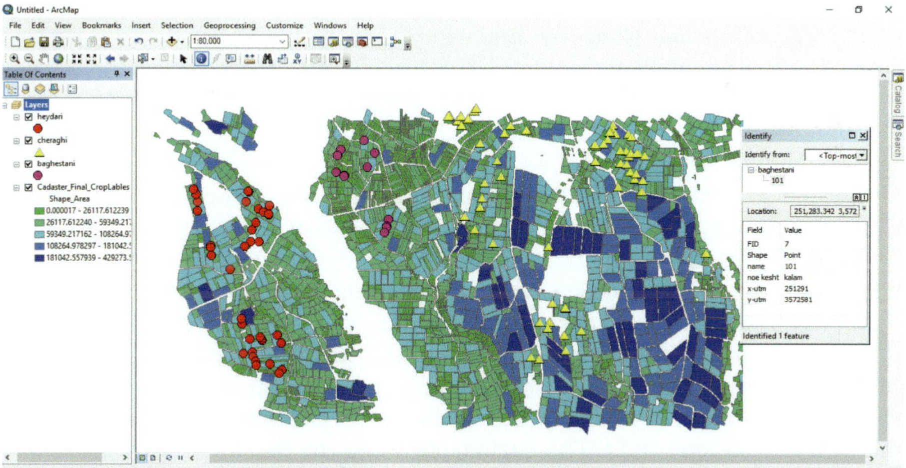

FIGURE 6.8
Agricultural cadastral mapping of Dezful, Iran

6.7 Conclusions and General Remarks

The world is shifting to the second data revolution and transformative technology by the means of geospatial information and geoanalytics utilizing computer vision and artificial intelligence. Therefore, this revolution will impact disaster risk reduction and mitigation, tremendously, by the means of data sharing via web and cloud-based platforms. There is a big change towards the Internet of Things (IoT), block-chain technology, various open sources, drones and many other technological tools in support of SDGs achievements. For this reason, to recommend building smarter cities, there is a need for people, intelligence and integrated technologies, including computer vision and AI for global challenges.

This chapter indicated that without applying the power of geospatial information technologies and integration with other technologies such as artificial intelligence and machine learning, computer vision, digital transformation, and multi-disciplinary research, there will be even greater challenges for future generations. Therefore, educating the young generation in innovative and cutting-edge technologies and research in geospatial disciplines through the Academic Network and understanding of the problems with the help of the private industry sectors at the UN-GGIM will help enable the SDGs implementation and monitoring. This chapter also presented examples of geospatial information and integration of technologies for earthquake disasters risk reduction and management to build a better world.

It is recommended that each and every country build and develop its National Geospatial Information Infrastructure (NGII) that can enable coping with disaster risk reduction and mitigation resilience. It is suggested that every country develop its architecture of GGIM with a strategic road map based on the global development policy of the 2030 Agenda and beyond. In addition educational resources should be allocated for emerging technologies of robotic and AI, enabling the making of maps and improving the cartographic workflow and the training of map makers. This is possible when all countries can access satellite images, needed for SDGs implementation and monitoring.

As discussed, in order to recognize humans in drone images using the integration of geospatial information and computer vision associated with artificial intelligence, it is necessary to describe objects using proper features. In this context, there are several challenges such as the existence of various objects and complex backgrounds. The flexible nature of the human body leads to numerous situations. Different viewing angles and different dimensions related to distance from the camera are other challenges that most algorithms face. The environment in which the image is taken also affects how the injured person appears on the image.

Finally, there is a clear lack of environmental and geospatial information laws and regulations. In order to move forward environmental and geospatial information laws need to be established or improved.

Acknowledgement

This chapter of the book is dedicated to the people at the United Nations who work on certain the new transformative 2030 agenda for SDGs 2030 and beyond. I would single out Dr. Amor Laaribi, the UN retiree and former UN-GGIM coordinator for reviewing the chapter and providing us with invaluable comments. Finally, the authors appreciate reviewers of this chapter for their remarkable comments.

Bibliography

[1] Rifaat Abdalla. Evaluation of spatial analysis application for urban emergency management. *SpringerPlus*, 5(1):2081, 2016.

[2] Jake K Aggarwal and Quin Cai. Human motion analysis: A review. *Computer vision and image understanding*, 73(3):428–440, 1999.

[3] M Alam, A Mahmoodzadeh, and S Pirasteh. A method for rapid evaluation of masonry buildings against earthquakes. *Int Disaster Adv J*, 2(3):15–23, 2009.

[4] SA Ali and S Pirasteh. Geological application of landsat etm for mapping structural geology and interpretation: aided by remote sensing and gis. *Int J Remote Sens*, 25(21):4715–4727, 2004.

[5] Mustafa Taghizadeh Ansari and Soodabeh Namadrian. The role and effect of freshwater on international law security and peace. *International Journal of Humanities and Cultural Studies (IJHCS) ISSN 2356-5926*, 3(1):264–274, 2016.

[6] M. Arshadi and S. Pirasteh. Determining lacks of the iranian laws in environmental protection and awareness. 2019.

[7] M Barazangi. Continental collision zones: Seismotectonics and crustal structure. *The Encyclopedia of Solid Earth Geophysics*, pages 58–75, 1989.

[8] Mesay Bejiga, Abdallah Zeggada, Abdelhamid Nouffidj, and Farid Melgani. A convolutional neural network approach for assisting avalanche

search and rescue operations with uav imagery. *Remote Sensing*, 9(2):100, 2017.

[9] Rodrigo Benenson, Mohamed Omran, Jan Hosang, and Bernt Schiele. Ten years of pedestrian detection, what have we learned? In *European Conference on Computer Vision*, pages 613–627. Springer, 2014.

[10] M Berberian. Natural hazards and the first earthquake catalog of iran, vol. 1: historical hazards in iran prior 1900. *IIEES report*, 1994.

[11] M Berberian. The first seismicity catalogue and the natural phenomenon of iran, first edition; natural hazards prior to twentieth century, (iiees). IIEES report, 1995.

[12] Paul Blondel, Alex Potelle, Pr C Pégard, and Pr R Lozano. How to improve the hog detector in the uav context. *IFAC Proceedings Volumes*, 46(30):46–51, 2013.

[13] A Cavallaro, A Ferraro, S Grasso, and M Maugeri. Topographic effects on the monte po hill in catania (italy). *Soil Dynamics and Earthquake Engineering*, 43:97–113, 2012.

[14] Navneet Dalal and Bill Triggs. Histograms of oriented gradients for human detection. In *international Conference on computer vision & Pattern Recognition (CVPR'05)*, volume 1, pages 886–893. IEEE Computer Society, 2005.

[15] R. Eltiyaminia and A. Hosseini. World domination of modernity and its relationship to the environmental crisis in the world. Journal of Humans and Environment. 13(2): 92-77., 2015.

[16] Mustafa Erdik, K Şeşetyan, MB Demircioğlu, U Hancılar, and C Zülfikar. Rapid earthquake loss assessment after damaging earthquakes. *Soil Dynamics and Earthquake Engineering*, 31(2):247–266, 2011.

[17] Tim Foresman and Ruth Luscombe. The second law of geography for a spatially enabled economy. *International Journal of Digital Earth*, 10(10):979–995, 2017.

[18] Paolo Franchin, Laura Ragni, Maria Rota, and Alessandro Zona. Modelling uncertainties of italian code-conforming structures for the purpose of seismic response analysis. *Journal of Earthquake Engineering*, 22(sup2):1964–1989, 2018.

[19] M Ghasemi. Implementation of a system for identifying injuries in natural disasters using uavs. Master's thesis, K.N. Toosi University of Technology, Iran., 2018.

[20] S Grasso and M Maugeri. The seismic microzonation of the city of catania (italy) for the maximum expected scenario earthquake of january 11, 1693. *Soil Dynamics and Earthquake Engineering*, 29(6):953–962, 2009.

[21] Salvatore Grasso and Michele Maugeri. The seismic microzonation of the city of catania (italy) for the etna scenario earthquake (m= 6.2) of 20 february 1818. *Earthquake Spectra*, 28(2):573–594, 2012.

[22] Saw Htwe Zaw, Teraphan Ornthammarath, and Nakhorn Poovarodom. Seismic reconnaissance and observed damage after the mw 6.8, 24 august 2016 chauk (central myanmar) earthquake. *Journal of Earthquake Engineering*, 23(2):284–304, 2019.

[23] Iunio Iervolino, Andrea Spillatura, and Paolo Bazzurro. Seismic reliability of code-conforming italian buildings. *Journal of Earthquake Engineering*, 22(sup2):5–27, 2018.

[24] Sadra Karimzadeh, Masakatsu Miyajima, Reza Hassanzadeh, Reza Amiraslanzadeh, and Batoul Kamel. A gis-based seismic hazard, building vulnerability and human loss assessment for the earthquake scenario in tabriz. *Soil Dynamics and Earthquake Engineering*, 66:263–280, 2014.

[25] Andrzej Kijko and MA Sellevoll. Estimation of earthquake hazard parameters from incomplete data files. part ii. incorporation of magnitude heterogeneity. *Bulletin of the Seismological Society of America*, 82(1):120–134, 1992.

[26] A. Komeijani. Overview of the formation of the world trade organization and analysis of the legal and economic aspects of iran's accession to the wto. Magazine-the Economy, to Research and Economic Policies, 6: 5-30., 1996.

[27] Khuzestan Water Plant Authority (KWPA). Agricultural cadastral mapping and crop identification of dezful, iran. 2019.

[28] E Levin, A Zarnowski, JL McCarty, J Bialas, A Banaszek, and S Banaszek. Feasibility study of inexpensive thermal sensors and small uas deployment for living human detection in rescue missions application scenarios". the international archives of the photogrammetry. *Remote Sensing and Spatial Information Sciences*, 8, 2016.

[29] Baozhen Liu, Hang Wu, Weihua Su, and Jinggong Sun. Sector-ring hog for rotation-invariant human detection. *Signal Processing: Image Communication*, 54:1–10, 2017.

[30] Ying Liu. Studies on seismic identification and reinforcement design of building structures. *Open Journal of Civil Engineering*, 8(03):292, 2018.

[31] K Maiti. Bhuj earthquake mapping surface features using remote sensing data. *M. Tech Dissertation submitted to Department of Earth Sciences, University of Roorkee, Roorkee (Uttaranchal)*, 185, 2001.

[32] S1 McClusky, S Balassanian, A Barka, C Demir, S Ergintav, I Georgiev, O Gurkan, M Hamburger, K Hurst, H Kahle, et al. Global positioning system constraints on plate kinematics and dynamics in the eastern mediterranean and caucasus. *Journal of Geophysical Research: Solid Earth*, 105(B3):5695–5719, 2000.

[33] N N. AMBRASEYS and K A. SIMPSON. Prediction of vertical response spectra in europe. *Earthquake Engineering & Structural Dynamics*, 25:401 – 412, 04 1996.

[34] David J. Maguire David W. Rhind Paul A. Longley, Mike Goodchild. *Geographic Information System and Scienc.* Wiley; 3 edition, 2011.

[35] Saied Pirasteh, J Li, and I Attarzadeh. Implementation of the damage index approach to rapid evaluation building resistance for earthquakes. *Earth Science Informatics. doi*, 10, 2015.

[36] Pirasteh Saied, Mahmoodzadeh Amir, and Mahtab Alam. Integration of geo information technology and survey analysis for development in mitigation study against earthquake: a case study for esfahan iran. *Disaster Advances*, 1(3):20–26, 2008.

[37] AK Saraf, P Mishra, S Mitra, B Sarma, and DK Mukhopadhyay. Remote sensing and gis technologies for improvements in geological structures interpretation and mapping. *International Journal of Remote Sensing*, 23(13):2527–2536, 2002.

[38] UNISDR. Roles, mandates and results of key un entities. disaster risk reduction in the united nations. United Nations, 2013.

[39] Shanshan Zhang, Rodrigo Benenson, Mohamed Omran, Jan Hosang, and Bernt Schiele. How far are we from solving pedestrian detection? In *Proceedings of the IEEE Conference on Computer Vision and Pattern Recognition*, pages 1259–1267, 2016.

[40] Zhenzhong Zhang and Lanmin Wang. Geological disasters in loess areas during the 1920 haiyuan earthquake, china. *GeoJournal*, 36(2-3):269–274, 1995.

7

Application of Unmanned Aircraft Systems for Coastal Mapping and Resiliency

Michael J. Starek
Conrad Blucher Institute for Surveying and Science, Texas A&M University-Corpus Christi, USA

Melanie Gingras
Conrad Blucher Institute for Surveying and Science, Texas A&M University-Corpus Christi, USA

Gary Jeffress
Conrad Blucher Institute for Surveying and Science, Texas A&M University-Corpus Christi, USA

This chapter provides an overview of UAS technology with focus on surveying and mapping. A case study on the use of UAS for coastal monitoring to aid community resiliency following a hurricane impact is also presented. The information and applications of UAS presented herein are applicable to a variety of UN SDGs including sustainable land use for "Life on Land" and sustainable agriculture for crop security and "Zero Hunger".

7.1 Introduction

Coastal zones are some of the most dynamic environments on Earth and some of the most threatened. According to the United Nations (UN) Atlas of the Oceans, 44% of the world's population (more people than inhabited the entire globe in 1950) live within 150 kilometers of the coast [2]. Growing population demand, impact from storms, climate change, and relative sea level rise puts coastal communities at the forefront of engineering and scientific efforts for sustainable and resilient development. As part of the UN Sustainable Development Goals (SDGs) for 2030, Goal 11 specifically identifies "Sustainable

Cities and Communities". In the context of coastal communities, resilience is a measure of the extent to which a coast is able to respond to external pressures without losing actual or potential functions [7]. Improving coastal resilience is considered to be a cost-effective approach to prepare for increasingly uncertain coastal environments. The ability to rebound more quickly can reduce negative human health, environmental, and economic impacts [5].

At a base level in building more resilient communities is the need for updated geospatial information. Coastal communities rely on adequate and timely geospatial data to guide decision-making in the event of a disaster, mitigate coastal erosion, and plan for sustainable development and growth. Emerging technology for the acquisition of spatially referenced data are rapidly transforming science, society, and decision-making. At the forefront of this revolution are technological advancements in unmanned aircraft systems (UAS), more commonly referred to as drones. UAS enable us to rapidly map and monitor our evolving world, with unprecedented detail, to tackle a range of problems in support of UN SDGs.

7.2 Overview of UAS Technology

UAS provide a new paradigm for aerial surveying and mapping. UAS are used to collect overlapping imagery, which can then be post-processed to derive two-dimensional (2D) and three-dimensional (3D) mapping products for geographic information. These data products can be used to characterize built and natural environments at a level of spatial detail previously unattainable or not practical with traditional remote sensing techniques. Spectral data can also be acquired from multispectral sensors onboard the UAS for performing traditional remote sensing tasks such as mapping vegetation health. Compared to traditional aircraft or satellite remote sensing, UAS provides certain advantages: rapid deploy capabilities, flexibility to target ideal weather conditions and for temporal repeatability, hyperspatial image resolutions, and cost-effectiveness at localized geographic extent [10, 12].

Application of UAS for surveying and mapping is primarily conducted with small UAS equipped with consumer-grade digital cameras and miniaturized sensors. Currently, the Federal Aviation Administration (FAA) of the United States classifies small UAS as weighing less than 25 kg (55 lbs) including payload. Often these platforms weigh only a few kgs with payload. UAS can be broadly classified into two types: rotary craft and fixed-wing (Figure 7.1).

Rotary craft typically provide more flexibility in sensor payload integration, enable more stabilized flight in windy conditions, provide vertical take-off and landing (VTOL) capability, and are better suited for inspection surveying (e.g. hovering), videography, and flying lower or slower to collect higher resolution imagery with reduced motion blur. In contrast, fixed-wing UAS

FIGURE 7.1
Examples of different types of small UAS mapping platforms operated by of the Conrad Blucher Institute for Surveying and Science at Texas A&M University-Corpus Christi. (a) Rotary platform called the DJI Phantom 4 Pro; (b) Fixed-wing UAS called the SenseFly eBee being hand-launched by the author; (c) Hybrid UAS called the Wingtra that is a vertical take-off and landing (VTOL) system which converts to fixed-wing during flight.

are generally more efficient at mapping larger areas due to more effective lift generation for a given payload thereby maximizing battery endurance [4, 9]. The majority of current hobbyist UAS activity is centered around small rotary platforms, such as the DJI Phantom series (see Figure 1a), due to their widespread availability, ease of use, and lower cost compared to commercial fixed-wing platforms designed for mapping. There are also hybrid platforms integrating rotary and fixed-wing design that provide the advantages of both, such as a vertical take-off and landing with fixed-wing flight (Figure 1c). With continued technological progression, the endurance and efficiency of UAS will improve and differences in performance capabilities between platform types will become more blurred.

Main system components of a UAS include the global navigation satellite system (GNSS) receiver for position, inertial measurement unit (IMU) for orientation, radio link for communication, onboard processor, and sensor/camera for data acquisition. Autonomous flight is performed using an autopilot that is fed by the onboard GNSS, IMU, and other sensors. UAS often include other navigation aiding sensors, such as magnetometers and pivot tubes, as well as sense-and-avoidance features with built-in artificial intelligence to reduce risk of collision. Sensors equipped on UAS can range from visual RGB digital cameras to multispectral and hyperspectral sensors, thermal imagers, and active sensors such as light detection and ranging (lidar). UAS appropriate for mapping include easy-to-use onboard autopilots for controlling image acquisition, which is important for image stitching and photogrammetric applications. The onboard radio is also essential because it dictates the strength and length of the command and control link between the operator's Ground Control Station (GCS or remote controller; typically, a laptop, tablet, or smart phone with mission control software for automated flight) and the UAS, which has implications for both data collection and safety.

7.3 Aerial Mapping with UAS

Two-dimensional image mosaics are the most commonly created products from imagery collected by a UAS. The simplest way to create a mosaic from aerial imagery is by using photo stitching software, which combines a series of overlapping aerial photographs into a single image called an image mosaic. However, this approach does not account for perspective distortion and camera lens distortion thereby resulting in non-uniform scale and inaccurate distance measurements for geospatial applications [6]. An orthomosaic is a mosaicked series of overlapping aerial images that have been geometrically corrected (orthorectified) using an elevation model of the terrain to give them a uniform scale. Orthorectification removes distortion from the images by adjusting for topographic relief, lens distortion, and camera tilt creating an aerial map. Orthormosaics are a fundamental geographic information system (GIS) data product used in a variety of planar mapping applications.

Another commonly created geospatial data product from UAS imagery is a digital surface model (DSM). A DSM provides a "2.5D" digital representation of the three-dimensional landcover and exposed ground. It is created from a dense set of georeferenced 3D point measurements called a point cloud derived from the overlapping UAS imagery using specialized photogrammetric processing. The DSM is a raster data type that can be easily ingested into a GIS for further analysis such as to measure the height of an object or estimate the volume of an object. If the data is tied to a vertical datum, a DSM can be considered a specialized type of digital elevation model (DEM) where raster cell values represent the elevation of the exposed ground and landcover. By generating DSMs from repeat UAS surveys across time, the change in elevation can be measured by raster differencing. For example, along a sandy beach, repeat UAS surveys can be used to generate DSMs and then differenced to measure change in elevation and estimate volumetric sediment loss or gain. The generated DSM can also be used by photogrammetric software to orthorectify the UAS imagery and create the 2D orthomosaic product described above. Furthermore, through automated ground point filtering, a bare-earth DEM (called a digital terrain model or DTM) can be created from the set of points identified as stemming from the ground. The accuracy and success of such a process depends highly on the density of landcover and the method implemented by the software. The sections below outline UAS mission planning, flight design, and photogrammetric processing to derive 2D and 3D mapping products.

7.3.1 Mission Planning

Planning a mapping mission with a small UAS entails a number of considerations. A first-order decision to be made is whether the flight will be conducted under autonomous control with mission planning and flight control software using GNSS waypoints or conducted manually. The decision of whether to use manual or autonomous control hinges on your application. As a general rule, autonomous control is more useful when one is trying to fly in a systematic pattern to acquire imagery for aerial mapping and photogrammetry purposes. In contrast, manual control is generally used for events that require reacting to information in real time such as inspection surveying (e.g. wind turbine) or video monitoring and streaming of events in real time (e.g. search and rescue operations). Both types of missions can be flown in either manner, or in a hybrid of manual and automatic control [6].

Prior to conducting any UAS flight operation it is important to analyze the area to be mapped before liftoff. The area should be visually inspected by walking or driving the premises or otherwise evaluated before the mission starts so as to identify obstacles such as power lines, water bodies, large trees, sensitive areas, or other potential pitfalls. It is good practice to use existing aerial or satellite imagery to inspect the area and plot out a flight before take-off. Pilots should know how to competently fly their UAS, even if they plan to use it primarily for autonomous mapping missions. This means understanding the launch and recovery procedure, flight control software, endurance limitations, trouble shooting, and safety procedures. UAS should remain within the visual line of sight of pilots and ground observers unless the pilots have special regulatory permission by their respective regulatory agency for beyond visual line of sight (VLOS) operation.

7.3.2 Flight Design

Designing a flight plan for autonomous mapping is done using mission planning and control software (Figure 7.2).

These are often proprietary software provided by the UAS manufacturers; however, there are open-source solutions. Mission Planner is a current open-source software package widely used by the hobbyist community that works with various compatible autopilots. There are also numerous "flight apps" designed to work off of smart phones and tablets for programming flight missions. The functionality for most of this software, regardless of modality, is generally similar. UAS aerial mapping missions are usually flown in a specific pattern of parallel lines, commonly described as transects, which are connected by a series of geopositional waypoints [6]. These flight patterns are a method of ensuring that the UAS captures an adequate quantity of images at the desired amount of image overlap. Overlap is essential for high fidelity photogrammetric processing to derive mapping products and described in more detail below. Once the mission plan is completed to satisfaction, it is uploaded to the UAS controller and the system is ready for launch. The operator can then launch the drone using the manufacturer's prescribed launch procedure.

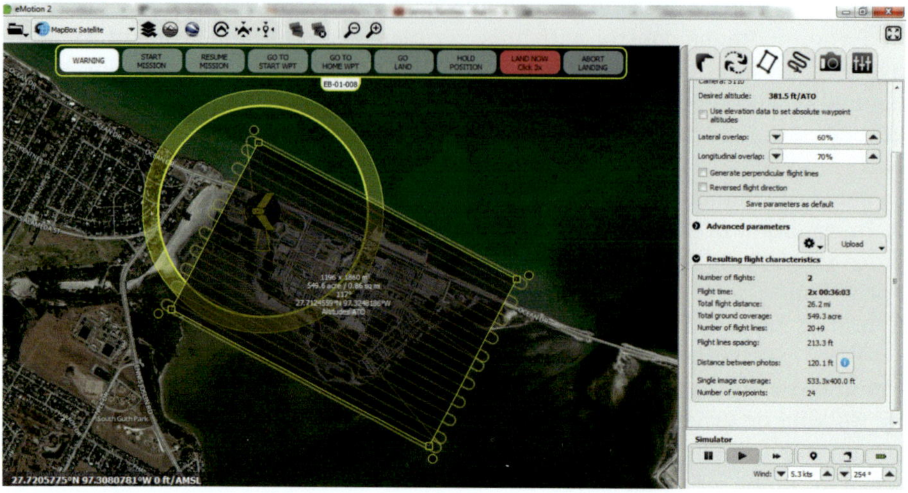

FIGURE 7.2
Example of a UAS flight design created with proprietary mission planning software called eMotion 2 by SenseFly. Parallel transects can be scene along with image overlap settings. The area shown is the Texas A&M University-Corpus Christi island campus.

Imagery collected during flight is typically stored onboard the platform using a SD memory card located in the camera or via other data storage means. There are services that aim to provide real-time wireless transfer of collected imagery to a cloud architecture during flight for downstream rapid processing using a cellular network or other wireless connection mode. Similar approaches can be implemented for wireless transfer of image data from platform to ground control station. But feasibility of these approaches will depend on bandwidth, available wireless connectivity, size and amount of imagery, and other factors. Presently, the most common operating mode for image acquisition is onboard storage and download of imagery after landing.

7.3.3 Image GSD and Overlap

To successfully perform UAS mapping, proper flight design is critical. Two important parameters to consider in designing any UAS survey are ground sample distance (GSD) and image overlap. GSD is the projected pixel width on the ground and is a function of the camera focal length, physical size of the individual sensor elements inside the camera (called pixel pitch), and flying height above ground level. Figure 7.3 shows this relationship.

GSD can be estimated as follows:

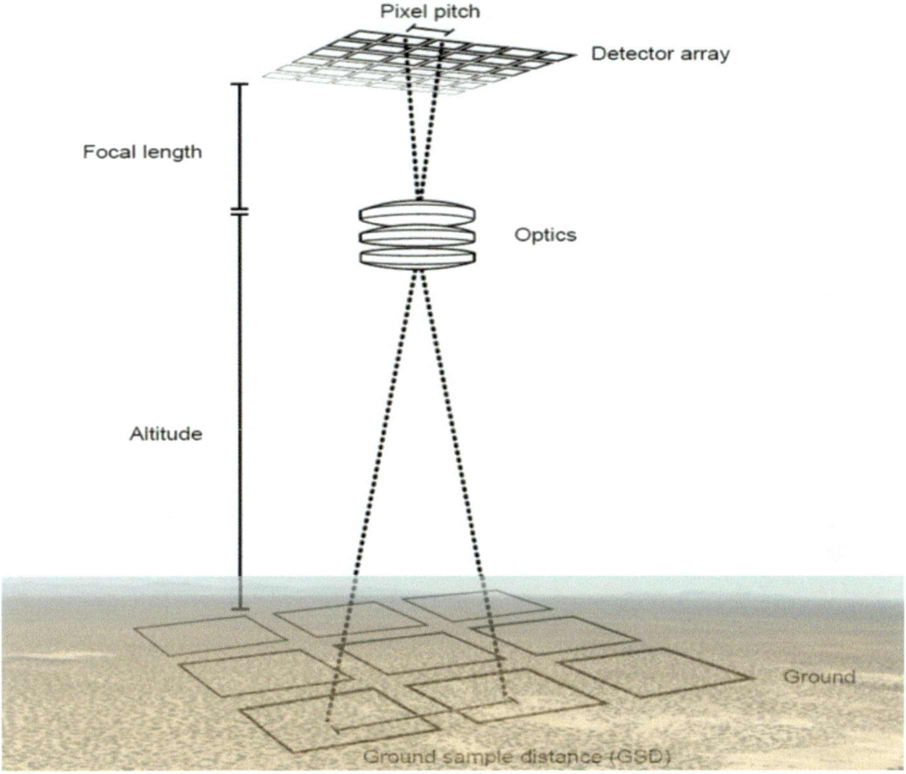

FIGURE 7.3
Relationship between ground sample distance (GSD), flying height above ground, camera pixel pitch and focal length. (Modified from [6]).

$$GSD = \frac{(p)(h)}{f} \tag{7.1}$$

where f = focal length of the respective camera, h = flying height above ground level, p = pixel pitch. During flight design, mission planning software will allow one to specify the specific camera model and focal length or it will be autodetected by the software. The software then allows one to adjust the nominal flying height above ground level to achieve a desired GSD for a given camera model. Dependent on project GSD requirements, the flight altitude may need to be adjusted during the mission to compensate for large variation in terrain elevation. Certain UAS can be programmed to use elevation models to follow the terrain for image acquisition.

The second primary consideration for flight design is image overlap. Sufficient overlap is crucial due to the type of photogrammetry employed to process the UAS imagery and reconstruct a 3D model of the imaged scene. In a typical scenario, images are acquired with at least 75% frontal overlap along the flight line and at least 60% side overlap between adjacent flight lines. For most scenarios, it is recommended that images are acquired with a regular grid pattern (Figure 7.4).

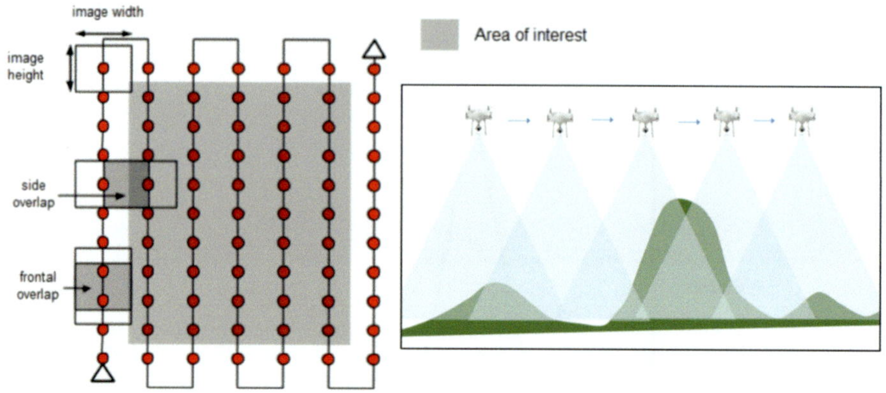

FIGURE 7.4
Grid image acquisition plan showing the meaning of frontal overlap and side overlap (Source of grid image: [1]). Inlay shows along track frontal overlap.

Over more complex environments, more overlap and different flight patterns may be necessary to achieve desired mapping results. For example, terrain covered by trees and dense vegetation or flat terrain with homogeneous texture pose challenges for photogrammetry software to auto-identify characteristic points (keypoints) shared between overlapping images. This correspondence is necessary for 3D reconstruction from overlapping images. Such cases may require more overlap in both directions and fling higher can sometimes improve results. For details on flight design for UAS mapping, the reader is referred to [1] and [3].

7.3.4 Structure-from-Motion Photogrammetry

Photogrammetry is the science of making measurements of objects from photographs. Common survey products include 2D orthorectified image mosaics and 3D models of the imaged scene. Traditional airborne photogrammetry utilizes large-format metric cameras precisely calibrated such that their interior properties, like focal length, are accurately known. However, metric cameras are expensive and not conducive for widespread use of UAS for mapping applications. In contrast, small UAS mapping typically employs consumer-grade digital cameras using a technique called Structure-from-Motion (SfM) photogrammetry. SfM exploits information from multiple overlapping images to extract 3D object information and perform self-calibration of the camera negating the need for precisely calibrated metric cameras. SfM derives three-dimensional structure from two-dimensional image sequences through movement of the camera thereby providing different perspective views of the scene. By using the UAS as the moving platform, SfM can be implemented with an onboard camera by acquiring images with sufficient overlap as previously explained. The SfM image processing workflow is summarized as follows [8]:

1. Image sequences are input into the software and a keypoint detection algorithm, such as the scale invariant feature transform (SIFT), is used to automatically extract features and find keypoint correspondences between overlapping images using a keypoint descriptor.

2. A "bundle block" adjustment is performed to minimize the errors in the correspondences by simultaneously solving for camera interior and exterior orientation. Based on this reconstruction, the matching points are verified and their 3D coordinates calculated to generate a sparse point cloud. Without any additional information, the coordinate system is arbitrary in translation and rotation and has inaccurate scale.

3. To further constrain the problem, ground control points (GCPs) can be used to transform the point coordinates to a real-world coordinate system and improve rectification. Similarly, initial camera positions from the onboard GNSS can also be used to constrain and georeference the solution.

4. Finally, the interior and exterior orientation for each image are used as input into a Multi-View Stereo (MVS) algorithm, which attempts to densify the point cloud by projecting every image pixel, or at a reduced scale.

The base data product output from UAS-SfM processing is a densified 3D point cloud of the imaged scene colored by the RGB values of the camera. UAS-SfM point clouds can be considered hyperspatial (densely-spaced points exceeding 1000 pts/m2) due to the high camera resolution (e.g. 20 MP+) and typical low altitudes at which data are collected (Figure 7.5).

The 3D point cloud can then be used by software to create a DSM of the imaged scene, which can subsequently be used to correct distortion in the imagery and generate a seamless orthomosaic. As a result, geospatial data products typically output from a UAS-SfM survey are the following: 3D point cloud, DSM, orthomosaic. Several commercial and open-source SfM software options are available to process UAS imagery. Widely used commercial software include Pix4D and Agisoft PhotoScan. Open Drone Map is a popular open source solution. Figure 7.6 summarizes the UAS-SfM processing workflow.

UAS are integrated with onboard GNSS receivers that can be used to geotag the acquired imagery with respective latitude and longitude coordinates thereby allowing them to be positioned on earth (georeferencing). However, the quality of GNSS receivers onboard these platforms can highly vary. Most

FIGURE 7.5
(left) RGB colored 3D point cloud of Little St. George Island, FL USA. (right) RGB colored 3D point cloud of a wetland study site located on Mustang Island, TX USA. These point clouds were generated from UAS imagery using structure-from-motion photogrammetry. They are very dense with >1000 pts/m^2

small UAS come equipped with low-accuracy GNSS receivers used for navigation only that provide 3D positional accuracies on the order of several to tens of meters whereas some systems come equipped with differential GNSS receivers capable of delivering horizontal and vertical positional accuracies down to a few centimeters. To improve positional accuracy of derived mapping products, ground control targets can be uniformly distributed throughout the study site and their positional coordinates accurately surveyed using a traditional land surveying approach, such as real-time kinematic (RTK) GNSS. These ground control points can then be input into the SfM processing workflow to more accurately locate the derived geospatial data products.

7.4 Regulations

Any UAS flight conducted for commercial or hobbyist use must ensure that it adheres to appropriate airspace regulations and that proper safety precautions are taken. For example, in the United States, commercial use of UAS for aerial surveying or other applications is regulated by the FAA's small UAS rule (called Part 107), that became effective on August 29, 2016. Current highlights of the operational rules include: aircraft must weigh less than 25 kg (55 pounds) and be registered with the FAA; operation is limited to Class G airspace, within visual line-of-sight, under 400 feet, daylight hours, and at or below 161 kilometers per hour. Part 107 provides an option to apply for a certificate of waiver, which allows for a small UAS operation to deviate from certain operating rules if the FAA finds that the proposed operation can be performed safely.

Finally, it is important to mention that at the time of this writing (2019),

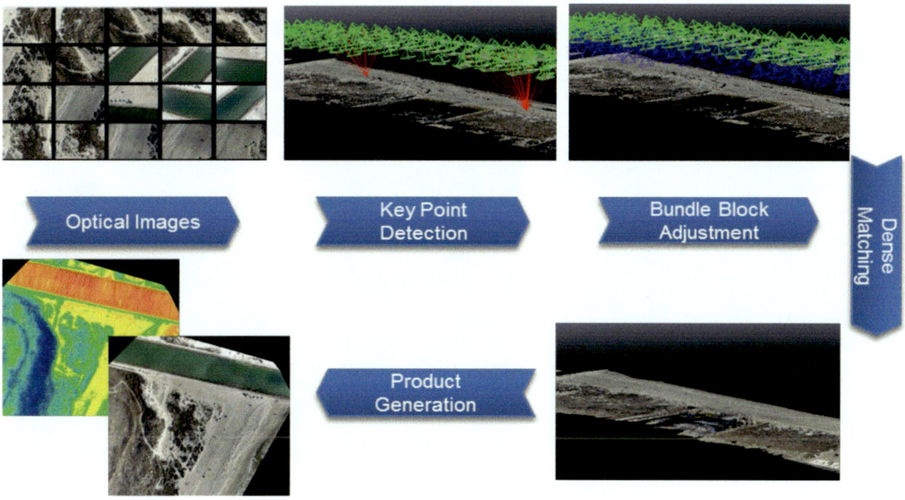

FIGURE 7.6
SfM workflow to process UAS image sequences into densified 3D point cloud, DSM, and orthomosaic. Example here is from Packery Channel; an open-water inlet located along the Texas Gulf coast

UAS technology and regulations are still emerging. The technology, regulations, and capabilities for surveying will rapidly evolve over the coming years. However, the sensor types, data products and general methods for data acquisition and processing, as discussed here, will likely remain similar.

7.5 Case Study: Hurricane Harvey Impact Assessment

Hurricane Harvey formed as a tropical storm over the Atlantic Ocean on August 17, 2017, weakened to a tropical depression as it crossed the Yucatan Peninsula, but rapidly intensified to a Category 4 strength hurricane in the Gulf of Mexico just before making landfall on the Texas coast 50km east of Corpus Christi on August 25. Hurricane Harvey caused severe wind damage in coastal towns, extensive flooding as it stalled over Texas from August 25–30, 2017, and unprecedented rainfall with totals in Cedar Bayou reaching 1318 mm (58.59″) in just three days [11].

Shortly after the hurricane, the Measurement Analytics (MANTIS) Lab with the Conrad Blucher Institute for Surveying and Science at Texas A&M University-Corpus Christi conducted small UAS surveys at various beach locations along Mustang Island, TX to assess coastal erosion. Study sites investigated included Gulf-facing beach sites at the Port Aransas jetty, which was exposed to the southern edge of the eyewall, and Newport Pass located about 25 km to the south (Figure 7.7).

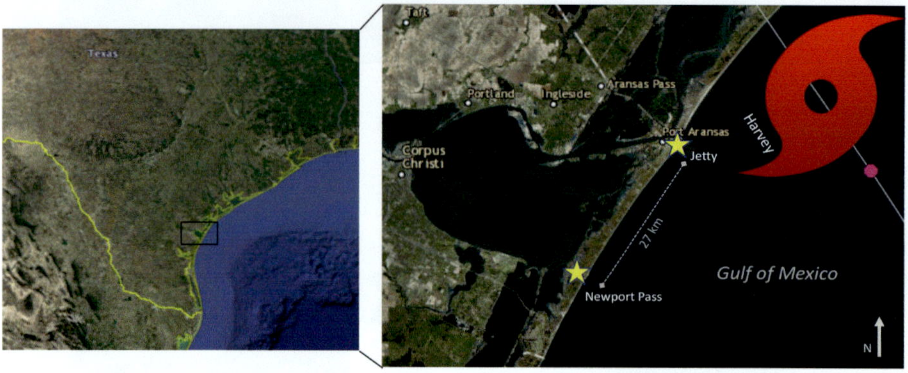

FIGURE 7.7
Hurricane Harvey Mustang Island, TX study sites

Change at the beach locations was distinguished by comparing post-hurricane UAS survey data with pre-storm airborne lidar data collected by the United States Army Corps of Engineers (USACE) National Coastal Mapping Program. The USACE airborne lidar survey was conducted in August-October 2016. The MANTIS lab collected post-impact UAS data in September-October 2017. Newport Pass study site was surveyed using a rotary DJI Phantom 4 Pro (Figure 7.1a), and Port Aransas study site was surveyed using a fixed-wing SenseFly eBee. Both systems were integrated with 20 MP RGB cameras. Ground control targets were used to ensure high-accuracy georeferencing and to tie the data to a vertical datum. All UAS imagery were processed using SfM to generate orthomosaics and DSMs. Because the UAS DSM raster cell values were referenced to a vertical datum for elevation, they represent elevation of the exposed ground and landcover and are referred to herein as DEMs.

Figure 7.8a is a traditional aerial image acquired prior to the hurricane at the Port Aransas South Jetty site, and Figure 7.8b shows a high resolution (3 cm) UAS orthomosaic produced from a survey a few weeks after the hurricane.

Results show that the most noticeable change was the beach erosion and scouring near the jetty and along the roadway where as much as 4 m was devoured by the large wave action and storm surge. The UAS-derived mapping products provided up-to-date information to the County to aid damage assessment of the jetty and guide their recovery efforts. At Newport Pass, there was perhaps the most noticeable change where a previously weakened dune blowout (Figures 7.9a and c) was completely breached by the storm surge during the hurricane and a temporary inlet to Corpus Christi Bay was formed (Figures 7.9b and d). This resulted in elevation changes upwards of 4 m losses where the channel cut through the dune, and gains where the flooded bay wa-

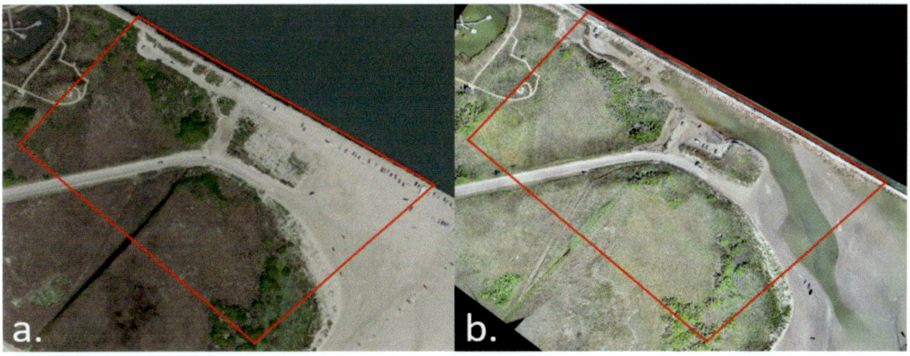

FIGURE 7.8
These images depict storm-related changes to the Port Aransas South Jetty site. (a) Google Earth aerial image from south of the Aransas Pass jetty before Hurricane Harvey; (b) High resolution UAS imagery from south of the Aransas Pass jetty after Hurricane Harvey.

ters and strong northwest winds from the southern eyewall pushed sediment and water against the backside (bayside) of the dune (Figure 7.9e).

In support of a different initiative related to Harvey, UAS surveys were conducted over a heavily damaged inland neighborhood near Rockport, TX where some of the strongest wind fields were observed during the storm. The effort was conducted as part of a volunteer emergency response effort in collaboration with engineers from the University of Notre Dame and the United States National Science Foundation (NSF) Geotechnical Extreme Events Reconnaissance (GEER) project. The purpose of the NSF GEER effort is to collect high-resolution remote sensing data along with information on structural damage recorded by reconnaissance teams on the ground to evaluate performance of structures during disaster events. Figures 7.10a and 7.10b show aerial images of the neighborhood prior to Hurricane Harvey and Figures 7.10c and 7.10d show parts of a UAS orthomosaic of the same area generated after the storm. Figure 7.10e is an oblique UAS image of six houses before the storm and Figure 7.10f is an oblique image of a 3D point cloud constructed from the UAS imagery using SfM. The collected UAS information from Harvey and other storms are being used by engineers to improve structural design and refine building codes for reducing impacts from hurricanes and other natural disasters. This effort will lead to more sustainable development and in return, more resilient coasts.

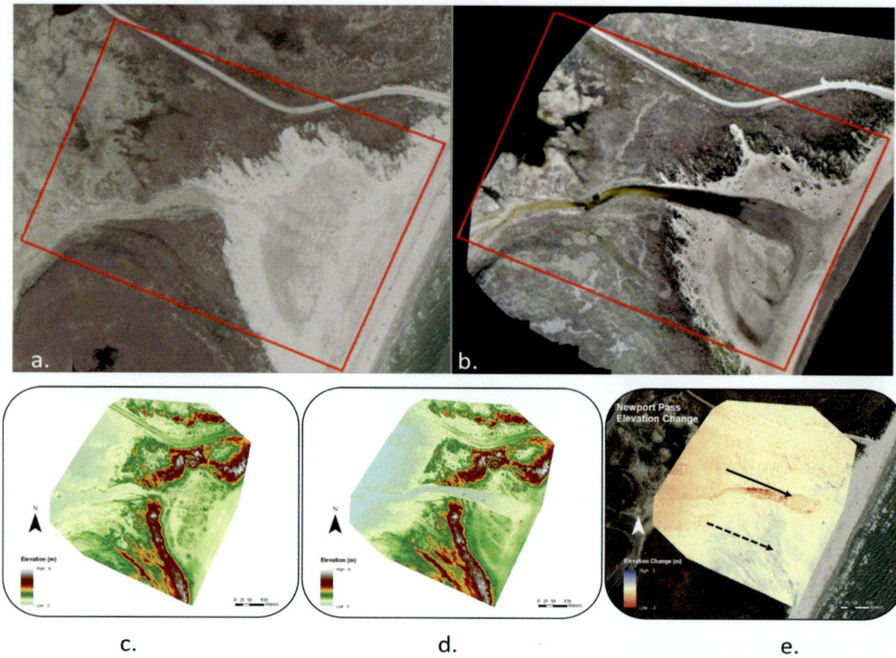

FIGURE 7.9
These images depict storm-related changes to the Newport Pass study site. (a) Aerial image south of Newport Pass before Hurricane Harvey; (b) UAS imagery of Newport Pass after Hurricane Harvey; (c) DEM of Newport Pass created from a 2016 USACE airborne lidar survey before Hurricane Harvey; (d) DEM of Newport Pass created from a UAS survey after Hurricane Harvey; (e) Computed elevation change of Newport Pass due to Hurricane Harvey (before DEM surface subtracted from after DEM surface). Solid arrow shows zone of large erosion stemming from the breach. Dashed arrow shows zone of deposition.

7.6 Conclusion

Engineering and scientific solutions for sustainable development of cities and communities requires updated geospatial information. For communities residing within the highly dynamic and vulnerable coastal zone, accurate and timely geospatial data is critical to aid disaster response, mitigate risks to coastal hazards, and plan for more sustainable and resilient infrastructure. UAS is transforming our ability to map and monitor our evolving world at unprecedented detail. As UAS technology continues to rapidly evolve, so will the speed at which we can attain increasingly accurate and detailed spatial

FIGURE 7.10
These images depict storm-related changes to a neighborhood near Rockport, TX: (a) Traditional aerial imagery of a Rockport neighborhood before Hurricane Harvey from Google Earth; (b) Zoomed in view of the aerial image showing the house before Hurricane Harvey; (c) UAS imagery of a Rockport neighborhood after Hurricane Harvey; (d) Zoomed in view of UAS imagery showing damage to the same house after Hurricane Harvey shown in the aerial image above; e) oblique imagery of a Rockport house before hurricane Harvey; (f) oblique view of a SfM derived 3D point cloud of the same house after hurricane Harvey.

information. With increasing reliance upon geospatial technology and data to inform our decisions, it becomes ever more important to understand the applications and limitations with such measurements and how to effectively apply them to better navigate our future world. It becomes ever more important to understand the accuracies associated with such measurements and how to effectively apply them to better navigate our future world.

Bibliography

[1] Designing the image acquisition plan, article 202557459, May 2018.

[2] United nations atlas of the oceans, October 2018.

[3] Jonathan L Carrivick, Mark W Smith, and Duncan J Quincey. *Structure from Motion in the Geosciences.* John Wiley & Sons, 2016.

[4] Ismael Colomina and Pere Molina. Unmanned aerial systems for photogrammetry and remote sensing: A review. *ISPRS Journal of photogrammetry and remote sensing*, 92:79–97, 2014.

[5] National Research Council et al. Disaster resilience: A national imperative, 2012.

[6] F Greenwood. How to make maps with drones. *Drones and aerial observation, July*, pages 355–47, 2015.

[7] Richard JT Klein, Marion J Smit, Hasse Goosen, and Cornelis H Hulsbergen. Resilience and vulnerability: coastal dynamics or dutch dikes? *Geographical Journal*, pages 259–268, 1998.

[8] Richard Slocum and Christopher Parrish. Simulated imagery rendering workflow for uas-based photogrammetric 3d reconstruction accuracy assessments. *Remote Sensing*, 9(4):396, 2017.

[9] Michael J Starek, Thomas Davis, Dan Prouty, and Jacob Berryhill. Small-scale uas for geoinformatics applications on an island campus. In *2014 Ubiquitous Positioning Indoor Navigation and Location Based Service (UPINLBS)*, pages 120–127. IEEE, 2014.

[10] Darren Turner, Arko Lucieer, and Christopher Watson. An automated technique for generating georectified mosaics from ultra-high resolution unmanned aerial vehicle (uav) imagery, based on structure from motion (sfm) point clouds. *Remote sensing*, 4(5):1392–1410, 2012.

[11] Geert Jan Van Oldenborgh, Karin Van Der Wiel, Antonia Sebastian, Roop Singh, Julie Arrighi, Friederike Otto, Karsten Haustein, Sihan Li, Gabriel Vecchi, and Heidi Cullen. Attribution of extreme rainfall from hurricane harvey, august 2017. *Environmental Research Letters*, 12(12):124009, 2017.

[12] AS Woodget, PE Carbonneau, Fleur Visser, and IP Maddock. Quantifying submerged fluvial topography using hyperspatial resolution uas imagery and structure from motion photogrammetry. *Earth Surface Processes and Landforms*, 40(1):47–64, 2015.

Part III

Supporting SDGs: Legal, Policies and Institutional Components and Capacity Building

8

Legal and Policy Paths for Effective Sustainable Development

Harlan J. Onsrud
University of Maine, USA

This chapter begins by reviewing basic economic and legal principles that have been used effectively by governments in supporting the marketplace in regard to traditional economic resources such as land, labor and capital. Because information and knowledge can and often do replace the need for traditional resources in contemporary societies, the chapter focuses as well on policies and laws promoting the growth of information economies to support SDGs implementation.

8.1 Introduction

Many of the UN Sustainable Development Goals (SDGs) are likely to be achieved most efficiently and effectively under legal and policy frameworks possessing certain basic legal and policy principles and frameworks [6]. If not already in place, recommended long-term principles and frameworks may be pursued concurrently or in tandem with short-term policy and legal adaptations that may be needed to address more immediate and pressing short-term SDG needs. Among SDGs that are unlikely to be addressed successfully without stable and well-reasoned long-term legal and policy frameworks in place include:

- **Goal 8.** Promote sustained, inclusive and sustainable economic growth, full and productive employment and decent work for all.
- **Goal 10.** Reduce inequality within and among countries.
- **Goal 16.** Promote peaceful and inclusive societies for sustainable devel-

opment, provide access to justice for all and build effective, accountable and inclusive institutions at all levels.

Most societies across the globe recognize the extreme importance of a vibrant marketplace as a prerequisite to sustainable development. They further recognize the strong role of government in creating appropriate legal bounds within which competitive markets in goods and services may thrive while concurrently providing safety-net services for the disadvantaged. That is, a nation with few natural resources but with a highly educated and innovative workforce may achieve a robust economy and sustainable development through primarily the development of information infrastructure and the provisioning of digital products and services. Yet, digital economies come as well with their own set of legal and policy challenges.

8.2 Fundamental Economic Policies Germane to Traditional Resources

Adam Smith, in his 1776 treatise on the Wealth of Nations, postulated that the ideal private good has the characteristics of *excludability, rivalry,* and *transparency.* Land, labor and capital exhibit these ideal marketplace characteristics but only fully through enforcement of appropriate laws.

In exploring these characteristics and the role of law in their support, we may use the illustrative and familiar example of land. *Excludability* in land is achieved by granting owners the ability to keep others from using or trespassing on their private property through action by government, typically through the judiciary and police enforcement. Land owners also have the legal right typically to exclude others using physical means, such as through the use of fences, buildings and other barriers.

Rivalry is the concept whereby my use of a resource deprives or affects your use of the same resource. If I use land to build a house on it, you can't build your house on it. Thus, land is far more rivalrous than a resource like digital information. Millions can benefit from an information resource without degrading or depriving anyone else's use of the same resource but this is not typical for land.

Transparency is achieved when a buyer has the ability to see the quality and prices of all competing goods in the marketplace. In land markets, our legal systems often support this characteristic by mandating that all land ownership records are publicly and readily accessible to all and an operational system is supported that guarantees or insures the validity of the title rights acquired.

Because land has the characteristics of a private good, the sale and trade of land parcels through the open marketplace has worked well and can continue to work well in creating wealth both for individuals and for national populations as a whole. The same holds true for other traditional economic resources possessing the classic characteristics of private goods. Information system goods and services, on the other hand, lack some of these characteristics and therefore must typically be transformed through action of law to exhibit the classic characteristics of private goods.

8.3 Role of Legal Controls

What then is the role of law in marketplace contexts? What should be the role of law in transitioning from least technologically developed cities and nations to smart cities and economically vibrant nations? Most western culture legal scholars, policy makers and economists argue that the priority of controls in society, in order, should be: (1) the marketplace, (2) private arrangements and (3) the law. Laws should be used only as a last resort in managing society's affairs. If the marketplace is truly free and open, the market itself is a far better means for setting prices and controlling quality than any state-imposed guesses enacted into law. If equally sophisticated private parties can work out pricing and quality issues for themselves through individualized contracts, that negotiation will give far more efficient results than conditions imposed by government through the law.

Thus, new laws are often NOT the best solution in addressing many societal challenges. When competitive markets exist, market forces are often a far better choice in establishing relations between parties than are detailed legal regulations. In Western culture, we often argue that the law should step in to regulate only where the free and open marketplace isn't working. In these instances, the primary role of the law should then be to:

(a) correct the marketplace to return it to being open, free, and competitive, and/or

(b) construct means to provide important goods and services desired by society that will not otherwise be produced by the marketplace.

In the latter case, lack of a marketplace to produce goods is due often to the inherent characteristics of the goods. That is, highly desirable goods such as street lighting or military defense are non-rivalrous and the benefits once supplied are difficult to exclude from others. It is inefficient for the marketplace to supply them. Thus, in these instances, governments often step in to produce the goods directly. Alternatively, government might convert a good by action of law such that the good then possesses rivalry and excludability. By example, copyright law, and its enforceable sanctions if violated, make creative works excludable. Copyright law provides an incentive to authors to make their knowledge, ideas, discoveries, and creative works widely available to the benefit of all in society. With the production and dissemination of the works of millions of authors and artists available in the marketplace, science and the useful arts are advanced. Without protection making the works legally excludable, such works often would be readily and widely distributed to millions with no compensation to the creators. This creates a strong disincentive for the sharing and distribution of creative works by authors and would make their sale, licensing and distribution far more burdensome.

While government by itself or through its agents would be unable to create the variety and extent of valuable goods and services provided by the marketplace, governments serve a strong role in establishing conditions to allow the marketplace to thrive. Among areas in which the law has a likely justifiable role in supporting both established and emerging economies include settling disputes, protecting citizens against excessive or unfair private power, protecting citizens against excessive or unfair government power, ensuring people an opportunity to enjoy the minimum decencies of life, and maintaining order.

8.4 Policies and Laws Germane to Digital Economy Resources

When appropriately bounded by laws to control negative behavior within the marketplace by private and even government parties, the marketplace has been very efficient and effective in promoting economic growth. However, economies falter when laws preventing skewing of the market and protecting human rights fail to be enacted and enforced. Because information technology advances and the digital resources they create continue to advance rapidly, it is very difficult for governments to respond quickly and effectively through government executive actions, legislation, and constitutional amendments to dampen or eliminate their adverse effects. A burgeoning information economy and developing cyberinfrastructure inevitably raise conflicts and legal challenges related to:

- intellectual property rights such as those involving copyright, patent, trademark, trade secret, and sui generis data laws,
- privacy rights supporting the ability of individuals to control the collection, dissemination and use of information about themselves and to prevent overly intrusive behavior by others,
- security, encryption, and data management practices,
- hacking and other cybercrimes,
- citizen access to government data, records, and online services,

- control of private power such as through anti-trust and anti-competition law,
- licensing and contract self-help mechanisms,
- liability in the provisioning of data and online services, and
- jurisdiction over disputes.

In many countries, resolving new conflicts within the context of digital economies are being left largely to emerge through processes carried out primarily in the private marketplace by competing business, consumer, and public interests. Government intervenes with new laws often only when technology and changed societal conditions have advanced to the point where problems and inequities have become clearly evident through the process of litigation in the courts. In other nations, government is far more proactive. Regardless of the timing of government intervention, there exist numerous conflict areas in which government may be justified in stepping in to resolve digital economy disruption, particularly in regard to failure of the market to achieve equitable results or protect human rights. The following paragraphs provide examples of illustrative instances.

8.4.1 Settling Disputes

Digital Tracking: Numerous disputes have arisen in modern contexts in the tracking of objects, humans, transactions by individuals, and interactions among individuals and their surroundings. These conflicts are continuing to grow as digital tracking becomes more pervasive. Tracked data, if stored somewhere, is subject to misuse and hacking. Law is often the best means of determining who should bear the costs of harms in such instances. For example, a breaching party such as a hacker may be inaccessible or unable to pay. The law might step in to establish rules for placing the damage burden on either victims, system developers, private operators, or government, or settle the dispute by distributing the burden among some or all of them.

Technology Gone Wrong: Assume that software in an autonomous vehicle in an emergency situation opts to hit three adult pedestrians in a crosswalk rather than one child and two pets located to the side of the travel lane. Laws enacted by elected representative government and enforced by the courts may be the best means for weighing and balancing the reasonableness of software algorithmic implementations and artificial intelligence decision-making, forcing corrections of technologies, and distributing the costs of harms.

8.4.2 Protecting People Against Excessive or Unfair Private Power

Actual Monopolies: Many areas across the globe have only a single supplier of Internet services because a sufficient market fails to exist to support more than one. Assume that the single digital service exists with non-negotiable conditions on a take-it or leave-it basis. The law might justifiably regulate such a service provider as a public utility to create a level playing field for all citizens within the jurisdiction. Alternatively, legislative action might be taken in appropriate cases to break up a monopolist into multiple smaller companies to create competitive choices for consumers.

Natural Monopolies: The information system with the greatest number of users often has the greatest utility which in turn attracts even more users. Assume that the corporate owner of a dominant information service offers the service on take-it or leave-it terms, such as giving up personal privacy or doing without the service. Because a competitive market fails to exist, the law may justifiably force a policy change or impose an alternative solution to restore an open, transparent, and competitive market. By example, it has been argued that Amazon dwarfs all private natural monopolies of the past and should be broken into at least two pieces through the action of anti-trust law. One company would operate its retail operation while a second separate company would operate its e-commerce operation which provides order and delivery services for thousands of companies [5].

Unfair Private Practices: Assume that consumers buy robots for answering questions and doing mobile chores around the home such as vacuuming floors. However, the company switches its privacy policy after capturing a large market share. That is, perhaps under the new policy the robot now records all voices all the time and photo documents all items in the home when no humans are present. A competitive market fails to exist due to sunk costs and thus the law is justified in stepping in to correct any unfair practices.

8.4.3 Protecting Citizens Against Excessive or Unfair Government Power

Controlling Government Collection: If government does not have access to certain aspects of your life, it can't control those aspects. Law is often used to ban data collection on citizens by government agencies in order to temper the power of the State. Thus, many governments have bans on domestic surveillance of citizens without first obtaining search warrants authorized by the courts under tightly controlled criteria and circumstances. However, if similar strong laws are not imposed on the commercial sector, technology has advanced to the point where personal data is able to be massively collected, retained, and exchanged by businesses and thereby also becomes much more accessible to the State.

Monopolistic Information: Because it can force the gathering of certain data by action of law, government often collects information to which only government has access and provides services that only government can provide. Open access and equal access laws are often enacted so that citizens can readily know what government is up to. In many democratic nations, citizens have a right to know what personal information about them is being held in government records, the right to inspect the records, and the right to make corrections or add explanations to those records. Numerous narrowly drawn exceptions to accessing government records also typically exist. These provisions typically might enable government agents to decline requests for information and data relating to police investigations, the national defense, confidential employee records, and similar matters that, if exposed to the general public, might have strong negative consequences for citizen rights or national interests.

8.4.4 Ensuring People an Opportunity to Enjoy the Minimum Decencies of Life

The education of young children and ensuring that they have enough to eat are viewed almost universally by governments as minimum decencies of life that all governments are justified in supplying and morally mandated to supply when needed. Although many taxpayers don't have children of their own, they benefit as well when they help support such societal needs. The investment supports the creation of a citizenry that is better able to take care of themselves and their families, engage in the future in the workforce, contribute to the economic wellbeing of society, participate actively in the functioning of government, and contribute to the advancement of knowledge, science, and innovations in society. Government support of at least a minimum education also diminishes the need for welfare services and generally strengthens the social fabric of the nation. Thus, government is often justified in passing laws that promote minimum standards of living for all.

In the digital age, the use of information technologies and growth of digital economies has become so prolific throughout the globe that those without access to at least a minimal level for communications, learning, and transactions are at a distinct disadvantage compared to others that do have this access. Yet, an estimated 4 billion people around the globe currently lack access to the internet [8]. Similar to rural electrification programs of the past and present, programs to provide access to information infrastructure and digital devices is viewed by many governments as a justifiable goal in meeting the minimum decencies of life for all citizens. Thus, if the marketplace is not adequately supplying such services, the government justifiably has a strong role to play in either addressing the issue directly or incentivizing the market to provide minimal equitable access to all.

8.5 Maintaining Order

Another justifiable role for government in passing new laws is for maintaining order within a nation. Typically, one might think of government relying primarily on the police and courts for maintaining order. In day-to-day dealing with cybercrimes or resolving digital economy conflicts among businesses, attorneys, police and the courts are indeed on the frontlines in processing and resolving such disruptions. In the case of a major cyberattack or persistent cyberattacks against a country, the defense and military branches of government likely also become involved. However, when the societal conditions of a nation are substantially disrupted such as by technological advancements or by radically changed global business models, legislative bodies may need to enact new laws in order to maintain order.

Across large sectors of the globe, spanning from very poor to very wealthy countries, economic inequality is growing. Large segments of the global population are being left behind by today's economies. Automation, robotics, and artificial intelligence supporting algorithmic decision-making are expanding and rapidly displacing many workers [3, 8]. Corporate business models have shifted radically by decentralizing many functions such that contractors at diverse locations compete with each other to supply parts, products, and services *just in time* using part-time employees that receive few benefits. These societal shifts are resulting in concentrated economic growth in each nation primarily where wealth is already concentrated. Discontent is growing among the poor and middle classes where job opportunities at reasonable pay are rapidly dwindling. The results to date of expanding digital economies are deepening dissatisfaction by large segments of populations in many countries. The inability of a nation's political and economic systems to address fundamental problems angers many, creates widespread mistrust in democratic institutions, and makes civil disobedience more likely. This is resulting in growing risks for democracies and challenges to the functioning of government.

A justifiable role of government in maintaining order under these circumstances, in addition to maintaining civil control through police and the courts, is to pass new laws creating more just and equitable distributions of the benefits arising from technological progress among the population. Such laws create a stronger societal and economic foundation for all and thereby advance the nation as a whole.

8.6 Open Access to Domestic Government Data

One policy area that has been particularly problematic for many nations has been with government agencies competing with private businesses in the marketplace. Laws are justified in preventing government agents from doing so. If a nation desires to grow a strong digital economy, it should not allow the government to claim intellectual property rights in domestic government data and then use those rights to compete with private businesses in providing goods and services to its population. Government should collect or create data, information, and records to meet its legislatively mandated purposes and then make these government works openly and freely available to all businesses and individuals as digital public goods. Government should not compete in data sales or services with the private sector. Rather, open access government data provides a public asset that all industries may mine and use in competing against each other in providing better services and products for citizens as well as increasing personal and corporate taxable income benefitting the nation.

Collective experience across the globe shows that the open use and sharing of scientific and technical data, stimulates economic growth, enhances accountability, and accelerates scientific discoveries [2]. Subject to a few narrowly drawn exceptions, such as those frequently set forth in national freedom of information acts, the recommendation to enact laws providing open access to government data applies to all domestic government data at all levels.

8.7 Correcting Unjust Laws and Policies Within Growing Information Economies

The wealthiest nations on Earth have begun creating an *information civilization* for the globe which is currently characterized by inestimable numbers of smart phones and personal digital devices, massively distributed sensors, rapid growth of automation in manufacturing and service industries, institutionalization of ubiquitous surveillance by the corporate and government sectors, pervasive data mining, machine learning, predictive analytics, algorithmic decision-making through artificial intelligence techniques, emerging deployment of autonomous vehicles, and burgeoning robot applications [10]. The spread of technological advancements and applications globally shows little sign of slowing down. This technological tsunami over the past few decades has resulted in untold benefits in increasing business and government efficiencies and delivering products and services at much lower costs to constituencies, clients, and consumers.

These information economy advancements have also caused massive problems within wealthy nations. In the United States, during a recent three-decade period of mind-boggling technological advancements, the average pre-tax income of the lower half of Americans when adjusted for inflation has remained virtually unchanged at just over $16,000 annually. Meanwhile, the average pre-tax income of the top one percent of the population has more than tripled up to $1.3 million in 2014 and increased more than sevenfold for the top .001 percent up to $122 million per year in 2014 [4].

The disparities have only increased since. Job creation and wages favor the high and low ends of the pay scale with middle class opportunities continuing to wane. Startups of small businesses have precipitously declined, severely narrowing long-standing paths to reach the middle class. While in the U.S. currently there may be sufficient numbers of jobs resulting in low unemployment, most of the emerging jobs are insufficient to provide a living wage, even for many college-educated citizens. "The system in America and around the world has been organized to siphon the gains from innovation upward such that the fortunes of the world's billionaires now grow at more than double the pace of everyone else's" [4]. The top ten percent of humanity now controls 90 percent of the planet's wealth [4].

A fundamental research question of our time is how information societies might correct or adapt to enable rich opportunities for all humans to more equitably share in the benefits of information technology advancements rather than funneling the benefits primarily upwards towards those individuals with greatest existing wealth [7]. Much of the problem appears to be with the inability of governments across the globe to effectively utilize the mechanisms described in the previous section.

A wide range of legislative actions have been proposed for addressing inequitable human and societal conditions brought about by technological advancements and digital economies. Many have already been tested in various forms, particularly in more affluent nations across the globe. Among the approaches for more equitable distribution of benefits have included universal health care, widespread implementation of paid family and parental leave, remunerating work of value to society that may not currently be credited in monetary terms such as for parenting, volunteering and mentoring, creating citizen accounts able to accrue the value of such contributions outside of jobs, as well as many additional similar approaches depending on national circumstances [8]. The list of suggested methods for paying for such programs, most of which would also need to be deployed through legislative action, is very long as well.

While numerous solutions have been proposed, their widespread implementation has been severely lacking to date. A combination of approaches for ensuring sustained worker income and benefits as well as government approaches for generating revenues to pay for them are likely to be enacted as societal disruption becomes more pronounced and governments are forced to respond.

The types of remedies just raised, however, will not be achieved or will fall far short in achieving objectives if foundational constitutional or other controlling governmental framework principles are misaligned with supporting core democratic principles. Foundational principles at the highest level may need to be adapted to current and emerging circumstances due to the vastly changed global landscape brought about by information technologies.

Many nations, including those that are wealthy, likely need to increase their focus on political reforms to strengthen democratic processes, provide equal voices for ordinary citizens, and reduce polarization in politics. Among foundational-shifting law-making actions suggested in higher wealth nations have included decreasing misalignments among population and representation in governmental bodies, instituting substantive campaign finance reform to lessen the effects of moneyed interests in elections, making the process of redrawing election districts less susceptible to political maneuvering, lessening political polarization through actions such as implementing widespread ranked-choice voting and requiring all citizens to vote such that even less politically fervent citizens vote, and strengthening, broadening and enforcing anti-trust law to protect citizens and businesses from the deleterious economic and political effects of historically large and national boundary crossing monopolies.

Leading legal scholars also have long argued that there is a fundamental need to strengthen the rights of individuals such that humans would have much stronger rights compared to the competing rights of corporations [9, 1]. Stronger human rights would enable humans to be much better able to control information exposure about themselves and potentially place them in a position to directly share in revenue streams partially based on use of their private personal data by businesses and other parties.

8.8 Conclusions

All nations struggle with selecting and constantly revising legal and policy paths that will allow the social, economic, and political well-being of their citizens to thrive and that will achieve long-term sustainability for the nation as a whole. Open competitive private markets within and among nations have key roles to play in generating wealth for each nation as well as for its' citizens. Governments have a major role to play in ensuring that markets provide a level playing field by passing and enforcing laws that keep them open, free, and competitive. Governments also have a role to play in providing important goods and service strongly desired or needed by society that that will not be produced otherwise by society. As a general proposition, particularly in response to continually emerging technological advancements, governments across the globe need to do a much better job of revising laws to enable more equitable spreading of economic benefits across far broader and much larger swaths of the population.

Bibliography

[1] Charles Lund Black. *A New Birth of Freedom: Human Rights, Named and Unnamed.* Yale University Press, 1997.

[2] Catherine Doldirina, Anita Eisenstadt, Harlan Onsrud, and Paul Uhlir. Legal approaches for open access to research data. *LawArXiv*, 2018.

[3] Hannah Fry. *Hello World: Being Human in the Age of Algorithms.* W. W. Norton Company, 2018.

[4] Anand Giridharadas. *Winners Take All: The Elite Charade of Changing the World.* Knopf; First Edition, First Printing edition (August 28, 2018), 2018.

[5] Stacy. Mitchell. Amazon doesn't just want to dominate the market – it wants to become the market. The Nation, 2018.

[6] United Nations. Sustainable development goals, 2015.

[7] Cathy O'Neil. *Weapons of Math Destruction: How Big Data Increases Inequality and Threatens Democracy.* Crown; 1 edition (September 6, 2016), 2016.

[8] Darrell M West. *The future of work: robots, AI, and automation.* Brookings Institution Press, 2018.

[9] Alan F Westin. Privacy and freedom. *Washington and Lee Law Review*, 25(1):166, 1968.

[10] Shoshana Zuboff. *The age of surveillance capitalism: the fight for the future at the new frontier of power.* Profile Books, 2019.

9

Developing a Framework for National Institutional Arrangements in Geospatial Information Management

Joep Crompvoets

KU Leuven, Belgium

Serene Ho

Royal Melbourne Institute of Technology (RMIT), Australia

Strong national institutional arrangements in geospatial information management are essential for successful implementation of the Sustainable Development Goals. This chapter presents a comprehensive framework based on a set of core instruments that has been developed to assist stakeholders. Examples of good practices in member states were collected for each instrument, enabling stakeholders to apply the framework in their decision making processes.

9.1 Introduction

The strategic importance of national institutional arrangements in geospatial information management was recognized by the United Nations Committee of Experts on Global Geospatial Information Management (UN-GGIM) at its third session in July 2013 when it identified the need for countries to examine institutional arrangements in geospatial information management, and thereby provide governments with options on how best to create national geospatial entities [12]. This need arose from earlier discussions at its second session in August 2012, when the Committee of Experts considered an inventory of issues that should be addressed in the coming years.

At its third session, the Committee of Experts further agreed that there was an urgent need to identify good practices related to national institutional arrangements for geospatial information management. A small Working Group on National Institutional Arrangements (WG-NIA) was established to continue the work with Member States and regional and international entities. WG-NIA aims to examine national institutional arrangements (NIAs) that support geospatial information management. The output of the WG-NIA identifies good practices of institutional arrangements in Member States to provide national governments with principles and guidelines for institutionalizing geospatial information management. Relevant institutional arrangements contribute to the strengthening of geospatial information management taking into account (technological) issues related to digitalization, geo-referencing, standardization, fundamental geospatial datasets. Moreover, new aspects such as volunteered geographic information and open data are also taken into account.

At its fourth session in August 2014, the Committee of Experts reiterated the strategic importance of national institutional arrangements, noting that Member States are at different stages of geospatial development, and that institutional and policy frameworks are dependent on these legal, fiscal arrangements and governance models, which are quite different across the globe [13]. At this session the WG-NIA proposed the following definition for institutional arrangements: "National Institutional Arrangements (NIA) for Geospatial Information Management (GIM) may be defined as formal and informal cooperation structures that supports and links public and private institutions and/or organizations and which are used to establish the legal, organizational and productive frameworks to allow for sustainable management of geospatial information, inclusive of its creation, updating and dissemination, thereby providing an authoritative, reliable and sustainable geospatial information base for all users."

At its fifth session in August 2015, the WG-NIA presented to the Committee of Experts an extensive analysis of the results of a set of questionnaires from Member States which provided evidence to the importance and complexity of national institutional arrangements, and which generated a valuable source of information to be used in the future [14]. The Committee, in its decisions, provided guidance on how the WG-NIA might evaluate the status of efforts on progress in national institutional arrangements, including providing additional clarity on the process and on the conclusions drafted.

At its sixth session in August 2016, the Committee of Experts recognized the complex and broad scope of the work that the WG-NIA was undertaking, noting that there is no single universal solution or model that fits all countries [15]. Reiterating the need to provide Member States with options on how best to create robust national geospatial institutional structures, the Committee encouraged the WG-NIA to continue its work and, in order to give the work greater focus, to report on its progress to the UN-GGIM Bureau prior to presenting it to the Committee at its next session.

Subsequent to the sixth session, the WG-NIA made presentations at two UN-GGIM regional meetings. Based on discussions coming out of these meetings and interventions from Member States representatives and the Bureau, it was agreed that the WG-NIA should focus on generic elements that provide

Member States with guidelines and principles with which to make decisions on their national institutional arrangements, and not delve into technical methods and detail.

The next step was the execution of a small project "To develop a framework and guidelines in support of national institutional arrangements in geospatial information management for Member States". The purpose of this small project was to support the WG-NIA by evaluating the work done to date, package, develop and deliver outputs that will satisfy the objectives of the Committee of Experts under the agenda item "trends in national institutional arrangements in global geospatial information management." The project was executed from the end of February until the beginning of June 2017. The results were presented by the chair of the Working Group and approved by the experts during the seventh session (August 2017) [16].

The approved framework consisting of NIA-instruments forms the foundation for the current WG-NIA activities focusing on the development of a Foundational Guide to NIA-instruments for Geospatial Information Management.

Before the development of the framework, key documents were reviewed, such as UN Economic and Social Council (2013; 2014; 2015; 2016) and WG-NIA meeting reports, and discussions with key persons of WG-NIA took place.

Important work has been carried out by WG-NIA providing a strong grounding in the mechanisms of many parts of national geospatial information management. For the purposes of supporting the WG-NIA's objectives, a method to better identify how various mechanisms combine to deliver effective geospatial information management was required. There was scope to consider those institutional arrangements related to setting direction(s) through prioritisation and decision making, and monitoring performance, compliance and progress against agreed-on direction and objectives. Consequently, strong case studies were welcomed so that a set of key examples of good practices of institutional arrangements in context could be identified.

The development of the NIA-framework would benefit from drawing from existing governance and/or institutionalization disciplinary expertise and literature dealing with relevant governance structures and business models. These generally provide frameworks for allocating tasks and resources and/or taking into account appropriate instruments for collaboration, regulation and/or market forces which extend to the various levels of government. Such frameworks also take into account the distribution of powers and responsibilities within the Member States (e.g. partnership building, legal frameworks, market regulations).

As a means to assist the WG-NIA and enable it to have a renewed and greater focus, it was strongly recommended to develop a relevant, specific and comprehensive framework for national institutional arrangements based on a set of core instruments, while taking into consideration that no single universal approach exists which fits all Member States of the United Nations. Such a framework needs to be able to logically generate generic elements and/or improving on the national institutional arrangements. In this context, the challenge was to develop a framework that is simple and straightforward in design, so that the key examples of good practices are logically borne out of the application. Another final challenge was that the framework had to be able to simply aggregate the findings at a global level while remaining relevant for individual Member States. The framework developed forms the foundation for identifying key examples of good practices of institutional arrangements.

This chapter is divided in five sections. This introductory section 1 included the review of the achievements of WG-NIA. The next section 2 "Framework development" introduces the framework for national institutional arrangements based on a set of core NIA-instruments followed by section 3 "Framework application" in which the followed application approach is presented. Section 4 'Key examples' presents the descriptions of key examples of good NIA-instrument practices. The reports ends with a short section presenting the main lessons learnt.

9.2 Framework Development

9.2.1 Introduction

To assist the WG-NIA, the development of a relevant overarching framework for NIAs based on a set of core instruments was recommended. While taking into consideration that no single universal approach exists which will fit all Member States of the United Nations, such a framework nonetheless needs to be able to generate generic elements that support delivery and/or improvement on current NIAs. There are two key challenges inherent to the development of a framework. First, the design should be simple and straightforward to enable key examples of good practices to be logically borne out of the application. Second, to support the ability to aggregate elements to be applicable at a more global level while remaining relevant for individual Member States.

9.2.2 Concepts

In this context, institutionalization is considered to be a process of creating 'appropriate' routines that become habitualized or internalized as legitimate behavior, and institutional arrangements provide instruments that governments can use to facilitate this (policy) process within and/or between organizations or programs. Institutionalization here refers to formal and informal structures that aim to enhance, frame or regulate the voluntary or forced alignment of tasks and efforts of organizations in the pursuit of geospatial information management. These instruments are used to create greater coherence and to reduce redundancy, lacunae and contradictions with and between policies, implementation or management [5].

Three mechanisms underpinning institutional arrangements (in the public sector) – with an emphasis on coordination – can be distinguished: hierarchies, markets and networks. Each of these mechanisms has something to contribute to understanding the causes of problems experienced in institutional arrangements, the gains to be achieved through institutional arrangements, and the mechanisms through which better institutional arrangements can be achieved. The distinction between hierarchies, markets and networks of institutional arrangements in social life is widely accepted [11].

In hierarchy-based institutional arrangements, patterns of interaction have two main drivers: authority, operationalized in administrative orders, rules and planning on the one hand, and dominance and authority as the basic control system on the other. Market-based institutional arrangements are based on competition, bargaining and exchange between actors. The price mechanism, incentives and self-interest of actors steer activities of different actors by creating an 'invisible hand'. Network-based institutional arrangements take the form of cooperation between actors, where inter-organizational relations are ruled by the acknowledgement of mutual interdependencies, trust and the responsibilities of each actor [5].

Each of these mechanisms illuminate different aspects of institutional arrangements, but each also has some important explanatory deficiencies. Although these mechanisms are introduced as alternatives to one another, in reality many attempts on the part of government to enhance institutional arrangements will involve more than one of these forms. Under certain circumstances, attempts to impose direct hierarchical control over an organization or set of organizations will work better if the institutional 'arrangers' can build a more cooperative network among the organizations involved or among lower-level employees in those organizations. On the other hand, attempts to embed institutional arrangements that are more bottom up will work better if hierarchy casts a deep, dark shadow on the participants. As well as providing an intellectual understanding of policy making and evaluation, these mechanisms are also closely related to a set of instruments that can be leveraged to deliver national institutional arrangements.

9.2.3 Instruments

The three mechanisms for institutional arrangements presented above are of a more general and abstract level. They refer to the basic processes which may underpin institutional arrangements (authority, price and competition or trust and solidarity) in a sustainability context. In turn, institutional arrangements rely on certain instruments, i.e. specific activities or structures, which may themselves refer to specific operational mechanisms.

Instruments can be either structural or managerial. Institutional arrangements may be realized by creating new or changing existing structures or

TABLE 9.1
Classification of NIA-instruments into structural and managerial instruments

Structural	Managerial
- S1. Establishment of coordinating functions or entities - S2. Reshuffling division of competences - S3. Establishment of a legal framework - S4. Regulated markets - S5. Systems for information exchange and sharing - S6. Entities for collective decision-making - S7. Partnerships	- M1. Strategic planning - M2. Financial management: input-oriented - M3. Financial management: performance-oriented - M4. Financial management: joined up working and cooperation - M5. Inter-organizational culture and knowledge management - M6. Capacity building

management forms within the government. Managerial instruments refer to procedures, incentives and values which plan, monitor and evaluate the use of resources (HRM, finance) or the implementation of policies. Relevant structural instruments in the context of NIA are: S1. Establishment of coordinating functions and entities, S2. Reshuffling of competencies, S3. Establishment of a legal framework, S4. Regulated markets, S5. Systems for information exchange and sharing, S6. Entities for collective decision-making, and S7. Partnerships. Relevant managerial instruments are: M1, Strategic planning, M2. Financial management: input-oriented, M3. Financial management: performance-oriented, M4. Financial management fostering joined up working and cooperation between public organizations, M5. Inter-organizational culture, knowledge management, and M6. Capacity building. Table 9.1 presents the classification of NIA-instruments into structural and managerial instruments types. Each of these instrument will be briefly introduced below.

S1. Establishment of coordinating functions or entities. This structural NIA-instrument refers to the creation of influencing lines of control with the establishment of new functions or entities (e.g. coordination body) with clearly allocated roles, or responsibility tasks. In this context, a coordinator, respectively an individual or unit whose only or main function is to coordinate the geospatial information management activities of the different organizations in an inter-organizational system, and a lead organization which has besides its coordinating function, some operational line functions. The exact position of the coordinating entity vis-à-vis other organizations may determine to what extent hierarchical authority and power as resource is available. Most common coordinating functions or entities within the public sector imply some hierarchical difference between coordinator and the coordinated organizations. Moreover, their coordinating power is mostly stipulated and enforced by laws and statutes. Their task is often to streamline, monitor and control the implementation of a centrally decided specific objective, goal or policy [5].

S2. Reshuffling of competencies. This structural NIA instrument contributes to new or changing structures and institutional forms in the context of the management of geospatial information. A well-known example is the reshuffling of competencies between ministries or departments in response to changing contextual pressures. NIA is enhanced by bringing related activities together by merging organizations or by separating them from other organizations with completely different activities. In addition, this instrument also takes into account the issue of (de)centralizing activities.

S3. Establishment of a legal framework. This structural NIA-instrument refers to the construction and adoption of a regulatory framework(s) for geospatial information management at different administrative levels and the associated legal conditions. Such a legal framework consists of a broad set of rules and regulations, aiming to organize a particular element in society (in this case the management of geospatial information). These rules and regulations are not necessarily developed specifically for a particular subject, but may have been created for other purposes in society and are now applied to the management of basic reference datasets. This can include legislation that deals with (digital) information, (open) data, standards or content, such as freedom of information, intellectual property rights or the protection of personal data. It can also involve legislation and policy with an even broader scope, such as tort liability and contract law, which apply to any kind of actor, situation or object falling within the field of application [6].

S4. Regulated markets. Another set of structural NIA instruments relates to the creation of regulated markets in order to create stimuli and sanctions that induce appropriate behavior by public organizations. The institutional arrangement of tasks and activities by different organizations is done through mechanisms of price and competition, offer and demand. Money and incentives are crucial. Providers of geospatial information are mainly funded through sales to their customers and purchasers, and their demand determines the activities of these providers. Such markets are generally created by government and, depending on the kind and number of users and providers, the kind and level of competition and the level of regulation, the market can be internal or external [5].

S5. Systems for information exchange and sharing. Applying the creation and maintenance of this structural NIA-instrument may induce organizations to take into account the actions of other organizations through processes of mutual adjustment. Through new or re-oriented flows and systems of information, decision-making organizations can be better informed about the latest developments and activities in line with those of organizations [9]. Through systems and arrangements for information exchange, information flows and exchange can be better organized. For example, the development of national geoportals as a key element of geospatial data infrastructures – which are web portals used to effectively find and access geospatial information and associated geospatial services (e.g. display, editing, analysis), are a good example of this instrument in the context of geospatial information management [3]. Information from various organizations can also be integrated in a government-wide information system, giving a strategic overview of government activities. The focus would be on both technical ICT systems as a basis for making information accessible as well as on the content of the information systems.

S6. Entities for collective decision-making. This structural NIA-instrument refers to entities that can make binding decisions [1] affecting multiple actors. Strategic decision-making boards are established consisting of senior officials of different organizations belonging to the policy domain of geospatial information management in order to collectively set out strategy and control the implementation of it. Such joint decision-making bodies enable joint planning and joint working more easily than weaker forms of cooperation.

S7. Partnerships. The most extreme form of cooperation is the creation of a partnership between two or more organizations leading to a common organization controlled by the different 'parent' organizations. This enables the achievement of which these organizations are collectively responsible for, or simply perform joint tasks. Applying this structural NIA-instrument obviously stimulates ownership and creativity, but also assumes substantial autonomy, a common vision, and sufficient goodwill and capacity at organizational level to make collaboration possible. Public partnership can take myriad forms, but can be broadly categorised into: government to government partnerships (G2G); government to business (G2B); and government to community or citizen (G2C).

M1. Strategic planning. This management NIA-instrument refers to the existence, implementation status and political support of strategy plans regarding geospatial information management in which activities of public organizations are aligned to a system of interconnected levels of plans, objectives and targets. NIA is fostered by giving individual organizations clear objectives and targets within a framework of broader inter-organizational or even government-wide goals. These different levels of plans are linked to one another to avoid duplication, gaps and to enhance the pursuit of overarching goals. These plans are monitored and evaluated, after which plans can be adjusted and fine-tuned.

M2. Financial management: input-oriented. This is the first NIA-instrument related to financial management system encompassing processes and instruments of budgeting, accounting and auditing. The set of instruments may entail budgetary guidelines, framework letters. Expenditure review committees, bilateral negotiations and conflict resolution processes, budgetary advice at the centre, formats, systems and provisions for accounting and audits [8, 5]. The hierarchical, input-oriented budget process defines clearly what resources related to geospatial information management should be spent on, and in great detail. There is not much autonomy for organizations to spend the budget as they see fit. Making savings are expressed as a multilateral demand, to which all organizations have to comply. Through the budget, policy priorities are set and communicated downwards.

M3. Financial management: performance-oriented. This second financial management NIA-instrument is result-oriented, with a heavy emphasis on organizational incentives for performance. The focus of the management system is on providing incentives to organizational units to improve their performance. The budget is linked to the expected or past performance (price times quantity: p*Q) of the organizations, and financial sanctions in case of underperformance are possible. Such budgeting is a pre-condition of creating (quasi)markets.

M4. Financial management fostering joined up working and cooperation. This third financial managerial instrument aims to join-up working and cooperation between public organizations. In such a perspective, the focus of the financial management system is on the consolidation of financial and performance information across organizations and policy fields. The emphasis is on information consolidation and exchange, new budget formats, geared towards horizontal policies (for example, outcome- or program-based budgets related to geospatial information management), as well as joined and exchangeable budgets in order to achieve cross-cutting objectives [9, 1, 5]. If organizational or individual incentives for collaboration are present in financial management systems, they are heavily geared towards joined-up activities and cooperation. Such financial management systems oriented towards collaboration will usually include great flexibilities for budget shifts between organizations and years, a limitation of input controls, as well as longer time-span.

M5. Inter-organizational culture and knowledge management. Another NIA-instrument relates more to human resources as an important resource. This managerial instrument aims to enhance institutional arrangements by fostering shared visions, values, norms and knowledge between organizations. As such, this set of NIA-instruments fosters the creation and growth of inter-organizational networks [7] and hence is predominantly linked to the network mechanism to institutional arrangement. This could be achieved by means of the development of cross-cutting skills among staff; common education or common training; management development; mobility of staff between organizations; and the creation of systems for inter-organizational career management [9]. The introduction of behavioral and ethical codes for relevant staff members may be another vehicle for creating and cultivating such common values and norms.

M6. Capacity building. Capacity building or development is defined by United Nations Development Program as the process by which individuals, organizations, institutions and societies develop abilities to perform functions, solve problems and set and achieve objectives [20]. Applied to the geospatial information management context, this means establishing effective strategies for capacity assessment, development, and promoting geospatial advocacy and awareness. For example, the development of a competency framework to articulate the skillsets and knowledge required to function in the geospatial

TABLE 9.2
Classification of NIA-instruments into structural and managerial instruments

Instruments	Hierarchy	Market	Network
Structural	S1. Establishment of coordinating functions or entities S2. Reshuffling division of competencies S3. Legal framework	S4. Regulated markets	S5. Systems for information exchange and sharing S6. Entities for collective decision-making S7. Partnerships
Managerial	M1. Strategic planning M2. Financial management: input-oriented	M3. Financial management: performance-oriented	M4. Financial management: joined up working and cooperation M5. Inter-organizational culture and knowledge management M6. Capacity building

industry could serve as a basis for capacity assessment and development. Facilitating education and skills training at all levels, from building basic awareness to the development of specialist skills could help to ensure a sustainable pipeline of talent for the geospatial information workforce.

The structural and managerial NIA instruments can be clustered into the underlying mechanisms allowing to guide the application of the key instruments for strengthening a specific NIA-mechanism (see table 9.2). The instruments clustered can be considered as complementary to each other and it is up to the decision-maker (and policy makers) which one and/or how to apply.

Being aware that no single universal NIA approach exists which will fit all Member States of the United Nations, it is important to note that some NIA-instruments may appear more relevant than others in a specific national context. It is up to the decision-makers (and policy makers) to decide which NIA-instrument is more relevant, feasible, efficient and/or effective.

9.3 Framework Application

9.3.1 Introduction

In order to apply the identified NIA-instruments in the context of geospatial information management, existing good practices in Member States were collected for each NIA-instrument and described in a standardized template. To better understand how these instruments can be used effectively to support geospatial information management in practice, examples of good practices of these instruments were sought mainly through WG-NIA members.

The notion of a 'good' practice is highly subjective: it is a consequence of any number of variables including political stability, resource commitment, effective governance and management structures, application of guiding principles for geospatial information management [19], etc. Therefore, instead of imposing a definition of what constitutes a 'good' practice of a national institutional arrangement, we have left the selection of examples up to the WG-NIA members (deemed as experts) and assumed that the reported practice represents an example of a 'good' practice because it demonstrates outputs or outcomes that facilitate geospatial information management in that country. A field in the standardized template for describing the good practices ('Good Practice Motivation') aims to provide justification to its selection for readers. All the completed templates were reviewed by the WG-NIA members as a validation process.

This contextual assessment implies that what might be a good practice for one country may not necessarily be transferable to another country. This underscores the UN-GGIM's initial statement that there will not be a one-size-fits-all solution. To overcome this specificity, section 5 will draw out key lessons based on the recurrence of these themes across the examples to develop generalized principles and guidelines.

The instruments of the overarching framework for NIAs are applied to Member States with geographical representations of the UN-GGIM's five regions (UN-GGIM Africa, UN-GGIM Americas, UN-GGIM Arab States UN-GGIM Asia-Pacific, UN-GGIM Europe) to present good practices for each NIA-instrument and to derive principles and guidelines from these practice presentations.

9.3.2 Application Approach

The structural and managerial instruments of the overarching framework for NIAs were applied to Member States to: 1) identify and describe good practice examples of institutional arrangements; and 2) elicit generic elements and lessons learnt, making partial use of the previous works executed by WG-NIA. In section 4 the key examples of each NIA-instrument are briefly presented.

The first steps in the identification of good practices for each type of NIA-instrument were the intensive reviews of existing key source materials and documents, as well as provision of good practices by members of WG-NIA. The Key source materials used were:

- Detailed answers of two questionnaires executed by WG-NIA Task Groups (TG) 1, 2 and 3 (2015) [18]. Questionnaire 1 was a shared questionnaire of the three TGs (TG1: Geospatial Reference Information production systems analysis; TG2: Geospatial Reference Information (GRI) funding structures, dissemination systems and data policy models; TG3 Role of Volunteered Geographic Information). Questionnaire 2, from TG3, was on the Structure of Geospatial Management Organization (2015) [17].
- Documents stored in the UN-GGIM Knowledge Base with descriptions of National Spatial Data Infrastructures (2015-2016), examples of geospatial information laws/directives/regulatory practices (2015-2016), Case studies/best practices (2013-2014), country reports (2011-2016), and Country profiles (2014-2016).
- INSPIRE Member States Reports[1] (2010-2016) [10].
- INSPIRE State of Play reports[2] [2].

The objective of the application was to collect a minimum of three good practices for each type of NIA-instrument. The collection criteria were the following: 1) Relevance of the practice example clearly showcasing the meaning of the application of the NIA-instrument; 2) Availability of information from reliable sources (e.g. policy documents, official websites, documents stored in the UN-GGIM Knowledge Base, etc.); 3) Currency – practices older than 10 years were considered to be outdated except those that very clearly showcase the meaning of the instrument; 4) Geographical representation of all the good practices as per the UN-GGIM's five regions; 5) Submissions of good practices provided by members of WG-NIA.

The collected examples of good practices were mainly described by members of the WG-NIA. Some practice descriptions include contributions of representatives of UN Member States who had detailed knowledge about a specific good practice (Brazil, Canada, Chile, Ecuador, Slovenia, Sweden). In total, 61 key examples of good practices of NIA-instruments were identified and described of which twenty are from the UN-GGIM region Europe, sixteen from Asia-Pacific, seventeen from Americas, five from the Africa, and three from Arab States.

---

[1]According to Article 21 of EU INSPIRE Directive, EU Member States shall send a report about the implementation progress of the directive every 3 years including issues related to institutional arrangements.

[2]These studies were executed by the Spatial Applications Division of KU Leuven on the status of national spatial data infrastructures across Europe. The studies began in 2002, and the reports were updated every year up to 2007. Further studies were carried out using the same approach in 2010 and 2011. These studies also referred to institutional arrangements of geospatial information management in the countries

The descriptions were based on a standardized template. This template was designed so that those who have an interest in good practices of NIA-instruments would find the content accessible and easily understood. The template contained the following topics: Title, Country, Type of NIA-instrument, Aim, NIA-instrument description, Background, Use, Good practice motivation, a visualization illustrating the NIA-instrument practice, and Reference. These topics together introduce a good practice of a specific NIA-instrument in one of the Member States. The applicability of the template was tested and approved by members of the WG-NIA.

9.4 Key Examples

9.4.1 Introduction

The instruments of the overarching framework for NIAs were applied for Member States to identify good practices of each type of NIA-instrument in Member States. This application was presented in the previous section 3.

The objective of this section 4 is to showcase key examples of good practices for each NIA-instrument. Based on the input of members of WG-NIA, a list of minimum three key practice examples per NIA-instrument was compiled reflecting the meaning and the applicability of each instrument.

9.4.2 Description of Key Practice Examples

Table 9.3 below presents an overview of all described key examples of good practices for each NIA-instrument as introduced in section 2 "Framework Development" followed by country and title of the good NIA-instrument practice. Annex 3 of the consultancy report [4] presents the full descriptions of all the collected key examples of good practices for each NIA-instrument. The descriptions are based on the standardized template as introduced in section 3 "Framework application"

TABLE 9.3: Overview of all described key examples of good practices for each NIA-instrument

NIA-Instrument	Country	Title
S1	Mexico	Coordination of the National Information System Statistical and Geographic
	New Zealand	A Clear Geospatial Governance Framework
	Panama	Coordinating structure of the National Spatial Data Infrastructure of Panama

	Spain	SIGPAC Coordination Board
S2	Belgium	Reshuffling of agencies in the Belgian region of Flanders
	Czech Republic	Governmental role clarification and the development of an SDI Coordination Structure
	Portugal	Reshuffling division of competences in the Portuguese Spatial Data Infrastructure within the broader governmental reform context
S3	Mexico	Legal Framework of the National Information System for Statistics and Geography
	The Netherlands	Legal Framework of the National Information System for Statistics and Geography
	Russia	Law on geodesy, cartography and spatial data
S4	Denmark	Open Standard Licensing
	Rwanda	Rwanda Open Data Policy
S4 + S5	United Kingdom	Open data platform data.gov.uk
S5	Canada	Federal Geospatial Platform
	Ecuador	Spatial data infrastructure facilitating emergency response in case of earthquakes
	France	National geoportal of the French administration
	Indonesia	Coordinating Data Sharing Through Indonesia's National Geospatial Information Networks
	Kenya	National land information management system
	Mexico	Digital Map of Mexico
	Morocco	Development of governmental geoportals
	New Zealand	LINZ Data Service
	Republic of Korea	Integrated Approach Towards Data Sharing through NIIS
	Rwanda	SpIDeRR: Spatial Information and Data Portal for Disaster Risk Reduction
	Singapore	Sharing Data, Delivering Services and Building Communities in GeoPlatforms

	Spain	Cadastral Electronic Site (SEC)
S6	Fiji	Fiji Geospatial Information Council
	Singapore	Joint decision-making committee with multiple Government agencies to drive geospatial development
	Slovenia	Slovenian coordination mechanism for infrastructure for spatial information
S7	Australia	Building National Datasets Through Intergovernmental Partnerships in PSMA Australia Limited
	Canada	Canadian Ocean Mapping Research and Educational Network (COMREN)
	Japan	GSI Maps Partner Network
	Mexico	National and international arrangement signed by INEGI
	Spain	Public Agreements of the Spanish National Plan for Land Observation (PNOT)
	Sweden	Data sharing model – The Swedish Geodata Cooperation Agreement
Combined S1 S2 S3	Ghana	Land administration project and subsequent reforms of the National Institutional Arrangements
M1	Australia	The Consultative Approach of Australia's 2026 Spatial Industry Transformation and Growth Agenda
	Brazil	Action Plan for the Implementation of INDE
	Denmark	Good Basic Data Everyone – A driver for growth and efficiency
	Former Yugoslav Republic of Macedonia	Strategy for National Spatial Data Infrastructure of the Former Yugoslav Republic of Macedonia
	Mexico	Programs of the National System of Statistical and Geographic Information (SNIEG or System)
	Namibia	Namibia National Spatial Data Infrastructure (NSDI): Strategy and Action plan 2015-2020
	Singapore	The Comprehensive Scope of the Singapore Geospatial Master Plan
	United Kingdom	Place matters: the Location Strategy for the United Kingdom

M2	Bahrein	Government Investment in Bahrein Spatial Data Infrastructure
	China	Financial investments in Chinese geospatial information Management
	India	NSDI Financial Strategy and Funding Models
	Mexico	Cadastral Modernization Program
M3	Germany	Automated performance procedure for German SDI Monitoring
	United Arab Emirates	Geomaturity Assessment of Abu Dhabi Spatial Data Infrastructure
	USA	Geospatial Maturity Assessment
M4	Australia/New Zealand	Australia and New Zealand Cooperative Research Centre for Spatial Information
	The Netherlands	Geonovum
	Norway	Digital Norway (NSDI) shared financing of basis geodata
M5	Canada	Federal Committee on Geomatics and Earth Observations (FCGEO) and Canadian Committee on Geomatics (CCOG) – Public Sector Geomatics Cooperation in Canada
	Canada	The Canadian Geomatics Community Roundtable and GeoAlliance Canada
	Japan	Enhanced cooperation among relevant stakeholders of geospatial information applications and services at local level
	Poland	Training cycle on INSPIRE Directive implementation
	USA	The COGO Report
M6	Brazil	Capacity Building in the National Spatial Data Infrastructure of Brazil (INDE)
	Chile	Regional training workshops for managing the National System on Territorial Information (SNIT)
	Singapore	Strengthening geospatial information capacity and the use of Geospatial Information, Science & Technology

In total, 61 key examples of good practices of NIA-instruments have been identified and described of which twenty are from the UN-GGIM region Europe, seventeen from Americas, sixteen from Asia-Pacific, five from the Africa, and three from Arab States.

Examples of good practices of NIA-instruments from 38 UN Member States were collected: Australia, Bahrein, Belgium, Brazil, Canada, Chile, China, Czech Republic, Denmark, Ecuador, Fiji, France, Germany, Ghana, India, Indonesia, Japan, Kenya, Republic of Korea, Former Yugoslav Republic of Macedonia, Mexico, Morocco, Namibia, New Zealand, The Netherlands, Norway, Panama, Poland, Portugal, Russia, Rwanda, Singapore, Slovenia, Spain, Sweden, United Arab Emirates, United Kingdom, and USA. These examples reflect practice across a range of UN Member States. More than one good practice is described for 13 UN Member States.

The collected examples demonstrate that some NIA instruments are fairly easy to identify and describe. This is seen in the diversity of practice applications for the structural NIA-instrument, S5. "Systems for information exchange and sharing", and the managerial NIA-instrument, M1. "Strategic Planning".

Conversely, it was difficult to obtain good practice examples of some NIA instruments such as S2. "Reshuffling division of competences" and M5.

"Inter-organizational culture and knowledge management". This is not necessarily an indication of an absence of these practices, rather an absence of available information on these practices as NIA-instruments.

9.5 Lessons Learnt

Some lessons learnt can be derived from the collected examples.

Emergence of a common model. The examples show there exists an array of institutional strategies to achieve good geospatial information management, but there are also commonalities. These commonalities have been abstracted and are shown as a possible roadmap for institutional design in Figure 1. This should not be read as the ideal model for implementing the NIA instruments, but simply as a way to support a user's understanding of how to commence use and implementation of the instruments. This needs to be done with sensitivity to contextual variables in the country (e.g. sources of legitimacy for decision-making, resources, number of agencies involved, pre-existing inter-organizational relationships, etc.).

Figure 1. Proposed model of function and relationship of NIA instruments

Clear trends. Examples from Member States demonstrate some clear trends: that geospatial information is now considered a national asset; that the publishing and sharing of geospatial information has socioeconomic benefits and as such, is gaining characteristics of a public good; that this represents challenges in terms of operations and funding structures.

The need for an integrated change process. Governments are cognizant of these emerging and/or established characteristics and are seeking to legislate to establish the appropriate facilitative governance structures. However, the examples also demonstrate that it often falls to managers to negotiate the operational challenges that these structural changes bring. Therefore, it is important that these NIA-instruments are considered in an integrated way as much as possible, and not perceived as a hierarchical change process.

The importance of a strategic plan. Many countries had an element of strategic planning, that was conducted as a first step to identify the vision, mission, aim and objectives of the geospatial information management initiative. This provided the direction for selecting the appropriate instrument for instigating a new structure. Whether this was more hierarchy- (S3) or networks-based (S7), is really a function of a contextual variables like where authority comes from, previous initiatives that may have worked or failed, resource flows, existing successful relationships, etc.

Catalysing institutional change. Legal frameworks were also often used to catalyse an institutional change process as it represents a coercive force and demands a mandatory shift in mental models and culture. Often the benefit of legislation is the provision of enforcement mechanisms to ensure that organisations comply with changes. However, the example from the Netherlands also shows that a consolidated legal framework is also a strategic mechanism that aligns the development, use and management of geospatial data with sustainable development principles – a strategy that can enhance the legitimacy for change.

The need for clarity. Regardless of the coordinating mechanism, it was apparent that in a multi-organisational, and multi-sectoral collaboration, clarity over who did what was necessary. This is reflected in the link to S2. S1 and S6 can be seen as potential outcomes of S2, and its operationalisation into a governance structure. For managers on the ground, the change trajectory marked by S1. Establishment of coordinating functions and entities, and S2. Reshuffling division of competencies needs to be considered carefully as this has implications for M5. Inter-organizational culture and knowledge management and M6. Capacity building.

Being open to 'open' data. It is strongly recommended that governments explore the possibilities of open data policies by making use of Creative Commons licenses as open standard licenses allowing providers of public sector (geospatial) data to publish their data without the need to develop and update custom licenses. However, issues related to accountability, transparency and sustainable financing need to be also taken into account. In order to have a strong regulated market, the main guideline is to establish a consistent pricing policy regarding the use of geospatial data and services.

Diverse business models. The three financial management NIA-instruments (M2. Input-oriented, M3. Performance-oriented, M4. Joined up working and cooperation) represent funding and business model options. Each have their own benefits and limitations, but it is evident that an initial injection of funds is necessary for getting an large-scale geospatial system up and running. There is a growing tension between the cost of geospatial data production and maintenance and the diffused economic benefits that accrue from facilitating its use and reuse. Norway provides a good example of the use of obligatory co-financing of basic data to manage this financial tension.

The challenge of culture and capacity. NIA-instruments M5. Inter-organizational culture and knowledge management and M6. Capacity building can be difficult instruments to apply in practice. The normal approaches, as seen in the examples, tend to be trainings and workshops. While these should not be discounted, they do not necessarily translate to the types of culture change and capacity building that is required to sustain new ways of working. Singapore's example of multiple approaches at different demographics provides a good example of an approach.

Acknowledgements

This paper is the result of the consultancy 'To develop a framework and guidelines in support of national institutional arrangements on geospatial information management for Member States' for the United Nations Committee of Experts on Global Geospatial Information Management. The authors are grateful for all the support given by the Working Group members on 'Trends in National Institutional Arrangements in global geospatial information management'. Moreover, to all persons who kindly provided us key examples of each NIA-instrument. Without their support it would have been impossible to achieve this document.

Bibliography

[1] Perri 6. Joined-up government in the western world in comparative perspective: A preliminary literature review and exploration. *Journal of Public Administration Research and Theory: J-PART*, pages 103–138, 2004.

[2] European Commission. Spatial data infrastructures in europe: State of play, 2002-2007, 2010-11.

[3] Joep Crompvoets. Geoportals. *International Encyclopedia of Geography: People, the Earth, Environment and Technology: People, the Earth, Environment and Technology*, pages 1–6, 2016.

[4] Joep Crompvoets and Serene Ho. To develop a framework and guidelines in support of national institutional arrangements in geospatial informa-

tion management for member states. united nations committee of experts on global geospatial information management. 2017.

[5] Koen Verhoest Geert Bouckaert, B. Guy Peters. *The Coordination of Public Sector Organizations: Shifting Patterns of Public Management.* Palgrave Macmillan; 2010 edition (February 24, 2010), 2010.

[6] Katleen Janssen and Joep Crompvoets. *Geographic data and the law: defining new challenges.* Leuven University Press, 2012.

[7] Erik-Hans Klijn and Joop FM Koppenjan. Public management and policy networks: foundations of a network approach to governance. *Public Management an International Journal of Research and Theory*, 2(2):135–158, 2000.

[8] 1999 OECD. *Integrating financial management and performance management*, volume PUMA/SBO (99)4 Final. Organisation for Economic Co-operation and Development, 1999.

[9] Christopher Pollitt. Joined-up government: a survey. *Political studies review*, 1(1):34–49, 2003.

[10] European Commission INSPIRE Member States Reports. Monitoring and reporting, 2010, 2013, 2016.

[11] Grahame Thompson. *Markets, hierarchies and networks: the coordination of social life.* Sage, 1991.

[12] UN. Trends in national institutional arrangements in global geospatial information management. *United Nations Economic and Social Council, E/C.20/2014/5/Add.1*, 2013.

[13] UN. Trends in national institutional arrangements in global geospatial information management, report on the third session. *United Nations Committee of Experts on Global Geospatial Information Management, Economic and Social Council, E/2013/46-E/C.20/2013/17*, 2014.

[14] UN. Trends in national institutional arrangements in global geospatial information management. *United Nations Economic and Social Council, E/2015/46-E/C.20/2015/17*, 2015.

[15] UN. Trends in national institutional arrangements in global geospatial information management. *United Nations Committee of Experts on Global Geospatial Information Management, Economic and Social Council E/C.20/2016/6*, 2016.

[16] UN. Trends in national institutional arrangements in global geospatial information management. *United Nations Committee of Experts on Global Geospatial Information Management, Economic and Social Council E/2017/46-E/C.20/2017/18*, 2017.

[17] UN-GGIM. Working group on trends in national institutional arrangements, questionnaire on structure of geospatial management organization. task group 3, 2015.

[18] UN-GGIM. Working group on trends in national institutional arrangements, questionnaires to inform the work of nia task groups 1, 2 and 3, 2015.

[19] UN-GGM. The statement of shared guiding principles for geospatial information management, 2015.

[20] A UNDP Primer United Nations Development Programme. Capacity development, 2009.

10

Considerations for Institutional Interconnectivity

Serene Ho

Royal Melbourne Institute of Technology (RMIT), Australia

This chapter opens by acknowledging the framing of sustainable development as a 'wicked' problem before overviewing the concept of institutions. The institutional challenges around coordination and collaboration in the public sector are then reviewed, first in the context of wicked problems followed by geospatial data management and spatial enablement. Finally, the chapter closes with a summary and a brief discussion on potential strategies for progressing the issue.

10.1 Introduction

There is growing recognition of the importance of geospatial data for the implementation and measurement of progress of the goals and targets identified under the United Nations (UN) 2030 Agenda for Sustainable Development [38]. With an orientation towards spatial enablement, much of this book is focused on addressing the challenge of SDG connectivity from a technical perspective, which is a significant challenge.

There are, however, also significant non-technical challenges. The SDGs seek to improve outcomes, often relating directly to the lives of the most vulnerable, whose agency in contributing and collecting data is almost always compromised. Bringing together data about their lived experiences in a way that is useful and truthful is not straightforward, as can be seen from the lack of prescribed methodologies for numerous targets and indicators. Bringing such data together also requires recognition of ethical data collection and integration, and is contingent upon different public, private and non-government organizations being able to collaborate and coordinate their efforts. Such scenarios may invite new ways of working together, inferring the need to develop and shape inter- and intra-organizational relationships which may or may not have precedence.

These non-technical challenges are often subsumed under the broad umbrella of 'institutional' considerations. Addressing these challenges is important and central to the progression of the SDGs [41]. But what exactly are institutions? The UN's own guidance note suggests that the word is used interchangeably with organizations, while the economist, Douglass North, famously espoused institutions as 'rules of the game' [31]. Regardless, the only consistency is that 'institutions' is a fuzzy word and can mean different things to different people - sometimes even to those within the same discipline.

This chapter therefore seeks to provide a general overview - a sort of primer - on institutional considerations for those who may be involved in attempting to drive coordination and collaboration to integrate geospatial data (with other administrative or statistical data) for SDG-related activities. As it is governments who have explicitly undersigned commitment to the SDGs, and who mostly still hold custodian roles over these data types, this chapter is oriented to the behavior of the public sector. This is not to say though that the discussion will not be relevant to other non-government stakeholders equally active in pursuing sustainable development. Accordingly, this chapter dips into literature from multiple domains such as public administration, sociology, economics, management and innovation studies to provide breadth, rather than depth, of coverage.

10.2 SDGs as a 'Wicked' Problem

The challenge of sustainable development is now broadly understood and accepted as a 'wicked' problem. The general concept of 'wicked' problems is attributed to Rittel and Webber's seminal work on design thinking to deal with the limitations they saw in (then) approaches to deal with complex planning problems. They defined a list of ten properties of wicked problems, invoked a non-linear, non-rational approach to designing which challenged the prevailing approach of the time, and inextricably and explicitly linked designer/design with the political context [36]. This theory has since been expanded upon in the design sphere, where wicked problems are now commonly characterized by the presence of fuzzy problem boundaries, unclear responsibilities, and a plurality (and contesting) of values, drivers, contexts, solutions, connectedness, and legitimacy [8, 21, 17].

Peters (2017) proposes that the term 'wicked' problem has become somewhat indiscriminately used, broadly cast to describe any difficult problem. He argues that, "*few problems facing governments in 2015 and thereafter are actually wicked problems in the full conceptual meaning of the term*" ([33],p.386). Indeed, we live in an age where complexity and complicatedness is the new normal and many problems facing governments these days are unlikely to be definitively resolved. Perceptions of progress are instead likely to be contingent on, and a consequence of, how outcomes are framed to stakeholders vis-á-vis context [5, 17]. Alford and Head (2017) argue that a loose application of the term has promulgated a "totalizing" ([3],p.399) perspective, whereby the problem is not fully analyzed as the sum of its constituent parts but instead attempts to solve the problem in its cumulative form . This enables a tendency towards problem avoidance (we cannot develop a strategy if we cannot distinguish a starting point), or conversely pressure to define the 'right' solution that is immediately implementable, and conveys (perhaps unintendedly) an implicit expectation that success is largely difficult to achieve.

Nonetheless, the notion of wickedness is still useful and latter studies have focused on drawing out more nuanced conceptualizations. Using examples from natural resource management, Nie, (2003) offers a dualistic perspective: wickedness in policy problems can be **by nature** or **by design** whereby "*the very nature and context of some cases and issues essentially promise political conflict – they are wicked by nature. But they are also wicked by design in that political actors, institutions and decision making processes compound them*" ([30],p. 308-309). Newman and Head (2017) show how assumptions of 'wickedness' can be epistemological: here, problems that are technical in nature are assumed to be more amenable to traditional problem-solving approaches, while social problems are viewed as more intractable, thereby demanding non-traditional approaches [29]. Alford and Head (2017) developed a typology of 'wickedness' to help analysts think through problem structure, and hence potential solutions [3].

By various definitions, it appears that there is broad agreement that the challenge of sustainable development is a wicked problem [34, 18, 43]. In part, this is attributed to the fact that sustainable development as an outcome is the sum of resolving a multitude of other wicked problems like climate change, which also occupy highly contested spaces that inhibits agreement and action. Subsequently, this infers that many networks of stakeholders need to be enrolled in problem-solving, which has led to a growing interest in a system-of-systems approach to address sustainable development challenges [28]. This is further complicated by the fact that in our knowledge economy, data is power - information sharing within government, and across public sector organizational boundaries, is recognized as a longstanding chronic challenge even if such activity advances organizational or public benefit [40]. For governments who have committed to the SDGs, this indicates challenges in terms of both public administration and public policy.

10.3 Institutions

In any situation, the way we behave will be consciously or unconsciously influenced by any number of rules and practices that prescribe how to act appropriately. These rules and practices are relatively stable and resilient to change, and are simultaneously being produced and changed by the social structures that we are embedded within.

These social structures that produce regular patterns of behaviour are what we refer to as **institutions** in this chapter: they evolve from local customs and beliefs, strengthening to form normative rule-based structures (both formal and informal) that prescribe or preclude behaviours in actions at all levels of society, and hence are instrumental in political order [32]. Core to the discourse of institutions is the assumption that institutions reduce risks and uncertainty (thus lowering transaction costs) by creating expectations around order and predictability of how others might behave in certain situations, i.e. propagating a logic of appropriateness [25, 47, 4]. Such behaviour is accepted and expected as optimal ways of acting.

We focus here on three main types of institutions as defined by Scott (2001): *regulative*, which is legally sanctioned; *normative*, which is morally governed; and *cultural-cognitive*, which is culturally supported [39]. These three types of institutions exert pressure to conform to expected behaviour in different ways: regulative institutions exert coercive pressure, i.e. we feel compelled (i.e. no choice) to act often with the threat of sanction; normative institutions exert normative pressure, i.e. we feel that we ought to act due to social pressure; cultural-cognitive institutions exert mimetic pressure, i.e. we imitate behaviour without necessarily having a conscious understanding of motivation [12].

There is often an assumption that stable institutions persist simply because they are right and a sense of 'historical efficiency' becomes associated with it [25]. However, institutions can often be less than appropriate not only because they are difficult things to change, but also because the cost of enacting change may be greater than the benefits that change might bring [11].

10.4 Wicked Problems and Institutional Challenges for Coordination and Collaboration in the Public Sector

The public policy and political science literature is rife with many examples of challenges posed to the administrative capacity of governments to coordinate action. As Bouckaert, Peters, and Verhoest (2010) note, "*Coordination is one of the oldest problems facing the public sector*" ([14],p.13). Institutions, as described above, play an important role in coordination and collaboration. Here, we briefly emphasize three common challenges related to wicked problems.

Multiple stakeholders. Coordination and collaboration is thought to be the most effective way of addressing transboundary problems that involve an array of stakeholders [17]. However, this attracts risks of under- or over-coordination. With many stakeholders involved, directing and forming a solution could be compromised if the problem does not fall within the explicit responsibility of any one organization to direct and coordinate; on the other hand, if too many stakeholders want to lead, then it becomes difficult to formulate a coherent approach [22]. In this context, institutions can be useful or limiting in the sense that social rules and practices directly structure opportunities and access to resources for actors that enhance or constrain participation [37].

Mechanisms for coordination. The Weberian view of bureaucracy has embedded a hierarchical approach to coordination as convention [45]. This has been an effective mechanism for vertical coordination enabled by regulative pressure, and therefore important functions of government (e.g. planning, budgets) remain effectively hierarchical. However, such rigidity can limit flexibility or access resources to facilitate coordination. With the increasing presence of non-government stakeholders in public service delivery more generally but also in tackling wicked problems, other mechanisms for coordination have become pertinent, primarily networked approaches, where collaboration is voluntary and fostered by shared values [20, 7]. In networks, coordination tends to be more horizontal, and is a consequence of negotiation amongst network members, which requires a culture of social trust to be established (Peters 1998; Considine 2005). Therefore challenges for coordination here may be cultural, but also related to communication, i.e. expressing and agreeing on rules and practices [2].

Institutional legacy of public sector reform. For many governments around the world, efficient and effective public service delivery is contingent on standardized, routinized models - and this is directly oppositional to the nature of wicked problems. The ability of the public sector to contend with such problems is compounded by public sector reforms in the 1980s and 1990s under the banner of New Public Management (NPM). This resulted in a shift away from coordinated, multi-purpose organizations towards streamlined, specialized units with more explicitly defined service objectives, which incentivized competition rather than collaboration [14]. Consequently, the negative - albeit unintended - impact these reforms had on overall public administration motivated another wave of changes in the 2000s that sought to redress fragmentation through an emphasis on 'joining up' government agencies for the delivery of public services [16, 6].

10.5 Challenges of Coordination for Spatial Enablement

The challenge of coordination related to spatial enablement could be seen as having parallels with the experiences of implementing spatial data infrastructures (SDI). SDIs are commonly expressed as collection of technology, data, policies and standards to facilitate access and sharing of geospatial data across a network of custodians and users [35]. The push for SDIs coincided with the second wave of reforms around joined-up government and this is likely to have positively impacted the take-up of the concept.

However, the implementation of SDIs in reality have often been troubled. The collection and provision of statistically significant geospatial data is often a central government task as this includes fundamental information for governing including cadastral data, addresses, physical planning, topographic information, etc. Indeed, many governments define foundational geospatial layers, e.g. in Australia, there are ten such layers[1] prescribed that are pronounced to be 'trusted' base datasets - i.e. collected and managed by government custodians. Early research demonstrated a range of institutional challenges including lack of incentives to encourage the development of new practices for sharing data [9], or an alternative perspective is that the opening up of data threatened existing power relationships [13]. Additionally, many SDI initiatives also have governance arrangements that are legacies of their origins stemming from specialist national mapping agencies, which were subsequently found to be inappropriate for delivering on the larger scale objectives of whole-of-government spatial enablement [27].

In response, SDIs appeared to move from a hierarchical mode towards a more networked mode of coordination [23, 44]. However, a recent study of institutional arrangements of SDIs across 37 UN Member States revealed that in reality, a hybrid approach seems to prevail where hierarchical structures still play an important role in coordinating SDIs but that a shift towards a networked approach also existed, especially at inter-organizational boundaries [10]. Additionally, issues of access and coordination have also been, to some extent, mitigated where strong open public sector information policies have been enacted and 'open by default' positions on government data have eased access to government spatial datasets [42].

There are also similar challenges confronting land administration systems (LAS). LAS are commonly framed as an institutional framework since it comprises structures defined by both social aspects (e.g. regulation and policies) and technology. Although LAS are defined as a core part of SDIs due to their administration of cadastral data [46], they are often tackled as distinct entities. This should perhaps not be surprising as in many parts of the world, geospatial data is not managed within the same government organization as

[1] Foundation Spatial Data Framework (see https://www.anzlic.gov.au/resources/foundation-spatial-data-framework).

cadastral data. Multiple ways of organizing this exists, commonly through multiple agencies (e.g. Greece, Spain) or distributed across different levels of government (e.g. Poland).

Consequently, issues such as overlapping administrative authority (e.g. Philippines), lack of harmonised legislation, and mandatory submission to extensive formal processes to effect change can all conspire to constrain efforts to coordinate and collaborate [19].

In line with shifts in public management approaches around the world, there is also a trend to modernize LAS. LAS have traditionally been focused on regulated data processes and hence, facilitated more of a transactional relationship between actors (e.g. regulator, data producer, user, etc.). This has tended to effect a more bilateral, authority-based type of governance.

Presently, given the emphasis on the knowledge economy and inhabiting a digital milieu, there are corollary impacts for LAS where data processes can now be construed as knowledge processes supporting the development of social capital, i.e. LAS are becoming knowledge-intensive industries. In these scenarios, trust becomes paramount as a public management strategy and multilateral governance becomes a more appropriate form of coordination [15, 1].

10.6 Institutional Considerations: Moving Forward

This chapter sought to provide an overview of institutional considerations for those who may be involved in attempting to drive coordination and collaboration of geospatial data for SDG-related activities. While oriented towards the public sector, the emergent lessons are equally valuable for stakeholders from other sectors who may be similarly involved.

Institutions, those social structures constituted of stable and resilient rules and practices that influence behavior, are central to any political structure. In the case of wicked problems like sustainable development, where numerous governmental and non-governmental organizations are involved, it becomes important to consider what impact existing institutions have, as these help to establish the appropriate structures for facilitating coordination (e.g. when a group of stakeholders do not always interact in a consistent way) and collaboration (e.g. when new relationships are required).

It also becomes important to consider the impact of existing institutions as these directly structure opportunities and access to resources, which can influence or limit the ability of stakeholders to participate. As such, institutions are also mechanisms for coordination, but many governments operate under a vertical (hierarchical) structure, whereas open, transboundary challenges like sustainable development often requires a more horizontal (networked) structure that is cultivated less by directives, and more through ongoing negotiation amongst stakeholders. This however, requires time and insightful strategizing to establish a culture of trust.

Shifting modes of governance is also a challenge experienced in the context of spatial enablement. Experiences over the last few decades of SDI implementation have revealed the limitations of institutional arrangements initially established to drive SDIs under the mandate of national mapping agencies. Similar to the findings in public management literature, it was thought that a networked approach would be a more appropriate mode of coordination but recent studies have shown that in fact, a hybrid arrangement seems to prevail. In reality though, it seems many governments are beginning to move towards a more hybrid approach.

For example, in Rwanda, institutional arrangements such as overarching policies, legal and regulatory framework and financing and capacity building programs are used as hierarchical instruments of vertical coordination. Their Open Data Policy helps to ensure that agencies follow consistent rules on data release, privacy safeguards, and use of an "open" license and technical standards. As well, departments are mandated to follow a directive to make available all for-public-consumption data online without charge. At lower levels though, network-type instruments are used such as partnerships for training and the establishment of a portal to facilitate exchange of information. Rwanda's use of such hierarchical instruments are fairly typical (e.g. strategic plan, coordinating government body, etc.), but there is increasing presence of both network arrangements, largely through partnerships (both formal and informal) or other collective decision-making model, and market arrangements (e.g. user-pay models).

Finally, at a time when wicked problems seem like the new normal, the legacy of previous public sector reforms that sought to create specialized functions of governments have left unintended consequences. Whilst these reforms achieved more efficient services, it also inadvertently resulted in the fragmentation of governments which has negatively impacted on their ability to collaborate. There is now a need to consider how to enact the necessary structural and cultural changes to mitigate and reverse the impact of these reforms. This however, might lie beyond the mandate of most geospatial and land organizations.

Albeit brief, the overview presented in this chapter has illustrated a variety of institutional considerations for how a public sector organization might approach and facilitate coordination to deliver the types of interconnected administrative response to wicked problems, such as those represented by sustainable development and advancing of the SDGs. An institutionalist perspective, which argues that action is driven by a logic of appropriateness (versus consequence, i.e. actors more motivated by rationality and self-interest), indicates that coordination is not simply a product of designing the right structural arrangements in terms of economic rationality; it also requires the cultivation of a common culture where norms and values are shared - a fundamental premise for success [24, 26]. This often requires strategies that tap into normative or mimetic pressures to encourage the requisite collective behaviour. However, these are not easy strategies to develop and requires that attention be paid on how the problem is legitimized to design the appropriate incentive structures to attract buy-in and resources.

The reviewed literature also emphasizes that wicked problems are a sum of many parts, and instead of tackling the problem in its entirety, it may be more productive to attempt to better understand the structure of the problem since this will help identify the type of organizational structure, and hence institutional arrangements, best required to enable more effective coordination and collaboration, i.e. institutional connectivity. This will improve both response and implementation of solutions and support progression of the SDGs.

Bibliography

[1] Paul S Adler and Seok-Woo Kwon. Social capital: Prospects for a new concept. *Academy of management review*, 27(1):17–40, 2002.

[2] E. Alexander. *How Organizations Act Together: Interorganizational Coordination in Theory and Practice.* Routledge; 1 edition (December 3, 1995), 1995.

[3] John Alford and Brian W Head. Wicked and less wicked problems: a typology and a contingency framework. *Policy and Society*, 2017.

[4] Jens Beckert. Agency, entrepreneurs, and institutional change. the role of strategic choice and institutionalized practices in organizations. *Organization studies*, 20(5):777–799, 1999.

[5] Pam Carter. Policy as palimpsest. *Policy & Politics*, 40(3):423–443, 2012.

[6] Tom Christensen and Per Lægreid. The whole-of-government approach to public sector reform. *Public administration review*, 67(6):1059–1066, 2007.

[7] Mark Considine. *Making public policy: institutions, actors, strategies.* Polity, 2005.

[8] Richard Coyne. Wicked problems revisited. *Design studies*, 26(1):5–17, 2005.

[9] W. J. Craig. Why can't we share data: institutional inertia, 1995. pp. 107-118.

[10] Joep Crompvoets and Serene Ho. To develop a framework and guidelines in support of national institutional arrangements in geospatial information management for member states. united nations committee of experts on global geospatial information management. 2017.

[11] P. Dimaggio and W. Powell. The new institutionalism in organizational analysis. The International Encyclopedia of Organization Studies: 144–145, 1991.

[12] Paul J DiMaggio and Walter W Powell. The iron cage revisited: Institutional isomorphism and collective rationality in organizational fields. *American sociological review*, pages 147–160, 1983.

[13] WH Erik de Man. Understanding sdi; complexity and institutionalization. *International Journal of Geographical Information Science*, 20(3):329–343, 2006.

[14] Koen Verhoest Geert Bouckaert, B. Guy Peters. *The Coordination of Public Sector Organizations: Shifting Patterns of Public Management.* Palgrave Macmillan; 2010 edition (February 24, 2010), 2010.

[15] Sumantra Ghoshal and Peter Moran. Bad for practice: A critique of the transaction cost theory. *Academy of management Review*, 21(1):13–47, 1996.

[16] John Halligan, Fiona Buick, and Janine O'Flynn. Experiments with joined-up, horizontal and whole-of-government in anglophone countries. *International handbook on civil service systems*, pages 74–99, 2011.

[17] Brian W Head and John Alford. Wicked problems: Implications for public policy and management. *Administration & society*, 47(6):711–739, 2015.

[18] Peder Hjorth and Ali Bagheri. Navigating towards sustainable development: A system dynamics approach. *Futures*, 38(1):74–92, 2006.

[19] Serene Ho, Abbas Rajabifard, Jantien Stoter, and Mohsen Kalantari. Legal barriers to 3d cadastre implementation: What is the issue? *Land Use Policy*, 35:379–387, 2013.

[20] Jan Kooiman. Governance. a social-political perspective. In *Participatory governance*, pages 71–96. Springer, 2002.

[21] Marshall W Kreuter, Christopher De Rosa, Elizabeth H Howze, and Grant T Baldwin. Understanding wicked problems: a key to advancing environmental health promotion. *Health education & behavior*, 31(4):441–454, 2004.

[22] Per Lagreid and Lise H Rykkja. Organizing for "wicked problems"–analyzing coordination arrangements in two policy areas: Internal security and the welfare administration. *International journal of public sector management*, 28(6):475–493, 2015.

[23] Bastiaan Loenen. *Developing geographic information infrastructures: The role of information policies.* IOS Press, 2006.

[24] James G March. *Primer on decision making: How decisions happen.* Simon and Schuster, 1994.

[25] James G March and Johan P Olsen. The logic of appropriateness. In *The Oxford handbook of political science.* 2004.

[26] James G March and Johan P Olsen. *Rediscovering institutions.* Simon and Schuster, 2010.

[27] Ian Masser, Abbas Rajabifard, and Ian Williamson. Spatially enabling governments through sdi implementation. *International Journal of Geographical Information Science*, 22(1):5–20, 2008.

[28] Scott M Meentemeyer, Brian Sauser, and John Boardman. Analysing a system of systems characterisation to define system of systems engineering practices. *International Journal of System of Systems Engineering*, 1(3):329–346, 2009.

[29] Joshua Newman and Brian W Head. Wicked tendencies in policy problems: Rethinking the distinction between social and technical problems. *Policy and Society*, 36(3):414–429, 2017.

[30] Martin Nie. Drivers of natural resource-based political conflict. *Policy sciences*, 36(3-4):307–341, 2003.

[31] Douglass C North. Institutions. *Journal of economic perspectives*, 5(1):97–112, 1991.

[32] Elinor Ostrom. *Understanding institutional diversity.* Princeton university press, 2009.

[33] B Guy Peters. What is so wicked about wicked problems? a conceptual analysis and a research program. *Policy and Society*, 36(3):385–396, 2017.

[34] Jonathan Pryshlakivsky and Cory Searcy. Sustainable development as a wicked problem. In *Managing and engineering in complex situations*, pages 109–128. Springer, 2013.

[35] Abbas Rajabifard, Mary-Ellen F Feeney, and Ian P Williamson. Future directions for sdi development. *International Journal of Applied Earth Observation and Geoinformation*, 4(1):11–22, 2002.

[36] Horst WJ Rittel and Melvin M Webber. Dilemmas in a general theory of planning. *Policy sciences*, 4(2):155–169, 1973.

[37] Fritz W Scharpf. *Games real actors could play: actor centered institutionalism.* Westview Press; 1 edition (September 5, 1997), 1997.

[38] Greg Scott and Abbas Rajabifard. Sustainable development and geospatial information: a strategic framework for integrating a global policy agenda into national geospatial capabilities. *Geo-spatial Information Science*, 20(2):59–76, 2017.

[39] W Richard Scott. *Institutions and organizations: Ideas and interests.* Sage, 2008.

[40] A.L. Stinchcombe. *Information and organizations.* Berkeley, CA: University of California Press., 2000.

[41] UNDP. Institutional and coordination mechanisms: guidance note on facilitating integration and coherence for sdg implementation, 2017.

[42] Bastiaan Van Loenen and MJM Grothe. Inspire empowers re-use of public sector information. *International Journal of Spatial Data Infrastructures Research, 9, 2014*, 2014.

[43] Annemarie van Zeijl-Rozema, Ron Cörvers, René Kemp, and Pim Martens. Governance for sustainable development: a framework. *Sustainable Development*, 16(6):410–421, 2008.

[44] Danny Vandenbroucke, Joep Crompvoets, Glenn Vancauwenberghe, Ezra Dessers, and Jos Van Orshoven. A network perspective on spatial data infrastructures: Application to the sub-national sdi of flanders (belgium). *Transactions in GIS*, 13:105–122, 2009.

[45] Max Weber. *The theory of social and economic organization.* Simon and Schuster, 2009.

[46] Ian Williamson, Stig Enemark, Jude Wallace, and Abbas Rajabifard. *Land administration for sustainable development.* Citeseer, 2010.

[47] Oliver E Williamson. The institutions of governance. *The American Economic Review*, 88(2):75–79, 1998.

11

Implementing SDGs in Smart Cities Beyond Digital Tools

Zhixuan Yang
Dongbei University of Finance and Economics, China

Abbas Rajabifard
Centre for SDIs and Land Administration, The University of Melbourne, Australia

The chapter aims to establish the theoretical framework by exploring the key components of smart city to observe the implementable structure and action of SDGs at the city level, particularly, the key components of smart city beyond digital tools towards SDGs.

11.1 Introduction

The process of smart transformation of cities is complicated and full of challenges, for example, the challenges of unprecedented demographic growth being projected to 9.8 billion as well as the continuous growth of urbanization reaching 68 percent by 2050 [86], which drives global cities to the limited capacity regarding sustainable economy, society and environment [10]. In the face of such challenges, smart cities establish the agglomeration hubs in terms of intelligent inhabitants, dense trade and business, interconnected chain industry, advanced technology and knowledge, positive policy and etc., which keeps reshaping the cities' mobility, growth pattern and ecological system. In fact, the complex future of smart cities is not foreseeable. The crucial doubt is its maintenance of sustainable development. Recently, the advancement of technological tools generates the possibility of capacity building in smart cities, which is expected to alleviate urban problems such as urban sprawl, waste control, air pollution, traffic congestion and etc. towards sustainability through the harness of digital tools and information.

With such background, the UN General Assembly issued the resolution A/RES/70/1 "Transforming our world: the 2030 Agenda for Sustainable Development" [87], which calls for the global attention of 17 Sustainable Development Goals (SDGs), covering the issues of elimination of hunger and poverty, life and prosperity, work and living conditions, social justice and partnership, environment and industry in the face of global challenges.

In fact, SDGs are to be implemented at the city level to reach the goals of global sustainability. Particularly, the goals are implementable in smart cities, such as G7, G8, G11 and G12. In addition, G16 and G17 point out that technology-driven framework of cities is essential in that regard, which coincides with the digital facilitation in smart cities. Therefore, the focus of implementing SDGs in smart cities is essentially on the focus of the facilitation of digital tools.

However, as the concern of smart city and its sustainability raises, people start to discuss that implementing SDGs in smart cities not only needs the enablement of technologies but also requires further strengthening the institutional frameworks. In another word, sustainable development requires a long-term transformation during which the enhancement of digital capacity is crucial [10], but the holistic smart city framework of implementing SDGs is still unknown.

The following sections are described as follows. The chapter starts by discussing the SDGs and means of implementation in smart cities. In this first section, the research focus is the SDGs and its implementation framework, including the general framework and implementable data and indicators. The second section is the smart city context. In this section, the research analyzes the smart city context by proposing the argument of smart city and sustainability, the measures of making cities smart and sustainable and the needs of digital tools and living labs. The third section is the key components beyond digital tools. In this section, the research proposes the key components beyond digital tools, particularly, networked infrastructure, knowledgeable community and intelligent governance. The fourth section is action agenda of smart city towards SDGs beyond digital tools. The research proposes the action agenda at the city level consisted of the integration of innovation capacity, transformation of smart growth, and evolvement of socio-economic ecosystem. And the last section is the discussion and conclusion. In this section, the research discusses and reviews the smart city frameworks of SDGs beyond digital tools. The research contribution is the establishment of holistic smart city frameworks of the implementation of SDGs beyond digital tools.

11.2 SDGs and Means of Implementation in Smart Cities

11.2.1 Three-Tier SDGs

The Sustainable Development Goals (SDGs) have been advocated by the UN since 2015 at the UN Conference on Sustainable Development in Rio de Janeiro [87, 86]. The general aim is to build up the global sustainable capacity in the face of uncertainties regarding economic, social, environmental and political challenges by 2030. Apart from Millennium Development Goals (MDGs), SDGs specially highlight the adoption of data and digital tool, thus SDGs are the call for the "data revolution" [68].

The SDGs consist of 17 goals and 169 sub-targets, that are, Goal 1 (G1)-no poverty, G2-zero hunger, G3-good health and well-being, G4-quality education, G5-gender equality, G6-clean water and sanitation, G7-affordable and clean energy, G8-decent work and economic growth, G9-industry, innovation and infrastructure, G10-reduced inequalities, G11-sustainable cities and communities, G12-responsible consumption and production, G13-climate action, G14-life below water, G15-life on land, G16-peace, justice and strong institutions, and G17-partnerships for the goals. The research finds that the 17 goals can be grouped into three tiers in one hierarchy, that are, fundamental tier, i.e. ecological balance and well-being goals (including G1, G2, G3, G4, G5, G6, G7, G10, G14 and G15), middle tier, i.e. industrialization goals (including G8, G9, G11 and G12) and top tier, i.e. higher-level targets (including G13, G16 and G17). The middle tier, the goals to be accomplished in cities, is the column bone supporting whole hierarchy, connecting higher-level targets with the fundamental tier (Figure 11.1). Therefore, the means of implementing SDGs in cities are crucial regarding the realization of SDGs.

11.2.2 Means of Implementation-Framework

Among the goals in the middle tier, the four goals, G8, G9, G11, and G12, form the implementation frame of implementing SDGs in smart cities in general, covering the interactions and connections of stakeholders in the process of city development moving towards more inclusive, resilient, harmonized, innovative, informative, technological, industrialized and sustainable future.

G11 is the most frequently mentioned in Urban Sustainable Development Goals (USDGs) aiming to ensure cities are "inclusive, safe, resilient and sustainable". It includes 7 sub-targets, covering the living conditions, environment, disaster, equalities of urban residents, determining to leave no one behind. Besides G11, the overall aim of G9 "Industry, Innovation and Infrastructure" is to build resilient infrastructure, promote sustainable industrialization and to foster innovation as the strong infrastructure provides the cornerstone of sustainable industrial development, supporting the technolog-

FIGURE 11.1
SDGs in three tiers

ical progress and shaping a more sustainable future, which fosters economic development and promotes social progress and well-being. It also includes targets to promote inclusive and sustainable industrialization, adopt clean and environmentally sound technologies, enhance scientific research and development, encourage innovation, and increase access to information and communications technology. Moreover, G8, "Decent work and economic growth", aims to promote sustained, inclusive and sustainable economic growth, full and productive employment and decent work for all. Furthermore, G12, "Responsible consumption and production", aims to support the commodity cycle in the supply chain using the relevant information and awareness for sustainable development and lifestyles in harmony with nature [87].

In addition, the overarching SDGs' implementation framework is highlighted in G16 and G17, those are, G16 "Promote peaceful and inclusive societies for sustainable development, provide access to justice for all and build effective, accountable and inclusive institutions at all levels" and G17 "Strengthen the means of implementation and revitalize the Global Partnership for Sustainable Development" [87]. As suggested by Risse, "SDGs have a strong focus on the means of implementation, including the targets of finance, capacity building, trade, policy, institutional coherence, multi-stakeholder partnerships, data, monitoring and accountability, as well as public governance and technology" [74].

11.2.3 Means of Implementation - Data and Indicators

The measurement of SDGs by data and indicators is a scope that practitioners and researches focus on regarding the implementation of SDGs in cities. Normally, the different scales of measurement are regarded as keys to open the black box of SDGs at both global and local levels. Therefore, the massive indicators are developed by different authorities, for example, the UN Habitats Urban indicator programme, Commission for Sustainable Development's Sustainable Development Indicators, EU's Urban sustainability indicators, European Common Indicators, OECD's Better Life Index, ISO 37 120 indicators (Sustainable development of communities), and the indicator of SDGs reported by World Council of City Data (WCCD) [91, 95].

Those indicators cover the range of economic, social and environmental sustainability mainly in the theoretic study. However, as the regard of the value in practice, Roland Zinkernagel (2018) finds among the seven frameworks of indicators, the most popular indicator, the UN Habitats urban indicator programme which was launched in 1993 and adjusted respectively in 1996 and 2001, is merely used by 200+ cities worldwide. Following that, the European Common Indicators launched in 1998, is tested only by 42 cities. Many of the other indicators are not truly appreciated by cities. The data and indicator analysis have been over-emphasized for some years by regional authorities as well as researchers, which misleads the researches to purely develop the indicator from the perspective of science rather than focus on the localized solutions. Therefore, in fact, the adoption rate of indicators is relatively low [95].

In the regard of indicators covering the range of smartness and sustainability, Hannele Ahvenniemi (2017) analyzes that the indicators of smart cities cover 20% environmental, 28% economic, and 52% social sustainability whereas the indicators of urban sustainability cover 43% environmental, 10% economic and 47% social sustainability. The fact reveals that the evaluation emphasis of smart cities is on social sustainability whereas urban sustainable development is on the environmental sustainability [2]. That brings in the dilemma of mismatch of implementation focuses of smart city and sustainability.

11.3 Smart City Context

11.3.1 Smart City Concept

The definition of a smart city was firstly introduced in the 1990s [23]. And focuses of the smart city extend gradually afterwards in accordance with the updated knowledge of effective and sustainable city development. For example, the focuses include ICTs' facilitation as a key to the smart city [41, 44, 48, 69], improvement of quality of life to achieve prosperity, effectiveness, and competitiveness [4], the role of stakeholders of influence on social, economic and environmental sustainability, including the six smart domains, those are, the smart economy, smart mobility, smart environment, smart people, smart living and smart governance [43, 66, 33, 67], utilization of technological network and digital infrastructure as the enabler of sustainable urban development [43] and the involvement of capital and business to form the integrated innovative ecosystem as a whole, network and linked system [26, 55].

In regard of what the smart city is by definition, there are diverse concepts, for example, technological and human demands' driven city [43], green and sustainable and informational city [89], collaborative and participative city [58, 1], multi-disciplinary and developmental city [6], creative and innovative city [40, 57, 90], governmental and infrastructural city [66, 93], lively ecosystem and service city [22, 81], data-driven and open data city [47, 56] and etc.

In fact, such vast coverage of focuses diminishes the core strategy of city development. For the most circumstances, the concept is accepted as a fuzzy idea to attract capital investment in city infrastructures. However, the truth is with the permeation of technologies into the business services and governmental policies, the smart cities form the capacity of the self-configuration, self-healing, self-protection, and self-optimization, which maintains the city growth on a higher level of sociology. But the question is, is the smart city sustainable?

11.3.2 Argument of Smart City and Sustainability

As the discussion of smart cities gets heated, the argument of the difference between smart city and sustainability is raised by practitioners as well as researchers. The common misunderstanding is that smart cities are sustainable. However, the smart cities can only be sustainable on conditions of holistic and integrated framework towards the goals of sustainability.

Currently, the overarching approaches of smart cities are focused on the technology-driven method (TDM) and human-driven method (HDM). The former regards that smart cities are networked places where deploying ICTs into each activity in the city would improve standards of living. It is further emphasized that the use of ICTs by communities will enable them to participate more fully in so-called knowledge societies [24]. However, ICTs alone would not contribute to achieving the desired improvements in living standards, and there exists a need for enhancing human capital and other forms of skill development among the citizenry [67]. The argument is that these dichotomies generate a critical knowledge gap because they suggest divergent hypotheses on what principles need to be considered when implementing strategies for enabling smart city development [65]. Margarita Angelidou (2017) also suggests that the smart and sustainable city landscape is extremely fragmented both on the policy and the technical levels. There is a host of unexplored opportunities toward smart sustainable development, many of which are still unknown [5]. Other proponents of smart cities emphasize the potential for promoting economic prosperity, ecological integrity and social equity which would advance the larger goal of urban sustainability [32, 55].

11.3.3 Making Cities Smart and Sustainable

The smart city is regarded as a strategy as well as the center of solutions to alleviate demographic pressure and urban problems [20, 83]. The connective network and interconnected system facilitated by massive amounts of digital data leverage the complex information to make a better decision and resolve the problems [45, 84]. There have been several schools of smart cities in the research field since the 1990s. The main focus of defining smart cities is on the digital (ICT) infrastructure development to improve the living standards of people, which can be observed from the papers published between 2010 and 2014 [9]. In the regard of digital facilitation, for example, cities are expected to sustain the development enabled by a virtual and interconnected environment [80, 52]. Therefore, the smart city is also possible to be an intelligent city, techno-city, well-being city, wired city, ubiquitous city, information city, knowledge city, learning city, green city, and sustainable city. Making cities smart and sustainable means the integration of development strategies regarding the local choice, combining ICTs and data infrastructure with the sustainability concerns, generating the economic, social and environmental influences on long-lasting development without the consumption of resources of the next generations. Cities are expected being "smart when investments in human and social capital and traditional (transport) and modern (ICT) communication infrastructure fuel sustainable economic development and a high quality of life, with a wise management of natural resources, through participatory governance" [17]. In addition, specific factors that influence the cities in regard of smart and sustainable are tested, for example, "the size" of a city is considered as a crucial driver towards sustainable economic development in the study of making smart cities sustainable. Considering the effect of agglomeration economies for productive use of resources at the moderate cost of living and expense of social and environmental degradation [31, 63, 75], small and medium-size cities are intended to be the places where creation and innovation happen towards a friendlier and more ecological environment comparing to large cities [30, 29, 16].

11.3.4 Needs of Digital Tools and Living Labs

The SDGs emphasize the importance of technological support and digital infrastructure to developing countries. for example, in G9, the highlights are the facilitation of sustainable and resilient infrastructure to support domestic technology development and increase the access to information and communication technology by 2020 in the sub-targets 9.a, 9.b and 9.c. Digital tools are regarded as an innovative instrument to boost the knowledge and innovation economy, which is crucial to driving the smart city discourse [3]. The advancement of technologies has a strong effect on the sustainable development of smart cities. The endowment of big data and digital tools can meet with the challenges that smart cities face regarding socio-economic development as well as the quality of life [79]. The common understanding of the use of digital tools is to transmit information from stakeholders to form the core of the public-private-people-partnership ecosystem and agglomerate the advanced competitiveness of citizens and business to create added-value of current circumstances.

The digital tools can also provide the opportunities of sharing-economy. On the basis of knowledge of communities, data and tools provide the resources for citizens to 'shape urban change' in smart cities [78, 39]. The communication of information connects the social network, particularly, with the aid of Internet of Things (IoT), the data-based commodity and service in the smart cities provide social mobility, which minimizes the social cost but maximizes the economic benefit.

The concept of digital-driven life is tested in the living labs for the experiment and validation of future smart cities. The living labs consume the labs to unite the participants and stakeholders and consider them as the bodies involved in the business ecosystem. Normally, the stakeholders include citizens, organizations and local governments in the cities. The benefit of labs is open access to the public information, data mobility anywhere, high-level of interactions, reshaping and operating the innovative social ecosystem [78]. The labs provide the information platform with the facilitation of digital tools, which self-involved and updates by the users[1]. Therefore, the living labs are the user-driven information ecosystem.

In the common understanding, one of the large advantages is the living labs provide real-time platforms engaging participants, particularly the stakeholders, to share the activity information targeting on co-creation, exploration, experimentation, and evaluation [7].

Researches also reveal that the living labs create an open ecosystem for either user-end or supplier-end, which brings the government, business and citizens together, engaging the willingness of all stakeholders in the economic and social activities. Furthermore, the living labs concept has formed a unique methodology regarding the innovative and collaborative interactions shaping the urban norms in a new form in the social ecosystem [77].

As the living labs rely on digital tools and technological networks, the adoption in smart cities is in accordance with the trend. In activities, living labs in the smart cities practice the innovative technology and connectivity to move the smart cities to provide the conditions and resources for citizens

[1]The successful cases of living labs the European Network of Living Labs (ENOLL), such as, Smart Santander (www.smartsantander.eu), Experimental Living Lab for Internet of Things (ELLIOT) (www.elliot-project.eu), Peripheria (www.peripheria.eu), Open Cities, EPIC, Apollon (www.apollon-pilot.eu). And only Europe has 319 living labs, which grows fast worldwide (European Network of Living Labs, 2013).

to shape urban change. In this way, "the smart city is an urban innovation ecosystem, a living laboratory acting as an agent of change" [78]. Particularly, the endowment of big data, smart mobility, and the internet of things facilitate the development of living labs in the informational ecosystems [39] .

In general, there are several groups of focuses on the key components of smart cities. For example, resource endowment [18], self-decisive system [34], digital facilitation such as ICTs, IoTs and big data [61, 23], participation of citizen and government [18] and optimization outcome through infrastructures [37]. Therefore, there are several key components of smart cities beyond digital tools undiscussed. The following section, the networked infrastructure, knowledgeable community and intelligent governance are to be focused.

11.4 Key Components Beyond Digital Tools

11.4.1 Networked Infrastructure

The data infrastructure enabled by digital tools is the foundation of an integrated platform supporting users' communication. The ICTs and IoTs are the special focus regarding the infrastructure. However, the separate data infrastructure is not well established regarding forming effective communications among stakeholders unless the well-connected infrastructure is set up. The networked infrastructure forms the foundation of informative society, which "improves economic and political efficiency and enables socio, cultural and urban development" [43].

The mobility of big data is vital regarding the establishment of networked infrastructure, which is enhanced by ICTs. With the advantage of data mobility, the provision of services and commodities become smart and convenient. Moreover, the interconnected infrastructure provides a sufficient channel to collaborate, stakeholders, particularly, the end users on the smart platform to communicate thoroughly of the needs and requirements. The feedback can also be timely reflected the counterparts in the communication, so that the prompt response and adjustment are well functioned during the process, which enhances the resilient capacity of the networked group of people. Therefore, the networked infrastructure offers the network instrument, and well-functioning infrastructure to provide the opportunities for reshaping the communication process, enabling the communities' inclusiveness and stimulating the resilience of infrastructure [27].

11.4.2 Knowledgeable Community

Improving communities living is the core of making cities smart and sustainable, particularly, in the process of smart city development. As discussed by Hollands, "a smart city is a city that aims at connecting the physical, IT, social and business infrastructures in order to leverage the intelligence of the city's community" [43]. A smart community is a broadly defined group of people of common or shared interest, "whose members, organizations and governing institutions are working in partnership to use IT to transform their circumstances in significant ways" [66].

The smart community is the end user and major benefit receiver, who is most likely to take the position of advocating smart cities. Without the support of a knowledgeable community, the smart city is merely a shell of technology and infrastructure. In that sense, the knowledgeable community is crucial in regard to implementing SDGs in smart cities.

However, recent practice in smart cities is not fully understood by local communities. There are several reasons for this. First, advanced digital tools are not close to communities' lives excepts for smartphones. The data infrastructure is mostly developed for the working environment of business and government but not close to people's daily lives. Second, the communities are not clear about the benefits that they can get through the tools and infrastructure. Therefore, the training and education processes through various channels are vital regarding delivering the information and technology to communities to make them knowledgeable.

On the basis of networked infrastructure, the knowledgeable communities can form the communication capacity at their choices to function the social system towards a more sustainable way.

11.4.3 Intelligent Governance

The data-driven revolution transforms the citizens' living style as well as governance structure in a great manner through various aspects. First, the data-driven manner transforms the city growth to knowledge-sharing and sharing-economy, involving communities into the decision process, which drives the decision smarter and closer to the end-needs. Second, the integration of disaggregated data improves the governments' decision-making process [64], which enables governance structure towards a more intelligent gesture. Third, the open data provided by the public sectors creates the transparency of information, and ensure the accountability of counterparts in the connections, monitoring the right role of decision-makers and actors in the well-informed partnership.

Governance is the central core of responsibility to connect citizens with businesses and the living environment to foster a culture of innovation and sustainable economic development. Particularly, governance networks are more or less stable patterns of social relations between mutually dependent actors, which form around public issues, and which are formed, maintained, and changed through interactions between the involved actors [53, 88]. In that regard, the governance needs to be intelligent to perceive the right timing for the proper decision, connecting both "top-down" and "bottom-up" communication process and integrating eight factors of good governance[2] with the digital facilitation [94].

The governance structure is more important than ever before regarding shaping the growth path of smart cities. The ICTs change the traditional governmental process to the network governance interconnecting the dependent actors due to social relations on one governance platform [53, 88]. The intelligent governance structure adjusts itself to a flat manner forming the service ecosystem in cities [21].

11.5 Action Agenda of Smart Cities Towards SDGs Beyond Digital Tools

11.5.1 Integration of Innovation Capacity in Smart Cities

The smart data generates the innovation capacity in smart cities. Specifically, the smart data and technology enable the communities to acquire the information of know-how. The wide use of ICTs forms digital infrastructure improving the quality of life as well as enhancing the efficiency of economic transactions by the manner of innovations. The SDGs sub-target 17.8 suggests the technology and innovation capacity-building mechanism, particularly for the least developed countries. Particularly, on the basis of UN Economic and Security Council 2016 Session, the "fostering statistical capacity-building, partnerships and coordination" is introduced as an optimism [13] that brings the knowledge to share and transform among the responsible stakeholders in the commitment of partnership, which drives the future direction of smart cities towards SDGs.

In fact, the recent proliferation of big data has contributed to smart city transformation regarding the establishment of innovative capacities [11, 14, 42, 50]. "Big data" generally refers to large and complex sets of data that represent digital traces of human activities and may be defined in terms of scale or volume, analysis methods[19], or effect on organizations [62]. The enhancement of big data improves the decision-making process by the integration of disaggregated data, fostering the intelligent governance, innovative business, and empowerment of participatory citizens (Higher Level Panel, 2013). Moreover, the use of big data also promotes transparency and accountability, enhancing the efficiency of SDGs' implementation at the local level [76, 85].

The general aim of capacity-building is to motivate smart cities know-how to become self-regulating and responsible upon certain and uncertain social interactions through participation as well as market-based relations. In the

[2]The particular emphasis is on the roles of government and community, aiming at the enhancement of transparency, responsiveness and effectiveness and efficiency by the methods of participation, rule of law, consensus, equity and inclusiveness, and accountability.

contrast to the data and indicators suggested in the previous section regarding the measurement of performance, innovation capacity is more related to the participants and stakeholders involved in the smart city ecosystem, therefore, it is the social-context capacity in the business environment beyond the technological capacity.

Phillips and Ilcan (2003) conceptualize capacity-building as a technology of neoliberal governance, an apparatus of rule that requires a diverse range of new rationalities that aim to 'grow' institutional frameworks, enhance the skills of people, and transfer knowledge through the formation of new partnerships for international development [71]. Such a building process enables the process of knowledge development and knowledge sharing, which is crucial for the enhancement of innovation capacities. In addition, the integration of innovation capacity in smart cities facilitate the diverse possibilities regarding the growth of economy, data mobility, friendly environment, knowledgeable community, better living standards, and effective governance of cities contributing to the implementation of SDGs [1, 18, 6, 89].

11.5.2 Transformation of Smart Growth in Smart Cities

The idea of smart city originates in the smart growth in the 1990s [15], pursuing the ideal solutions of urban problems such as urban sprawl, traffic congestion, air pollution, loss of open space and etc. due to the ill-planned and ill-coordinated development [66]. Particularly, smart growth is indicated as an ideology facing the problems that urban sprawl brings in due to the fast urbanization. The smartness is paid attention to the economic growth especially aiming to achieve policy performance and success. Meanwhile, the smart growth builds up the community group involving stakeholders in the urban planning phase, seeking the applicable measures to achieve the smart decision. The concept of smart growth initiates the use of smart tools and technology in urban planning as well as the rethinking of governance necessities in the face of the smart community.

However, the argument of smart growth never ends as the complicated situation change via ages in cities. Wey and Hsu (2014) argue about the new urbanism and smart growth concept to deal with city problems especially environmental, housing and citizens' well-being [92]. European Parliament (2014) argues the aim of smart growth is to create a smart city "where the traditional networks and services are made more efficient with the use of digital and telecommunication technologies, for the benefits of its inhabitants and businesses" [54]. In general, the smart city transforms the growth pattern to a smarter choice, particularly in terms of the intelligent governance structure and strong partnership in social aspects [11, 14, 42, 50, 73].

The SDGs are universal on the main focuses of 5 key elements: people, planet, peace, prosperity, and partnership. The partnership unites the communities, business, government and other interest-related organizations in the involvement of sustainable development in smart cities.

As discussed earlier, the dilemma of dichotomous understanding of smart cities relies on the divert directions of the technology-driven method (TDM) and human-driven method (HDM). The integration of TDM and HDM needs the focus of partnership existing in the stakeholders who have a different emphasis on benefits. The partnership is first of all the awareness of the importance of a holistic approach to applying smart instrument in the united group including the public sector, private sector as well as communities on the integrated platform.

Moreover, the global partnership is a well-established framework for the collaboration of local governments, private organizations, social communities, academic institutes, and other participants. The partnership is a mechanism to implement SDGs in the more rooted grass field on the national and local levels with the cooperation of all the participants and stakeholders, which is the core driver of means of implementation [46].

The SDGs form the arena of global partnerships, which enables the multi-stakeholder getting involved in the partnership, ranging from the provision, supervision, evaluation to the analysis of information, covering all the participants with the interest, keeping the direction of sustainable development of smart cities. Particularly, with the facilitation of techno-partnership, the people are well-connected and kept informative to reinforce the long-lasting collaboration.

11.5.3 Evolvement of the Socio-Economic Ecosystem in Smart Cities

Smart cities are regarded as intelligent digital ecosystems installed in the urban space [67, 72, 25, 92, 60, 33]. Particularly, the ecosystem consists of complex interactions and inter-dependencies, which keeps evolving with the progress of digital advantages. As suggested by Angelidou (2015) "A smart city is a multi-agent ecosystem comprised of kinds of societal actors like public sectors, private companies, non-profit organizations, and citizens; it also represents a multidisciplinary field constantly shaped by advancements in technology and urban development" [6]. Komninos (2008) also suggested that "smart cities are the consequence of a dense innovation ecosystem that creates value through the use and reuse of information that may come from many different social connections and highly skilled human capital" [51].

The metabolism of such an ecosystem is crucial regarding the sustainable movement of the system. The released data of a smart city nurtures a lively ecosystem composed of agents (i.e., groups of companies and non-profit organizations) that create innovative products and services [22, 81].

Therefore, the smart city ecosystem is generally operated and maintained by the diverse businesses, promoting the growth of the economy, harmonizing the social community, sustaining the built environment so that the ecosystem maintains the self-organizing and self-evolving continuity particularly with the enablement of digital technologies. The so-called metabolism is actually through the service mobility to generate the advantage of economic agglomeration. And by the effect of agglomeration, the large cities become the hubs of business attracting intelligent labors to reside in and contribute to the city growth.

The merits of the ecosystem, such as interaction, balance, loosely coupled actors with shared goals, self-organization [36] are enlarged by the adoption of ICTs in the business environment, reinforcing the information flows and forming the digital business ecosystem [8]. The digital business ecosystem fosters the demands of citizens regarding the services and end-needs, facilitating the service deliveries out of the strains [28]. That means the larger digital human ecosystem is enabled by the integrated business environment through the information flow.

In the environment of alliances of smart business and knowledgeable community, cities attract the important edge of innovation [70, 69], which is also recognized as a biological system[12] as it has the capacity to self-regulate and self-evolve. Such advancement promotes the adjustments the socio-economic goals towards the SDGs.

11.6 Discussion and Conclusion

The basic understanding of smart cities is the facilitation by digital tools in the city development although the concept of smart cities is multi-faceted and multi-disciplined. However, the smart city concept can be distinguished from other similar ideas such as the digital city or intelligent city in that it focuses on factors such as human capital and education as drivers of urban growth, rather than singling out the role of ICT infrastructure. [59].

In recent years, the digital infrastructure is so much developed that the mass investment is not fully recognized. In addition, the argument is raised regarding the relationship between smart cities and sustainable goals, however, the path of achieving the sustainability of smart cities is unknown. Therefore, the gap exists between the smart cities and sustainability, particularly, the implementation framework of SDGs in smart cities is not clear.

Implementing SDGs in smart cities is a complicated issue. The reasons are as follows. First, cities are complex ecosystems, relying on inputs of recourses, generating outputs of commodities; second, cities divert due to different urban forms and features; third, cities have to achieve smart growth to maintain the sustainability; fourth, cities are the places where agglomerations happen; government, business and community influence the direction of cities. Therefore, to achieve SDGs, it requires a long-term transformation in cities. It is impossible to find an instant solution to all nor universal panacea to solve all problems.

In the regard of making smart cities sustainable, the possibility exists, because smart cities are the arena of implementation of SDGs at local levels not only because the common innovation of technology forms the digital ecosystem of the both, but also the future direction of city development coincides the concept of each other. However, it is also can be impossible if smart development is not towards SDGs.

To re-adjust the sustainable path of smart cities, researches during the period of 2017 and 2018 transits from the surface talk of the definition to the deeper insights of the truth of smart cities. For example, the paper was written by Maria Kaika (2016) strongly rejected the hypothesis of smart cities being sustainable and resilient [49]. This relatively simplistic imaginary of the smart city has been roundly critiqued on a number of fronts, especially around the entangling of neoliberal ideologies with technocratic governance and the dystopian potential for mass surveillance [83, 35, 38, 43, 50, 82, 89].

The inner goal of smart cities is to improve the living quality of people with the facilitation of digital solutions, such as ICT, big data, Internet of Things (IoT). The initiatives of smart cities are mostly encouraged by local governments in regard to ICT infrastructure development. Under such conditions, the fast effect of enhancement of digital living manners, such as mobile phone communication, broadband construction, online trade and service and etc. is duplicated preliminarily globally. However, the consideration of the accomplishment of smart cities is not directly linked to sustainability. There exists a vast gap in the ideology of both terminologies. Particularly, the experience of the implementation shows the smart city strategy cannot lead to sustainability if the emphasis based on digital development. Thus, the recent researches show the doubts of simply linking the smart cities with sustainable cities. In the face of such a dilemma, a few types of research start to dig into the institutional instruments of smart cities beyond digital tools. Therefore, the updated understanding is that smart cities can achieve the SDGs on conditions of understanding the key components to form a holistic implementation framework.

The chapter focuses on the non-financial instruments as well as the local implementation particularly in smart cities under certain jurisdiction and level of operation as the means of implementation are in response to local needs upon global imperatives.

Upon the framework of implementing SDGs in smart cities beyond digital tools, the research delivers a holistic implementation framework, including deployment of key components and action agenda of SDGs at cities levels.

In the contrast to the smart city initiative framework as suggested by Hafedh Chourabi et al. (2012) "management and organization, technology, governance, policy context, people and communities, economy, built infrastructure, and natural environment" [20], the chapter proposes three key components of smart cities beyond digital tools, those are, networked infrastructure, knowledgeable community and intelligent governance. The networked infrastructure is the foundation of networking facilities, enabling the inclusiveness of stakeholders on the integrated platform and enhancing the efficiency of

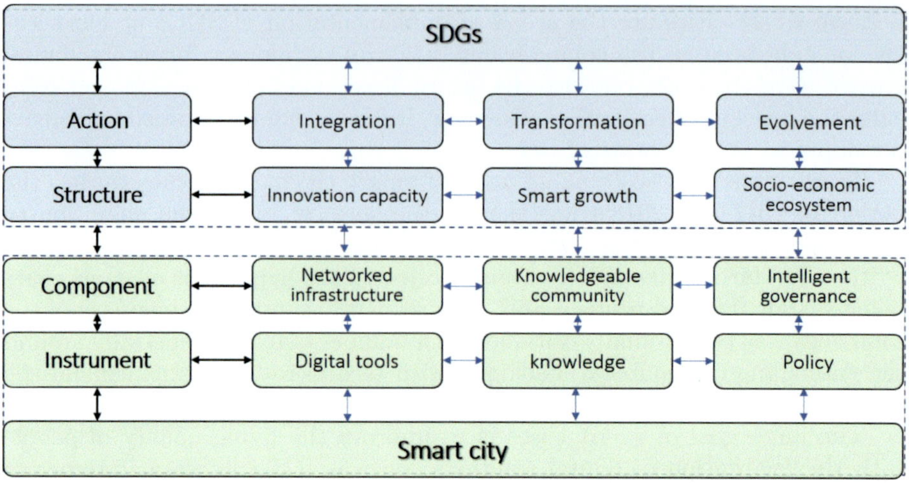

FIGURE 11.2
Framework of implementing of SDGs in smart cities beyond digital tools

communication. The knowledgeable community is information-rich end users in the smart cities, participating the communication in the spectrum of city development and building up the capacity to improve the decision-making process. The data-driven intelligent governance forms the inter-connected institutional structure among government, business, and citizens, driving the governance to a networked and demographic process.

The smart city action agenda is sophisticated in the combination of key components and structure regarding envisioning smart cities in the implementation of SDGs. The chapter proposes the action agenda consisting of three layers, those are, integration of innovation capacity, the transformation of smart growth engine and evolvement of the socio-economic ecosystem in smart cities.

The following-up research will focus on the deep insights into the three-layer implementation framework of SDGs in smart cities.

Bibliography

[1] Alberto Abella, Marta Ortiz de Urbina-Criado, and Carmen De-Pablos-Heredero. A model for the analysis of data-driven innovation and value generation in smart cities' ecosystems. *Cities*, 64:47 – 53, 2017.

[2] Hannele Ahvenniemi, Aapo Huovila, Isabel Pinto-Seppä, and Miimu Airaksinen. What are the differences between sustainable and smart

cities? *Cities*, 60:234–245, 2 2017.

[3] Gountaras-N.-& Tarani P. Angelidou, M. Engaging digital services for the creation of urban knowledge ecosystems: the case of thermi, greece. *International Journal of Knowledge-Based Development.*, 3(4):331–350, 2012.

[4] M. Angelidou. Smart city policies: A spatial approach. *Cities*, 41:S3–S11, 2014.

[5] M Angelidou. The role of smart city characteristics in the plans of fifteen cities. *Journal of Urban Technology*, 24(4):3–28, 2017.

[6] Margarita Angelidou. Smart cities: A conjuncture of four forces. *Cities*, 47:95 – 106, 2015. Current Research on Cities (CRoC).

[7] Psaltoglou-A.-Komninos N. Kakderi C. Tsarchopoulos-P. Panori A. J. J. o. S. & Management T. P. Angelidou, M. Enhancing sustainable urban development through smart city applications. *JOURNAL OF SCIENCE AND TECHNOLOGY POLICY MANAGEMENT*, 9(2):146–169, 2018.

[8] L. G. Anthopoulos. Understanding the smart city domain: A literature review. in transforming city governments for successful smart cities. *Springer International Publishing Switzerland*, pages 9–21, 2015.

[9] Tuba Bakıcı, Esteve Almirall, and Jonathan Wareham. A smart city initiative: the case of barcelona. *Journal of the knowledge economy*, 4(2):135–148, 2013.

[10] Werner R Balogh, Luc St-Pierre, and Simonetta Di Pippo. Towards a results-based management approach for capacity-building in space science, technology and applications to support the implementation of the 2030 agenda for sustainable development. *Acta Astronautica*, 139:385–389, 2017.

[11] Sarah Barns. Mine your data: open data, digital strategies and entrepreneurial governance by code. *Urban geography*, 37(4):554–571, 2016.

[12] Michael Batty. The size, scale, and shape of cities. *science*, 319(5864):769–771, 2008.

[13] Lauren Berlant. Thinking about feeling historical. In *Political Emotions*, pages 243–259. Routledge, 2010.

[14] Simon Elias Bibri. The iot for smart sustainable cities of the future: An analytical framework for sensor-based big data applications for environmental sustainability. *Sustainable cities and society*, 38:230–253, 2018.

[15] David Bollier. *How smart growth can stop sprawl: a fledgling citizen movement expands.* Washington, DC : Essential Books; First Edition edition (1998), 1998.

[16] Kamila Borsekova and Peter Nijkamp. Smart cities: A challenge to research and policy analysis. *Cities*, 78, 03 2018.

[17] Andrea Caragliu, Chiara Del Bo, and Peter Nijkamp. Smart cities in europe. *VU University Amsterdam, Faculty of Economics, Business Administration and Econometrics, Serie Research Memoranda*, 18, 01 2009.

[18] Andrea Caragliu, Chiara Del Bo, and Peter Nijkamp. Smart cities in europe. *Journal of urban technology*, 18(2):65–82, 2011.

[19] Hsinchun Chen, Roger HL Chiang, and Veda C Storey. Business intelligence and analytics: From big data to big impact. *MIS quarterly*, 36(4), 2012.

[20] Hafedh Chourabi, Taewoo Nam, Shawn Walker, J Ramon Gil-Garcia, Sehl Mellouli, Karine Nahon, Theresa A Pardo, and Hans Jochen Scholl. Understanding smart cities: An integrative framework. In *2012 45th Hawaii international conference on system sciences*, pages 2289–2297. IEEE, 2012.

[21] James JT Connolly, Erika S Svendsen, Dana R Fisher, and Lindsay K Campbell. Networked governance and the management of ecosystem services: The case of urban environmental stewardship in new york city. *Ecosystem Services*, 10:187–194, 2014.

[22] Nuno Ferreira Da Cruz and Rui Cunha Marques. Revisiting the determinants of local government performance. *Omega*, 44:91–103, 2014.

[23] Renata Paola Dameri and Annalisa Cocchia. Smart city and digital city: twenty years of terminology evolution. In *X Conference of the Italian Chapter of AIS, ITAIS*, pages 1–8, 2013.

[24] Gert de Roo, Allen Creedy, Christian Zuidema, and Geoff Porter. *Towards Liveable Cities and Towns - Guidance for Sustainable Urban Management.* 01 2007.

[25] Kevin C Desouza and Trevor H Flanery. Designing, planning, and managing resilient cities: A conceptual framework. *Cities*, 35:89–99, 2013.

[26] Susanne Dirks and Mary Keeling. A vision of smarter cities: How cities can lead the way into a prosperous and sustainable future. *IBM Institute for business Value*, 8, 2009.

[27] Aksel Ersoy. Smart cities as a mechanism towards a broader understanding of infrastructure interdependencies. *Regional Studies, Regional Science*, 4(1):26–31, 2017.

[28] Kumar H.-& Sharma T. J. M.-K. P. Fingar, P. Enterprise e-commerce, 2000.

[29] Richard Florida. *Cities and the creative class.* Routledge, 2005.

[30] Susanne A Frick and Andrés Rodríguez-Pose. Average city size and economic growth. *Cambridge Journal of Regions, Economy and Society*, 9(2):301–318, 2016.

[31] Susanne A Frick and Andrés Rodríguez-Pose. Big or small cities? on city size and economic growth. *Growth and Change*, 49(1):4–32, 2018.

[32] David Gibbs, Rob Krueger, and Gordon MacLeod. Grappling with smart city politics in an era of market triumphalism, 2013.

[33] Rudolf Giffinger, Christian Fertner, Hans Kramar, Evert Meijers, et al. City-ranking of european medium-sized cities. *Cent. Reg. Sci. Vienna UT*, pages 1–12, 2007.

[34] Rudolf Giffinger and Haindlmaier Gudrun. Smart cities ranking: an effective instrument for the positioning of the cities? *ACE: architecture, city and environment*, 4(12):7–26, 2010.

[35] Adam Greenfield. *Against the smart city (The city is here for you to use Book 1).* Do projects; 1.3 edition (20 December 2013), 2013.

[36] Ulrike Gretzel, Chulmo Koo, Marianna Sigala, and Zheng Xiang. Special issue on smart tourism: convergence of information technologies, experiences, and theories. *Electronic Markets*, 25(3):175–177, 2015.

[37] Robert E Hall, B Bowerman, J Braverman, J Taylor, H Todosow, and U Von Wimmersperg. The vision of a smart city. Technical report, Brookhaven National Lab., Upton, NY (US), 2000.

[38] Orit Halpern, Jesse LeCavalier, Nerea Calvillo, and Wolfgang Pietsch. Test-bed urbanism. *Public Culture*, 25(2):272–306, 2013.

[39] Hoon Han and Scott Hawken. Introduction: Innovation and identity in next-generation smart cities. *City, culture and society*, 12:1–4, 2018.

[40] Minako Hara, Tomomi Nagao, Shinsuke Hannoe, and Jiro Nakamura. New key performance indicators for a smart sustainable city. *Sustainability*, 8(3):206, 2016.

[41] Colin Harrison and Ian Abbott Donnelly. A theory of smart cities. In *Proceedings of the 55th Annual Meeting of the ISSS-2011, Hull, UK*, volume 55, 2011.

[42] Ibrahim Abaker Targio Hashem, Victor Chang, Nor Badrul Anuar, Kayode Adewole, Ibrar Yaqoob, Abdullah Gani, Ejaz Ahmed, and Haruna Chiroma. The role of big data in smart city. *International Journal of Information Management*, 36(5):748–758, 2016.

[43] Robert G Hollands. Will the real smart city please stand up? intelligent, progressive or entrepreneurial? *City*, 12(3):303–320, 2008.

[44] Hung-Nien Hsieh, Chiu-Yao Chou, Chung-Chih Chen, and Yuan-Yu Chen. The evaluating indices and promoting strategies of intelligent city in taiwan. In *2011 International Conference on Multimedia Technology*, pages 6704–6709. IEEE, 2011.

[45] IBM. Ibm 2010 annual report. Online, 2010.

[46] Tomasz Janowski. Implementing sustainable development goals with digital government–aspiration-capacity gap, 2016.

[47] Thorhildur Jetzek, Michel Avital, and Niels Bjorn-Andersen. Data-driven innovation through open government data. *Journal of theoretical and applied electronic commerce research*, 9(2):100–120, 2014.

[48] Robertas Jucevičius, Irena Patašienė, and Martynas Patašius. Digital dimension of smart city: critical analysis. *Procedia-Social and Behavioral Sciences*, 156:146–150, 2014.

[49] Maria Kaika. 'don't call me resilient again!': the new urban agenda as immunology... or... what happens when communities refuse to be vaccinated with 'smart cities' and indicators. *Environment and Urbanization*, 29(1):89–102, 2017.

[50] Rob Kitchin. Big data, new epistemologies and paradigm shifts. *Big data & society*, 1(1):2053951714528481, 2014.

[51] Nicos Komninos. *Intelligent cities and globalisation of innovation networks*. Routledge, 2008.

[52] Nicos Komninos, Charalampos Bratsas, Christina Kakderi, and Panagiotis Tsarchopoulos. Smart city ontologies: Improving the effectiveness of smart city applications. *Journal of Smart Cities*, 1(1):31–46, 2015.

[53] Joop Koppenjan and Erik-Hans Klijn. *Managing uncertainties in networks: Public private controversies*. Routledge, 2004.

[54] Harish Kumar, Manoj Kumar Singh, MP Gupta, and Jitendra Madaan. Smart neighbourhood: A tism approach to reduce urban polarization for the sustainable development of smart cities. *Journal of Science and Technology Policy Management*, 9(2):210–226, 2018.

[55] Rama Krishna Reddy Kummitha and Nathalie Crutzen. How do we understand smart cities? an evolutionary perspective. *Cities*, 67:43–52, 2017.

[56] Andrew Kusiak and Chun-Yu Tang. Innovation in a requirement lifecycle framework. In *Proceedings of the 5th International Symposium on Intelligent Manufacturing Systems, IMS*, pages 61–67. Citeseer, 2006.

[57] Angeliki Kylili and Paris A Fokaides. European smart cities: The role of zero energy buildings. *Sustainable Cities and Society*, 15:86–95, 2015.

[58] Gwanhoo Lee and Young Hoon Kwak. An open government maturity model for social media-based public engagement. *Government information quarterly*, 29(4):492–503, 2012.

[59] Jung Hoon Lee, Robert Phaal, and Sang-Ho Lee. An integrated service-device-technology roadmap for smart city development. *Technological Forecasting and Social Change*, 80(2):286–306, 2013.

[60] Jungwoo Lee and Hyejung Lee. Developing and validating a citizen-centric typology for smart city services. *Government Information Quarterly*, 31:S93–S105, 2014.

[61] Chiehyeon Lim, Kwang-Jae Kim, and Paul P Maglio. Smart cities with big data: Reference models, challenges, and considerations. *Cities*, 82:86–99, 2018.

[62] Andrew McAfee, Erik Brynjolfsson, Thomas H Davenport, DJ Patil, and Dominic Barton. Big data: the management revolution. *Harvard business review*, 90(10):60–68, 2012.

[63] Patricia C Melo, Daniel J Graham, and Robert B Noland. A meta-analysis of estimates of urban agglomeration economies. *Regional science and urban Economics*, 39(3):332–342, 2009.

[64] Tae-Heon MOON and Jung-Hun CHO. Road networks and crime occurrence multi-agent simulation for smart safe city. *Journal of the Korean Association of Geographic Information Studies*, 18(2):120–134, 2015.

[65] Luca Mora, Mark Deakin, and Alasdair Reid. Combining co-citation clustering and text-based analysis to reveal the main development paths of smart cities. *Technological Forecasting and Social Change*, 142:56–69, 2019.

[66] Taewoo Nam and Theresa A Pardo. Conceptualizing smart city with dimensions of technology, people, and institutions. In *Proceedings of the 12th annual international digital government research conference: digital government innovation in challenging times*, pages 282–291. ACM, 2011.

[67] Paolo Neirotti, Alberto De Marco, Anna Corinna Cagliano, Giulio Mangano, and Francesco Scorrano. Current trends in smart city initiatives: Some stylised facts. *Cities*, 38:25–36, 2014.

[68] Higher Level Panel. A world that counts: Moblising the data revolution for sustainable development, 2013.

[69] Sotirios Paroutis, Mark Bennett, and Loizos Heracleous. A strategic view on smart city technology: The case of ibm smarter cities during a recession. *Technological Forecasting and Social Change*, 89:262–272, 2014.

[70] Krassimira Antonova Paskaleva. The smart city: A nexus for open innovation? *Intelligent Buildings International*, 3(3):153–171, 2011.

[71] Lynne Phillips and Suzan Ilcan. 'a world free from hunger': Global imagination and governance in the age of scientific management. *Sociologia Ruralis*, 43(4):434–453, 2003.

[72] Giuseppe Piro, Ilaria Cianci, Luigi Alfredo Grieco, Gennaro Boggia, and Pietro Camarda. Information centric services in smart cities. *Journal of Systems and Software*, 88:169–188, 2014.

[73] Chirag Rabari and Michael Storper. The digital skin of cities: urban theory and research in the age of the sensored and metered city, ubiquitous computing and big data. *Cambridge Journal of Regions, Economy and Society*, 8(1):27–42, 2014.

[74] Nathalie Risse. Implementing the 2030 agenda and its sdgs: Where to start? *Policy Brief*, 2016.

[75] Stuart S Rosenthal and William C Strange. Evidence on the nature and sources of agglomeration economies. In *Handbook of regional and urban economics*, volume 4, pages 2119–2171. Elsevier, 2004.

[76] Anselm Kamperman Sanders, Tatiana Falcão, Alexander Haider, Jenna Jambeck, Cara LaPointe, Chris Vickers, Nicolas Ziebarth, et al. World economic and social survey 2018: Frontier technologies for sustainable development. 2018.

[77] Hans Schaffers. The future of smart cities: Open issues and research challenges. *Scienze Regionali, Italian Journal of Regional Science*, (1/2018):123–128, 2018.

[78] Hans Schaffers, Nicos Komninos, Marc Pallot, Miguel Aguas, Esteve Almirall, Tuba Bakici, Jean Barroca, Dave Carter, Michel Corriou, Joana Fernadez, et al. Smart cities as innovation ecosystems sustained by the future internet. 2012.

[79] Hans Schaffers, Nicos Komninos, Marc Pallot, Brigitte Trousse, Michael Nilsson, and Alvaro Oliveira. Smart cities and the future internet: Towards cooperation frameworks for open innovation. In *The future internet assembly*, pages 431–446. Springer, 2011.

[80] Jochen Schiewe, Alenka Krek, Irene Peters, Harald Sternberg, and Karl-Peter Traub. Hcu research group "digital city": developing and evaluating tools for urban research. *Digital earth summit on geoinformatics. Hamburg University, HafenCity*, 2008.

[81] Johannes M Schleicher, Michael Vögler, Schahram Dustdar, and Christian Inzinger. Enabling a smart city application ecosystem: requirements and architectural aspects. *IEEE Internet Computing*, 20(2):58–65, 2016.

[82] Richard Sennett. No one likes a city that's too smart. *The Guardian*, 4, 2012.

[83] Taylor Shelton, Matthew Zook, and Alan Wiig. The 'actually existing smart city'. *Cambridge Journal of Regions, Economy and Society*, 8(1):13–25, 2015.

[84] Anthony M Townsend. *Smart cities: Big data, civic hackers, and the quest for a new utopia.* WW Norton & Company, 2013.

[85] UN. Trends in national institutional arrangements in global geospatial information management. *United Nations Economic and Social Council, E/2015/46-E/C.20/2015/17*, 2015.

[86] UNDESA. Revision of world urbanization prospects. in: Un department of economic and social affairs, 2018.

[87] UNGA. Transforming our world: the 2030 agenda for sustainable development, 2015.

[88] Ingmar Van Meerkerk and Jurian Edelenbos. The effects of boundary spanners on trust and performance of urban governance networks: findings from survey research on urban development projects in the netherlands. *Policy Sciences*, 47(1):3–24, 2014.

[89] Alberto Vanolo. Smartmentality: The smart city as disciplinary strategy. *Urban studies*, 51(5):883–898, 2014.

[90] Nils Walravens. Qualitative indicators for smart city business models: The case of mobile services and applications. *Telecommunications Policy*, 39(3-4):218–240, 2015.

[91] WCCD. World council on city data, 2018.

[92] Wann-Ming Wey and Janice Hsu. New urbanism and smart growth: Toward achieving a smart national taipei university district. *Habitat International*, 42:164–174, 2014.

[93] Alan Wiig and Elvin Wyly. Introduction: Thinking through the politics of the smart city, 2016.

[94] Zhixuan Yang, Arbind Tuladhar, and Abbas Rajabifard. 3d city governance: Towards an integrated sustainability. *FIG 2015*, 2015.

[95] Roland Zinkernagel, James Evans, and Lena Neij. Applying the sdgs to cities: Business as usual or a new dawn? *Sustainability*, 10(9):3201, 2018.

12

Spatial Enablement to Facilitate the New Urban Agenda Commitments for Sustainable Development

Soheil Sabri
Centre for SDIs and Land Administration, The University of Melbourne, Australia

Abbas Rajabifard
Centre for SDIs and Land Administration, The University of Melbourne, Australia

In the context of SDGs, this chapter talks about spatial enablement to facilitate the New Urban Agenda commitments for Sustainable Development. It discusses historical account of how global began to be considered a threat and how it ended up as a potential development tool for the future generation established by the New Urban Agenda. The Chapter introduces the elements of action framework for the implementation of a selected case studio in Singapore. The chapter will come to an end with describing the future opportunities in research and capacity development in support of evidence-based and data-driven urban policy and planning.

12.1 Introduction: Background and Driving Forces

In light of rapid urbanisation worldwide, the complexity of cities is ever increasing, which complicate urban planning and management tasks. One of the major challenges is the limited capabilities offered by conventional approaches to urban and regional development and management [12]. Addressing current challenges created by urbanisation requires cutting-edge, interoperable tools and expertise, localised for each country and domain of application adopted and adapted from best practices worldwide to meet international standards (e.g. Sustainable Development Goals) [17].

Over the last four decades, several international organisations have developed standard indicators to foster nations to set, measure, monitor and evaluate development policies. Examples for these standards are Millennium Development Goals (MDGs) and Sustainable Development Goals (SDGs) adopted by the United Nations and member states in 2000 and 2016 respectively [18, 8]. Accordingly, other international and national organisations have developed initiatives, indicators and indices such as the New Urban Agenda, ISO 37120 urban sustainability indicators, and the City Prosperity Index for ensuring quality of life challenges are linked to sustainable development policies, strategies, and decision making [6, 12, 20].

In 2016, the United Nations Conference on Housing and Sustainable Urban Development (Habitat III) in Quito, Ecuador, adopted the New Urban Agenda (NUA). For the first time, Habitat III acknowledged cities as potentially the source of solutions rather than the cause of challenges the world is facing with. The NUA declared that, if well-planned and well-managed, urbanisation can be a powerful tool in achieving the sustainable development status in developing and developed countries [17]. Five main pillars of NUA implementation lay out the standards and principles for planning, development, construction, improvement, and management of urban areas:

1. National urban policy; *"the National Government is the level that holds the sovereignty of the nation, and it establishes the rules and functions of the subnational and local governments."*

2. Urban legislation and regulations; *"good urbanisation cannot be conceived without a good regulatory framework."*

3. Urban planning and design; *"urban planning and design is an essential technical part of the urbanization process and it refers to the physical layout of buildable plots, public space, and their relationship to one another. In line with the NUA, UN-Habitat believes that urban planning of design is a fundamental priority to achieving sustainable urban development."*

4. Local economy and municipal finance; *"one of the novelties of the urban paradigm shift of the NUA is the contribution of urbanization to the national economy. Urbanization should be approached not as a cost, but as an investment, because the cost of urbanization is minimal compared to the value that it can generate."*

5. And local implementation; *"as an important action plan, this pillar encourages spatial development strategies that take into account, as appropriate, the need to guide urban extension prioritising urban renewal by planning for the provision of accessible and well-connected infrastructure and services, sustainable population densities, and compact design and integration of new neighbourhoods into the urban fabric, preventing urban sprawl and marginalisation."*[17]

There have been extensive works to adopt the above-mentioned initiatives, standards and indicators to local policy and development strategies. As an example, UN-Habitat drafted an Action Framework for Implementation of the New Urban Agenda (AFINUA), which aimed to guide the implementation of the NUA, with necessary ingredients to lead each, identifying methods of measurement and their link to the provisions of the NUA. There are 35 key elements that group into the above-mentioned five main pillars of NUA [16].

However, recent studies indicate a weak connection between some of the indicators and government policies and decision making. Deng et al., (2017) highlight that city managers have been slow to adopt the urban sustainability indicators in their decision-making process due to the lack of unique interpretation by the urban experts [4]. Other barriers in adopting such indicators are the cost of adaptation, specifically for smaller governments, and lack of perceived benefit from international exposure and comparison. Other studies indicate a problem that arises in indicator-based comparisons, as the comparison might be invalid due to inconsistencies in the data used to derive them [19]. These studies highlight that the international standards should not only be considered for benchmarking but also be valued for the opportunity they present for comparative learning. In addition, data provenance, reliability, and consistency of analysis need to be evaluated for standards. It is significant and timely to investigate the role of spatial technology and underpinning frameworks such as hierarchy of Spatial Data Infrastructures (SDIs) [7] for supporting the implementation of sustainable development standards in different levels of government.

The crucial role of spatial technologies in capacity building across different levels of governments is emphasised in Articles 159 and 160 of AFINUA. While there are some initiatives for spatial enablement of governments in implementation of SDGs and NUA, there is limited understanding of how these technologies will be helpful. As such, this chapter explores the 35 key elements established by AFINUA and highlights the current spatial enablement initiatives that provide opportunity to implement these elements. This chapter is for decision-makers and urban and regional planners at different levels of government in local, state and national scales.

This chapter continues with providing a historical account of how global urbanisation began to be considered a threat and how it ended up as a potential development tool for the future generation established by the New Urban Agenda. The next section introduces the elements of action framework for the implementation of NUA. The principles of spatial enablement and some of the international initiatives will be presented in section three. The fourth section draws links between spatial enablement concepts and principles and the key elements of AFINUA. Finally, the last section of the chapter discusses some implications of implementing the spatial enablement framework for successful achievement of sustainable development. This chapter will come to an end with describing the future opportunities in research and capacity development in support of evidence-based and data-driven urban policy and planning.

12.2 Urbanisation; From a Threat to an Opportunity

The United Nations Conferences on Human Settlements, where a significant number of city governments and urban management officials are in attendance, have sought approaches to improving the urban quality of life. In the early 1970s, the global strategy to address challenges of human settlements was sought and, at the same time, the United Nations Resolution on Housing, Building and Planning was produced. However, one of the major issues raised was rapid urbanisation, which required a response to the emergence of large unmanageable slums with poor access to basic services including water and sanitation.

Accordingly, in 1976 the Vancouver Declaration on Human Settlements highlighted critical actions and requirements to improve the quality of life in all human settlements. This was the time that the United Nations Conference on Human Settlements (Habitat I) elaborated on issues of unplanned urbanisation, which resulted in overcrowded cities without corresponding capacity to provide basic services. Addressing these challenges, the UN General Assembly established UN-Habitat as a focal point for human settlements action. In addition, a year after the Vancouver declaration, the UN established its Center for Human Settlements (UNCHS/Habitat). However, despite these advances, poor living conditions in human settlements remained a persistent issue.

The global urban population soared dramatically and by 1990 42.5% of global population was urban; during this time ten megacities emerged with populations of 10 million or more. This led to the 1996 Istanbul Declaration on Human Settlements and a strengthening in the role and capability of the UN Center for Human Settlements (UNCHS). UN-Habitat II was held at the same time and HABITAT AGENDA was adopted. This agenda particularly emphasised promoting the provision of adequate shelters and basic services to familiarise sustainable human settlements.

In 2000, eight Millennium Development Goals (MDGs) were adopted by the United Nations. These goals, which targeted for 2015, range from reducing extreme poverty rates by 50% to fighting against the spread of HIV/AIDS and providing universal primary education. The MDGs set to respond to the challenges of urban inequality, slums, poverty, and environmental degradation. As a result of implementing MDGs, between 2000 and 2014, more than 320 million people living in slums gained access to improved water sources, improved sanitation facilities, or durable or less crowded housing. However, in 2015 more than 880 million people were estimated to be living in slums, compared to 792 million in 2000 and 689 million in 1990.

The implementation of MDG targets worldwide had been recorded with some progress. Several countries included the Millennium Development Goals (MDGs) into their national and sub-national development plans and strategies, and adopted specific measures with the aim of achieving the associated targets [18]. However, the achievements have been uneven across regions and countries, leading to significant gaps [21]. According to the UN report on MDGs in 2015, millions of disadvantage people had not leveraged these goals due to factors such as geographic location, gender, age, ethnicity, and disability [8].

Reaching to the deadline of MDGs in 2015, the discussion around the results and barriers of implementation led to formulating the Sustainable Development Goals (SDGs). Given that the MDGs failed certain people and geographical locations, the SDGs' 2030 Agenda sets out to "*reach the furthest behind first*" and concludes with a pledge that "*no one will be left behind*" [18].

Accordingly, In October 2016, the New Urban Agenda was unanimously adopted at the United Nations Habitat III, with the aim of serving as a new vision for cities and municipalities worldwide for the next 20 years [17]. UNDP "*demonstrated its full support to the implementation of the New Urban Agenda with the official launch of its Sustainable Urbanization Strategy*". This was the first time urbanisation was considered an opportunity for achieving sustainable development, which addresses SDGs, especially Goal 11 on sustainable cities and communities.

In the reviews for implementing NUA the community consultations and inputs from two important expert group meetings held in Surabaya, Indonesia, in July 2016 and New York in April 2017 led to formulating an action framework for implementation of the NUA (AFINUA) [16]. This framework articulated that while the process of achieving the Sustainable Development Goals are important, it cannot replace the outcome. In fact, the specific attention of AFINUA is enumerating desired urban outcomes. As such, the framework set out to assist the local authorities, major groups and relevant stakeholders to measure and monitor the targets they are aiming for.

The AFINUA key elements are also connected to other indicators identified by UN-Habitat's City Prosperity Initiative (CPI). The CPI is a global initiative that aims to turn the data to information and knowledge for cities to measure their performances and establish an evidence-based policy dialogue among decision-makers. The CPI aims to facilitate a higher accountability in the implementation of the SDGs and NUA [19]. This initiative and associated tools have been used individually or in combination with other indexes in evaluating several cities worldwide in monitoring sustainability performance [23], multi-scale sustainability evaluation [22], environmental quality [1], and urban resilience [13].

Like AFINUA, there have been other action guidelines established for localising the implementation of SDGs. In Europe, the Association of Flemish Cities and Municipalities (VVSG) worked with local authorities to translate the SDGs at the local level [5]. They published several tools and guidelines to help local authorities in monitoring and exploring the ways to generate a broader SDG policy in their respective legislation. VVSG transferred the focus of achieving SDGs to taking specific actions and raising awareness of this global ambition among residents, in government, and in industry. In the next section, the principles of the AFINUA and CPI are described in terms of how they are set for global action.

12.3 AFINUA and Its Relation to SDGs and CPI

The action framework for implementing new urban agenda, groups 35 key elements into the NUA's five pillars: (1) national urban policies, with six key elements, (2) urban legislation, rules and regulations, with nine key elements, (3) urban planning and design, with eight key elements, (4) urban economy and municipal finance, with six key elements, and (5) local implementation, with six key elements. While the NUA is exclusive to SDG Goal 11, other goals and targets provide urban-critical sectoral and cross-cutting areas. Some examples are the food security and urban-rural linkage relevant to Goal 2; health as a critical urban factor in Goal 3; education and culture is considered in Goal 4; gender equity reflected in Goal 5; water challenges in Goal 6; energy as a major concern in cities reflected in Goal 7; employment and GDP indicated in Goal 8; sustainable consumption and production in Goal 12; and climate change impacts in cities included in Goal 13. As such, these relations are sought in formulating the AFINUA elements [16].

In addition, the connection of AFINUA to CPI is through six dimensions set by the UN-Habitat:

a. Productivity (CPI-P):*This dimension measures the average achievements of the cities in terms of creating wealth and how it's shared, or cities contribution to economic growth and development, generation of income, provision of decent jobs and equal opportunities for all.*

b. *Infrastructure Development (CPI-ID):The Infrastructure dimension measures the average achievement of the city in providing adequate infrastructure for accessing clean water, sanitation, good roads, and information and communication technology - in order to improve living standards and enhance productivity, mobility and connectivity.*

c. *Quality of Life (CPI-QoL):The quality of life dimension measures the cities' average achievement in ensuring general wellbeing and satisfaction of the citizens.*

d. *Equity and Social Inclusion (CPI-ESI):The Equity and Social inclusiEquity and Social Inclusion (CPI-ESI):on dimenion measures the cities' average achievements in ensuring equitable (re)distribution of the benefits of prosperity, reduces poverty and the incidence of slums, protects the rights of minority and vulnerable groups, enhances gender equality, and ensures equal participation in the social, economic, political and cultural spheres.*

e. *Environmental Sustainability (CPI-ES):The Environmental Sustainability dimesion measures the average achievement of the cities in ensuring the protection of the urban environment and its natural assets. This should be done simultaneously while ensuring growth, pursuing energy efficiency, reducing pressure on surrounding land and natural resources and reducing environmental losses through creative and environment-enhancing solutions. [15].*

Figure 12.1 summarises the connection of AFINUA elements with SDGs for National Urban Policies. The details of these connections and their local actors can be found in [16].

FIGURE 12.1
The Connection of AFINUA Elements for National Urban Policies Domain of NUA

One of the major requirements for ensuring implementation of items set by AFINUA is a robust data infrastructure. This is significant because the implementation needs ongoing control, measurement, and monitoring of the AFINUA-related SDG and CPI indicators. The next section briefly explains the overall components of a reliable data infrastructure for this purpose.

12.4 Spatial Data Infrastructure Advancements and Opportunities

The MDG report in 2015 indicated the importance of sustainable data for sustainable development [8]. This report regarded the data as an *"indispensable element of the development agenda"*. In particular, this report indicated how the local data is important for measuring and monitoring subnational performances. The demand and policy making are regarded as two significant drivers for data improvements, and while there have been several initiatives for improvements of data collection worldwide, the critical data for policy making was still lacking [18].

In transformation from MDGs to SDGs, one of the main issues was the lack of quality data to enable regular monitoring and support evidence-based decision making. As such, several suggestions were made by international entities including UNDP, World Bank Group, and UN-Habitat to support evidence-based decision making. These suggestions include using real-time data, adopting geospatial data, strengthening statistical capacity, utilising new technologies, changing the methods of data collection and dissemination, developing global standards for integrated statistical systems, and promoting open data [8, 18].

However, several studies reported that many governments worldwide lack awareness, realising the importance of geospatial information and related technologies in enabling the implementation of SDGs. This lack of awareness is particularly at the policy and decision-making level, which hinders enabling robust tools such as National Spatial Data Infrastructures [14].

While the national policy makers are yet to fully implement the data policy in support of sustainable development, the international entities have provided several frameworks and standard guidelines, which have been used for developing system architectures as enablers for deriving several city and regional indicators in an ad hoc fashion.

At the United Nations level, several initiatives including UN-Global SDG Database [1] is available for access to data, which is compiled through the UN System in preparation for the Secretary-General's annual report on "Progress towards the Sustainable Development Goals". However, this data is not spatially enabled and remains at a national level, which provides only a limited understanding of the local performances. In addition, recently an Open SDGs Data Hub [2] was developed to fully implement and monitor progress on the SDGs. This platform aims to support decision makers everywhere, who need accurate and timely data and statistics. So far, however, limited countries have committed to this platform and the available data has limited capability for local governments.

[1] https://unstats.un.org/sdgs/indicators/database/
[2] http://www.sdg.org/

In the international technology standard community, Open Geospatial Consortium (OGC) formulated an enterprise framework to derive indicators for sustainable development and resilience of communities (ISO 37120) [9]. The OGC framework is considered to be based on cloud computing. This framework, which is called the OGC Smart Cities Spatial Information Framework, incorporates four layers of sensing (real-time data), data (access and quality checking), business (analysis and visualisation), and application (e.g. health, education, public safety and security, and urban planning). Figure 12.2 shows the components of OGC enterprise framework for smart cities.

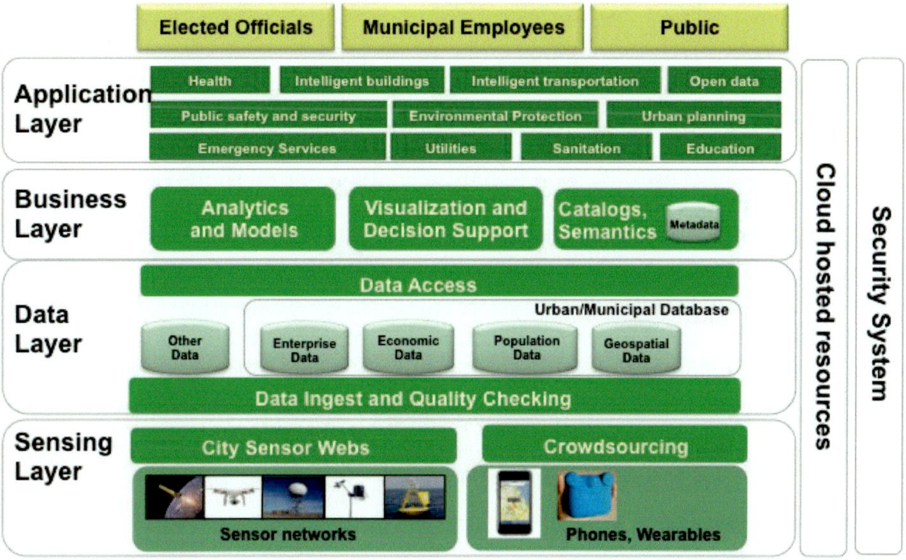

FIGURE 12.2
OGC enterprise framework for Smart Cities adopted from [9].

Several initiatives worldwide have implemented spatial data infrastructures addressing the components of the OGC framework. The examples are Australian Urban Research Infrastructure Network (AURIN), which is an e-infrastructure to support urban and built environment research in Australia. AURIN facilitates access to more than 2500 datasets across Australia, enabling researchers and data providers to integrate data, analyse and visualise several different urban and built environment data [11]. Another example is the Urban Big Data Centre (UBDC), an initiative by the UK Economic and Social Research Council at the University of Glasgow, in partnership with six other UK universities. The role of UBDC is to manage, link and analyse massive amounts of multi- sectoral urban open and authorised data in a portal allowing diverse users to conduct research and analysis [3]. Similarly, the

[3]http://ubdc.ac.uk/

Urban Centre for Computation and Data (UrbanCCD) at the University of Chicago has developed a platform called Plenarion to facilitate urban data discovery, exploration, and application of open city data [2]. However, most of these systems lack real-time data analysis as well as a semantic enablement layer for harmonising the fragmented and heterogeneous spatial data [10]. In addition, some of these infrastructures are yet to be used in deriving the SDG and CPI indicators as well as providing reliable sources for local, subnational, and national policy makers. A potential explanation for this might be that the sustainable development standards and indicators are too broad to be derived in smaller geographical boundaries than states and national levels. In addition, the national and international data policies are not reflecting the requirements of SDG and CPI.

Moreover, the lack of semantic enablement layer is an important one to address as AFINUA elements and associated SDG and CPI indicators are comprised of multi-disciplinary domains and fragmented data sources with heterogenous structures [3]. The semantic enablement methods and associated architectures will be presented in other chapter on enabling tools and technical components.

As such, while there are several national and international initiatives, standards and frameworks for spatial data infrastructures, they are not fully committed to enable the implementation, measuring and monitoring of the sustainable development indicators and localised NUA initiatives that set the future agenda for research and policy making.

12.5 Conclusion and Discussion

This chapter is intended to highlight the localising steps taken for implementation of SDGs in urban and territory environments. In addition, it highlighted several technological and spatial data initiatives worldwide that can potentially act as enablers for implementing, measuring, and monitoring the localised Sustainable Development Goals.

The literature and evaluation reports on SDGs indicated the weak connection between decision making and government policies and the standard indicators. This limited connection was attributed to the lack of robust data and various interpretation on the standards and indicators, and as a result the roles of Spatial Data Infrastructures and new technologies have been highlighted in several reports and studies.

While there is an emphasis on the critical role of spatial technologies for implementing the New Urban Agenda (articles 159 and 160 of AFINUA), this chapter highlights the lack of data standards and policy to address the AFINUA requirements. As such, two major aspects need to be addressed for local implementation and monitoring of New Urban Agenda and associated

Sustainable Development Goals. First, the broadness of the sustainable development standards and indicators and their definitions, which limits them to be derived in smaller spatial levels such as statistical areas or suburbs. Second, the national and international data policies need to be revisited to reflecting the requirements of sustainable development indicators such as SDG and CPI.

Bibliography

[1] Marino Bonaiuto, Ferdinando Fornara, Silvia Ariccio, Uberta Ganucci Cancellieri, and Leila Rahimi. Perceived residential environment quality indicators (preqis) relevance for un-habitat city prosperity index (cpi). *Habitat International*, 45:53–63, 2015.

[2] Charlie Catlett, Tanu Malik, Brett Goldstein, Jonathan Giuffrida, Yetong Shao, Alessandro Panella, Derek Eder, Eric van Zanten, Robert Mitchum, Severin Thaler, et al. Plenario: An open data discovery and exploration platform for urban science. *IEEE Data Eng. Bull.*, 37(4):27–42, 2014.

[3] Yiqun Chen, Soheil Sabri, Abbas Rajabifard, and Muyiwa Elijah Agunbiade. An ontology-based spatial data harmonisation for urban analytics. *Computers, Environment and Urban Systems*, 72:177–190, 2018.

[4] Derui Deng, Shiyao Liu, Lawrence Wallis, Elizabeth Duncan, and Phil McManus. Urban sustainability indicators: how do australian city decision makers perceive and use global reporting standards? *Australian geographer*, 48(3):401–416, 2017.

[5] Bart Van Herck, Valentijn Vanoeteren, and Katrin Janssen. Local indicators for the 2030 agenda (sustainable development goals), 2019. Brussels, Belgium.

[6] Robert W Marans. Quality of urban life & environmental sustainability studies: Future linkage opportunities. *Habitat International*, 45:47–52, 2015.

[7] Ian Masser, Abbas Rajabifard, and Ian Williamson. Spatially enabling governments through sdi implementation. *International Journal of Geographical Information Science*, 22(1):5–20, 2008.

[8] United Nations. *The Millennium Development Goals Report 2015.* 2015.

[9] George Percivall, C Ronsdorf, S Liang, D McKenzie, and L McKee. Ogc smart cities spatial information framework. *OGC Internal reference*, (14-115), 2015.

[10] Abbas Rajabifard, Serene Ho, and Soheil Sabri. Urban analytics data infrastructure: Critical sdi for urban management in australia. *SMART WORLD*, page 95, 2016.

[11] Soheil Sabri, Christopher J Pettit, Mohsen Kalantari, Abbas Rajabifard, Marcus White, Oliver Lade, and Tuan Ngo. What are essential requirements in planning for future cities using open data infrastructures and 3d data models. In *Proceedings of the 14th International Conference on Computers in Urban Planning and Urban Management, Cambridge, MA, USA*, pages 7–10, 2015.

[12] Soheil Sabri, Abbas Rajabifard, Serene Ho, Mohammad-Reza Namazi-Rad, and Christopher Pettit. Alternative planning and land administration for future smart cities [leading edge]. *IEEE technology and society magazine*, 34(4):33–73, 2015.

[13] Holger Schlör, Sandra Venghaus, and Jürgen-Friedrich Hake. The few-nexus city index–measuring urban resilience. *Applied energy*, 210:382–392, 2018.

[14] Greg Scott and Abbas Rajabifard. Sustainable development and geospatial information: a strategic framework for integrating a global policy agenda into national geospatial capabilities. *Geo-spatial Information Science*, 20(2):59–76, 2017.

[15] UN-Habitat. The six dimensions of urban prosperity — city prosperity initiative, 2016.

[16] UN-Habitat. Action framework for implementation of the new urban agenda, 2017.

[17] UN-Habitat. The new urban agenda, adopted by the united nations on 20 october 2016, quito, ecuador, 2017.

[18] UNDP and World Bank Group. Transitioning from the mdgs to the sdgs, 2016.

[19] Yetian Wang and Mark S Fox. Consistency analysis of city indicator data. In *International Conference on Computers in Urban Planning and Urban Management*, pages 355–369. Springer, 2017.

[20] Cecilia Wong. A framework for 'city prosperity index': Linking indicators, analysis and policy. *Habitat International*, 45:3–9, 2015.

[21] Zofia Wysokińska. Millenium development goals/un and sustainable development goals/un as instruments for realising sustainable development concept in the global economy. *Comparative Economic Research*, 20(1):101–118, 2017.

[22] Tan Yigitcanlar, Fatih Dur, and Didem Dizdaroglu. Towards prosperous sustainable cities: A multiscalar urban sustainability assessment approach. *Habitat International*, 45:36–46, 2015.

[23] Bohong Zheng and Komi Bedra. Recent sustainability performance in china: Strength-weakness analysis and ranking of provincial cities. *Sustainability*, 10(9):3063, 2018.

13

The Geospatial Capacity Building Ecosystem - Developing the Brainware for SDI

Josef Strobl
University of Salzburg, Austria

In the context of SDGs, this chapter aims at arguing for an ecosystem view of lifelong learning at the core of building and maintaining the brainware for geospatial information systems supporting our livelihoods.

13.1 Introduction

'Brainware', the human capacity and competence to manage Spatial Data Infrastructures (SDI) and related geospatial information frameworks is widely considered a bottleneck in generating decision support for societies, economies and environments [11]. This not only applies to technical skill sets for operating various systems of record, but even more so in leveraging the power of spatial thinking approaches towards reaching Sustainable Development Goals, most of which cannot be approached and monitored without a geospatial perspective.

Traditionally, higher education institutions (HEI) were and still are considered the places for academic capacity building, preparing and qualifying graduates for designing, implementing, maintaining and leading complex architectures like SDIs. Very few qualifications, though, today last a professional lifetime due to still accelerating cycles of innovation, technologies and disruptive changes. Capacity building today is far from a linear process, requiring a multitude of interventions, re-inventions, re-qualifications and actions by multiple stakeholders.

13.2 Status

Academic education in geospatial technologies, methods and applications have parallel roots in several spatially oriented disciplines [13]. Geography adopted quantitative methods and computer cartography as a pathway towards GIS. Surveying morphed into Geomatics through positioning technologies and automation. Remote Sensing emerged as an effective data acquisition and earth observation technology. Numerous application domains like planning, resource management, transportation or business intelligence took advantage of these emerging technologies (including 'Open' approaches – see [1]) and refined their respective business processes.

This 'transversal', methodology-oriented approach to geospatial education required the creation of novel curricular pathways, implemented through a sequence of 'core curricula' initiatives [4, 14, 16]. 'Transversal' also refers to the need for integration of geospatial competences in the full range of spatially oriented disciplines, ranging from A like Archeology to Z like Zoology. One approach to 'spatialize' disciplines is through the option of including a 'minor' in one's academic coursework [3]. This can be aligned with a 'spatial turn' currently experienced in a variety of disciplines, including the boost for Digital Humanities.

All these curriculum and (partly) learning media developments did not fully succeed with satisfying the wider industry needs of qualified experts, though. Only few dedicated study programs(e.g. in Geoinformatics) have been implemented, with a majority of geospatial methods courses embedded in traditional 'spatial' and application discipline programs.

While these take care of educating competent users of SDI elements, managing core SDI architectures suffers from shortages of qualified staff bringing the right mix of computational, architectural and geospatial knowledge to task.

In addition, due to the dynamic evolution of SDI technologies and standards, initial cycles of academic qualification will not sufficiently support a lifetime of professional leadership in such a complex domain. Continuing education requires multiple actors supporting different modes and facets of 'lifelong learning'. While dedicated online study programs like UNIGIS [12] facilitate in-service development of competences, the latter will not be sustained without involvement of a range of stakeholders.

13.3 Mix of Actors in an Education Ecosystem

An ecosystem, as defined by Encyclopedia Britannica, is understood as a 'complex of living organisms, their physical environment, and all their interrelationships in a particular unit of space.' Obviously, this concept is used metaphorically in non-biological domains, e.g. when referring to digital ecosystems as 'digital counterparts of biological ecosystems' [2].

To develop educational ecosystems, first the 'organisms', i.e. actors need to be identified before exploring potential relationships and interactions among these:

- Higher Education Institutions: universities continue to fulfill the roles of initial undergraduate and graduate education, continuously adjusting curricula, syllabi and pedagogical frameworks to evolving technologies and application demands [9, 10]. While in many cases undergraduate studies serve as an entry point into the geospatial domain, 'feeding' motivated youngsters into suitable academic tracks is an important task long neglected. Outreach initiatives like GIS Days, interaction with schools and participative activities increasingly are driven by HEIs. In addition, (some) HEI increasingly focus on mid-career continuing education – definitely a growth area with substantially more demand than current supply [15].

- Industry, composed of technology vendors as well as domain-oriented services, are addressing their need for talent by taking a stronger role in capacity building. This includes offering internships and dual-track study-while-working schemes, sponsoring of students, lecturing within academic programs, providing technologies and guidance for emerging innovations.

- Professional associations offer stimuli (like certification), sometimes arrange short courses and networking opportunities as well as conference services. These societies create professional identities which sometimes are limiting, but also help with motivating personal development.

- Media play an indispensable role not only as actors providing current updates on technologies and professional practice through web portals, magazines and conferences, but perhaps even more importantly as connecting facilitators enabling relationships among 'organisms' and individuals.

While considering these institutional actors, we need to focus on individuals as 'carriers' of brainware, developed while moving through the operational learning instances of this ecosystem. These students – in all stages of their professional lives – are exposed to these instances as opportunities for learning: academic courses and programs, short intensive courses and trainings, literature items and webinars, MOOCs and online trainings, mentoring, internships and many more.

Clearly, most of these learning instances cannot easily be provided by a single actor, therefore it is essential to establish a collaborative framework collectively generating the opportunities for personal development within an 'ecosystem'.

Individual learners can follow many different pathways through this ecosystem. While these pathways in many cases will start from undergraduate studies, progressing through various continuing education opportunities, they also can start from professional experience before turning towards formalized education. Pathways definitely are not only linear but also loop through re-tooling by continuing education, and also 'change sides' by swapping the roles of instructor and learner, moving towards a community-oriented common learning and support experience in later stages.

13.4 Case Study: the Copernicus Master in Digital Earth

Starting from 2019, the Universities of Salzburg (Austria), Olomouc (Czech Republic) and South Brittany (France) offer a joint international MSc program (https://www.master-cde.eu) with all students completing a first year in Austria before specializing in Geo Data Science (France) or Geovisualisation (Czech).

This kind of programme (based on experiences from [6, 7]) only is feasible within the context and with the support of organisations like Copernicus Academy, UNIGIS, Eurogi etc, numerous industry actors providing internships, professional placements and technologies, and geospatially oriented media outlets creating the required visibility. Supported through a generous European scholarship scheme, this MSc programme will lead the way towards qualifying global experts for the trend towards online geospatial technologies supporting societies, economies and environments, and indispensable for monitoring spatially distributed SDG indicators.

The European *'Copernicus Master in Digital Earth'* is underpinned by a strong emphasis on international perspectives and mobility. SDGs and SDIs cannot be sensibly bounded by national boundaries, and reaching across national entities requires experience in different countries, languages, cultures, technological environments, industries and institutional settings. Programmes like this therefore will be leading the way towards introducing excellent students into the future geospatial communities of practice.

13.5 Educational Ecosystem Services

As demonstrated in the previous section, sustained positive outcomes from capacity building measures will rarely be achieved by single organisations providing specific study programs or learning instances. Curriculum development has to be a multi-stakeholder effort, shall reach beyond initial education cycles to cover lifelong learning perspectives, and primarily address fundamental concepts adaptable to technological evolution and updated methods.

Employing the concept of 'ecosystem services' [5], the joint and collaborative contribution of stakeholders to geospatial capacity building is considered an essential service to SDI brainware development, which cannot be rendered by any individual institution alone.

Only a multitude of services ranging from traditional educational programmes to targeted short courses and trainings, MOOCs, certificate-induced learning, informative updates through magazine articles and blogs, individual mentoring and social learning frameworks etc. will ultimately succeed with qualifying and maintaining a workforce able to support the needs of SDI-for-SDG monitoring and evaluation.

13.6 Conclusions

Geospatial technologies and methods have been identified as indispensable for measuring and monitoring many or most of the 232 indicators for SDGs. Access to these metrics serving as KPIs for SDGs is facilitated through online services integrated with SDIs. Competences to design, develop and maintain SDIs on one side, and to work with these geospatial services from an application domain perspective on the other hand thus are indispensible for any operational approach to SDGs. Above we outline an argument why capacity building towards these required competences has to be a community effort [8]. Any singular approach, like a dedicated study programme, will not be able to fulfil the long term qualification goals critical for contributing the brainware components of making the SDG framework successful across human societies.

This chapter therefore serves as a call to action, addressing all stakeholders identified above. A collaborative and concerted effort is needed to succeed, it can not only be the HEIs responsibility to supply talent to 'the geospatial industry', but rather everyone involved will need to contribute towards strengthening the critical element of human capacity in this field.

Bibliography

[1] Mariana Belgiu, Josef Strobl, and Gudrun Wallentin. Open geospatial education. *ISPRS International Journal of Geo-Information*, 4(2):697–710, 2015.

[2] Gerard Briscoe and Philippe De Wilde. Digital ecosystems: evolving service-orientated architectures. In *Proceedings of the 1st international conference on Bio inspired models of network, information and computing systems*, page 17. ACM, 2006.

[3] Adrijana Car and Josef Strobl. Minor programs in the geospatial domain–referencing the giscience & technology body of knowledge. *GI_Forum 2016,*, 4:336–354, 2016.

[4] DeMers M. Johnson A. Kemp K. Luck A. T. Plewe B. & Wentz E. (Eds.) DiBiase, D. *Geographic Information Science & Technology Body of Knowledge (1st edition).* Assn of Amer Geographers; 1 edition (June 30, 2006), 2006.

[5] Stephen C Farber, Robert Costanza, and Matthew A Wilson. Economic and ecological concepts for valuing ecosystem services. *Ecological economics*, 41(3):375–392, 2002.

[6] A. Nazarkulova and Strobl J. 'gem' – international geoinformatics ms in central asia. proceedings of gisca-12., 2012.

[7] A Nazarkulova and J Strobl. From central asia to europe: Erasmus mundus" gsmart" student mobility. *International Journal of Geoinformatics*, 12(4), 2016.

[8] J. Strobl Nazarkulova, A. and M. Belgiu. Building an sdi as a community project – challenges in emerging economies. Proceedings GSDI-12 Conference, Ottawa., 2010.

[9] Uwe Schulze, Detlef Kanwischer, and Christoph Reudenbach. Essential competences for gis learning in higher education: a synthesis of international curricular documents in the gis&t domain. *Journal of Geography in Higher Education*, 37(2):257–275, 2013.

[10] A. Nazarkulova Strobl, J. and B. Markus. Collaborative components in blended learning curricula. *Proceedings of the IGU Congress on Geography Education.*, 2011.

[11] J. Strobl. "digital earth brainware". in: Geoinformatics paves the highway to digital earth(gi-reports@igf), 2008.

[12] Josef Strobl. Unigis–networked learning over a distance. *Teaching Geographic Information Science and Technology in Higher Education*, 2012.

[13] Josef Strobl. Geographic information science and technology: Educational directions. *International Encyclopedia of Geography: People, the Earth, Environment and Technology: People, the Earth, Environment and Technology*, pages 1–8, 2016.

[14] Danny Vandenbroucke and Glenn Vancauwenberghe. Towards a new body of knowledge for geographic information science and technology. *Micro, Macro & Mezzo Geoinformation*, 2016(6):7–19, 2016.

[15] Gudrun Wallentin, Barbara Hofer, and Christoph Traun. Assessment of workforce demands to shape gis&t education. *Transactions in GIS*, 19(3):439–454, 2015.

[16] John P Wilson. Geographic information science & technology: Body of knowledge 2.0 project. *Final Report: University Consortium for Geographic Information Science*, 2014.

Part IV

Enabling Tools and Technical Components

14

The Role of Geospatial Information Standards for Sustainable Development

Denise McKenzie

Open Geospatial Consortium, United Kingdom

Mathias Jonas

International Hydrographic Organization (IHO), Germany

Serena Coetzee

University of Pretoria, South Africa

Chris Body

Standards Consultant , ISO/TC211, Australia

Margie Smith

Geoscience Australia, Australia

Marcus Blake

Australian Bureau of Statistics, Australia

Joseph Abhayaratna

PSMA Australia Limited, Australia

Michael Judd

Land Information New Zealand, New Zealand

Marna Roos

AfriGIS, South Africa

The purpose of this chapter is to illustrate how standards for geospatial information facilitate the implementation of United Nations Sustainable Development Goals (SDGs).

14.1 Introduction

The UN Millennium Development Goals Report from 2015 states under the heading "Geospatial data can support monitoring" the following phrase:

> *"Knowing where people and things are and their relationship to each other is essential for informed decision making. Comprehensive location-based information is helping Governments to develop strategic priorities, make decisions, and measure and monitor outcomes. Once the geospatial data are created, they can be used many times to support a multiplicity of applications."*

The notion of this sentence is not to be challenged but raises the very practical question: How exactly can this multitude of data uses be technically enabled? The obvious answer is standardization, more precisely, a standardization which enjoys universal acceptance and application. Section 14.2 explains why standardization is inseparably associated with the age of digitization. In section 14.3, background on international standardization for geospatial information is provided. Section 14.4presents the ecosystem of international standards. In section 14.5, several case studies are briefly presented to demonstrate how suitable geospatial information based on common standards supports the overarching aspiration of sustainability.

It should be noted that in 2015 the OGC, ISO/TC211 and IHO presented the first version of the Guide to the role of standards in Geospatial Information Management and its companion document to the UN Global Geospatial Information Management Committee of Experts. This chapter on geospatial standards for the SDGs should considered in relation to the fundamental principles and best practice guidance in these documents. The authors strongly recommend that anyone embarking on work that utilizes geospatial data for the SDGs should ensure they have familiarized themselves with this document.

Excerpt from the decisions of the Fifth Session of the United Nations Committee of Experts on Global Geospatial Information Management (UN-GGIM). Held from 3-7 August 2015 at the United Nations Headquarters in New York.

5/108 Implementation and adoption of standards for the global geospatial information community The Committee of Experts:

(a) Welcomed the report by the Open Geospatial Consortium (OGC), Technical Committee 211 of the International Organization for Standardization (ISO/TC 211) and the International Hydrographic Organization (IHO), and thanked them and their many experts for their collaborative efforts in producing and finalising the Standards Guide and Companion Document.

(b) Adopted the final published "Guide to the Role of Standards in Geospatial Information Management" and the "Technical Compendium" as the international geospatial standards best practice for spatial data infrastructure, and encouraged all Member States to adopt and implement the recommended standards appropriate to their countries' level of spatial data infrastructure (SDI) maturity.

(c) Encouraged Member States to continue to work in cooperation with the international standards bodies, including participation, as appropriate, in the work programmes of the OGC, ISO/TC211 and the IHO, and requested the standards organisations to consider mechanisms to facilitate wider training programmes and to ensure the access to standards on reasonable terms, especially for developing countries.

14.2 Digitization Forces Standardization

With regard to standardization the age of digitization evolves tremendous changes along the whole chain of information processing. Starting with the acquisition and storage of so far unseen massive data amounts, continuing with the processing to normalize and interpret the gathered information through correlation and quality assessment up to the visual presentation and treatment through artificial intelligence procedures. Digital standardization has become the elementary requirement for all modern data appliances. Compared with the analogue treatment of information in the past, the digital handling of information as data enables the combining of information originally delivered from different knowledge domains. The phrase coined for this capability is "interoperability" and can be effectively achieved in a digital environment only. Interoperability is based in the first place on a consistent abstraction about a set of associated entities by means of attributed objects and their interrelations applicable across domains -the data modelling. Consequently, data modelling itself is a relevant subject of standardization- known as semantic standards.

Once such universal data model exists, a standardized form of digital encoding of the semantic information associated with the modelled entity can be applied. The result are machine readable data sets describing a situation generically. Encoding schemes does not need to be identical for all data sets as long as they are based on the same modelling paradigm, however, broad use of the same encoding eases the processing of the resulting data sets enormously and led to the dominance of a small number of such schemes. Well known ones are XML and its derivation for geospatial information named GML. These standards belong to the world of technical standards.

The semantic and technical standardization deliver ample foundation to create consistent data sets assumed there is prescriptive coding guidance to assign the raw information to a designated object and to form a inter object relations. Again, authoritative standardization is required here -at least within the knowledge domain to result in consistent data sets of the same theme but from different producers.

Once produced, the data sets have to be streamed to their users. Internet communication standards are ready to provide this carriage for geospatial information. There are basically two well-known concepts: the client-server architecture or the transfer of a copy of a data set toward the user to replicate the originating data base locally in parts or completely.

Today digital geospatial data sets manly contain pure information only, which means that different interpretations can be applied to. The most popular interpretation is visual presentation. Based on a consistent data model and universal encoding standards, customized presentation rules can be applied to the data set. End users software, commonly named *Geo Information Systems - GIS* - provide additional functionality to generate varying presentations of the same information content but customized to the task at hand. Like the data modelling and the encoding, presentation rules are subject to standardization as well since this supports identical interpretation through the user.

The visual presentation is still the most popular human-machine-interface for the provision of geospatial information. However, natural voice command and response gain more and more acceptance. The machine "reads and tells" the user the facts on his/hers vocal request. Even here standards in vocabulary and pronunciation are the prerequisite. But procedural interpretation of the transmitted information with little or without human intervention is about to become the mightiest tool to treat complex geospatial information-familiar to anybody who ever planned a route to drive by means of car navigation.

Semantic Standards and Technical Standards Combined

Over the last two decades, many information communities have learned the importance of data coordination and have learned how to do it. Information communities who depend on sharing information often put in place data coordination committees and processes for creating and maintaining standard data models and metadata content standards. The data model used by an information community is their standard way of describing spatial information. It provides a data dictionary and related details necessary for the sharing, aggregation and comparison of data within the community. Metadata associated with a data set includes the data model along with other data about the data - date of collection, person or organization responsible for the collection, etc. Data model development proceeds as a part of an information community's metadata standards development effort. Such standards are often referred to as "semantic standards". Because of these standards, different information systems used within the community can "speak the same language". Different data sets that use the same data model can be aggregated or compared. Semantic standards also facilitate communication between information communities: When each community's data model is published and relatively stable, translation between different data models is easier and more precise, despite some inevitable loss of information.

Data models necessarily evolve as information communities evolve, and so this data coordination process within and between domains is an ongoing activity. Data modelers working with other data modelers are key standards developers for the Anthropocene. Geospatial standards are important for environmental work because virtually everything in our environment has a spatial component and because interactions between environmental features and phenomena depend on proximity. In the geospatial world, an "information community" is an industry, profession, academic discipline or other domain that shares a set of spatial information communication requirements. Because the geospatial element is so important, many data coordination efforts have begun in efforts to create "spatial data infrastructures".

14.3 The Framework of International Standardization for Geospatial Information

There are three key international organizations with the objective of developing open standards for geospatial information.

ISO/TC 211, Geographic information/Geomatics, is a technical committee of the International Organization for Standardization (ISO). It works towards establishing a structured set of standards for information concerning objects or phenomena that are directly or indirectly associated with a location relative to the Earth. The Open Geospatial Consortium (OGC) is an international not for profit standards organization. The focus of OGC work is to define, document and test implementation standards for use with geospatial content and services. The International Hydrographic Organization (IHO) is an intergovernmental consultative and technical organization established in 1921 to support safety of navigation and the protection of the marine environment. Among its main objectives, IHO is to bring about the greatest possible uniformity in nautical charts and documents (i.e. standardization). The provision of hydrographic and nautical chart services is one of the obligations of coastal State signatories to the International Convention for the Safety of Life at Sea (SOLAS) under the responsibility of the International Maritime Organization (IMO).

Members of the international standards organizations represent government, industry, research, and academia, and develop standards through consensus. Together the standards developed by these organizations form an integrated ecosystem, e.g. OGC and IHO standards leverage the abstract standards defined by ISO/TC 211. This ecosystem facilitates the publication, discovery, access, maintenance and use of geospatial information across a range of applications, systems and business enterprises.

FIGURE 14.1
O Ostensen, D McKenzie & R Ward - Standards Report to UN-GGIM 2015

DGIWG	Defence Geospatial Information Working Group
WMO	World Meteorological Organization
ICAO	International Civil Aviation Organization
W3C	World Wide Web Consortium
OASIS	Organization for the Advancement of Structured Information Standards
IETF	Internet Engineering Task Force
IEEE	Institute of Electrical and Electronics Engineers
IEC	International Electrotechnical Commission
OMG	Object Management Group
ISO/IEC JTC 1	ISO/IEC Joint Technical Committee 1-Information Technology

To achieve interoperability, standardization in the field of geospatial information covers system heterogeneity (hardware, operating systems, communication systems, etc.), syntactic heterogeneity (physical representation of information), structural heterogeneity (concepts and the relationships between them) and semantic heterogeneity (meaning of concepts). In addition, service standards define interfaces for geospatial information functionality provided by a server, and procedural standards provide specifications for accomplishing a particular task.

Geospatial information standards rely heavily on general purpose information communication and technology (ICT) standards. Similarly, domain-specific geospatial information standards rely on generic geospatial information standards, as depicted below.

14.3.1 Technical Standards Link Environmental Standards to IT Innovations

Over the last two decades, the members of this integrated standardization ecosystem have developed policies and procedures for working together to develop consensus-based open interface and encoding standards that provide a way for any two computer systems to request and return any kind of spatial data. These "technical standards" are broadly useful within all spatial data information communities. They support inter-community communication and they are also essential for convergence and integration of different kinds of spatial technologies, such as 2D/3D/4D imaging, vector GIS, surveying, CAD, tracking, etc.

14.3.2 Standardization Driven by Innovation and Technical Evolution

Standardization has always an aspect of consolidation, i.e. freezes technical progress for a distinct duration in favor of uniformity. Modern concepts of digital standards try to overcome this paradigm. The ISO 19200 series for geospatial information standards introduced the registry concept which means that adaptations and new model items can be applied to a web based register at any time. This register is basically open to the interested public but administered by designated and acknowledged institutions. Once registered, the enhanced model items can be exported to form the basis for the future coding of a data product. The data product "hooks up" its most recent model for implementation at the user's device after delivery. This mechanism guaranties the application of the most recent data model at any time.

Like semantic standards, technical standards evolve. The fundamental domain-neutral spatial technology standards framework is now in place, but rapid advances in technology require that this foundation needs continual attention. Such industry-wide advances force revision and rethinking of established technical standards. Discussions about revision invariably run into the issue of backwards compatibility, a standard's lifetime of usefulness, and the importance of stability to both technology providers and technology users who have made investments based on the standard. These are difficult but important issues. Mature standards development organizations and their long-term members have experience in negotiating these issues. They also have a keen awareness of the costs and risks associated with letting market leaders establish proprietary standards outside of an open consensus process. Industry market leaders work in standards organizations because they, like their competitors and despite their natural desire to "lock in" customers, have business reasons to implement and help develop open standards. Technical standards are in place that can provide access control, security and certain privacy protections, but development also needs to address other issues such as geospatial data rights management and data quality. Much work remains in the broad area of technical standards for geospatial interoperability, despite the fact that a mature domain-neutral open spatial technology standards framework is already largely in place.

14.3.3 New Information Communities Emerge

One reason work remains is that technology is advancing so rapidly. Another reason is that new information communities keep appearing.

In climate science, as in many domains, new disciplines arise, and they are new information communities. Their data models differ, but they need to share data and communicate. Communities in relationship need interoperability. The OGC Geography Markup Language (GML) Encoding Standard and other OGC standards can be used to develop international domain-specific encoding standards that bring semantic standards and technical standards together. This is a key cyber infrastructure innovation for environmental science, business and policymaking. A domain that develops a domain-specific data encoding standard based on OGC standards and on the domain's semantic standards gives domain participants much fuller access to developments in the mainstream digital technology world: Web searches, chained computer models, full use of cloud infrastructures, Big Data, data analytics, data fusion, management tools for open data, heterogeneous sensor webs and much more.

14.4 Case Studies

Each of the case studies in this section identifies geospatial information standards from the ecosystem and describes how they contributed or can contribute to achieving one or more SDGs.

Traditionally, addresses were used for delivery to individuals and organizations. Today, there are many more possibilities. Addresses are widely used as a locational reference for all kinds of information, such as information about people, buildings, organizations and services [2]. This makes it possible to spatially analyze and visualize the different pieces of information on a single map in support of planning, management and decision-making.

Addresses are a key element for delivering policies at national and international levels in support of the sustainable development goals (SDGs), specifically "with regard to governance, rule of law, poverty reduction, disease prevention and the provision of basic services such as electricity, sanitation and water". Also, without an address, an individual does not have a legal identity, does not have equal opportunities to finding employment and is not socially integrated [6]. There are thus direct links to at least 9 of the 17 SDGs and others are supported more indirectly through the linking of information to addresses.

A variety of address standards and/or specifications are in use around the world. They are typically well integrated into various operational processes and, in some cases, legally enforced. At the same time, some countries are rationalizing their addressing system or creating a new one. Addresses are also increasingly used to reference new geographic objects (e.g. road furniture) and are integrated in new technologies, such as in-vehicle navigation, for which digital interoperability is essential. The ISO 19160 series of standards on addressing facilitates the entire address lifecycle, from planning and assignment of addresses to using, changing and retiring addresses. The case studies from Australia, New Zealand and South Africa exemplify some of the benefits of standardized addresses.

14.4.1 Australia

Addresses in Australia are managed under the National Address Management Framework underpinned by two standards: the Australian/New Zealand Standard (AS/NZS) 4819 *Geographic information – Rural and Urban Addressing* for address creation and the Australian Standard (AS) 4590 *Interchange of client information* [1]. AS 4590 contains the data element requirements for digital address collection, interchange and storage. This standard references ISO 19160-1. Both standards are published through the Standards Australia IT-004 Committee, which mirrors the ISO TC-211.

Addresses in Australia are first created by the (537) local governments in Australia using AS/NZS 4819, which are maintained by a cross-jurisdictional Permanent Committee on Addressing [1]. This address information is aggregated by each state and territory governments and then contributed to a standardised, authoritative, national product - the Geocoded National Address File, or G-NAF®, which is made publicly available through the Commonwealth Government's open data portal [2] [5]. G-NAF is produced and maintained by PSMA Australia Ltd, an independent and self-funded company that is owned by the nine governments of Australia [4].

Standardised address data underpins Australian governments' services to its citizens. For example:

- The national addressing system, as described above, is central to the Australian national statistical process that is maintained and implemented by the Australian Bureau of Statistics (ABS). The ABS maintains an internal Address Register sourced from G-NAF® and internal address

[1] icsm.gov.au/what-we-do/permanent-committee-addressing
[2] data.gov.au

datasets. The Statistical Spatial Framework and National Address Management Framework (NAMF) [see Figure 14.1] provide the framework for the Address Register, which contains information on all addresses (e.g. their location, residential/non-residential status, etc.) and is processed to create residential dwelling frames for the five yearly Australian National Census of Housing & Population and most ABS social surveys.

Accurate census and survey information underpins key national population statistics. These statistics are critical to enabling governments to meet public health and educational outcomes, as well as providing the infrastructure that supports economic growth and sustainable communities.

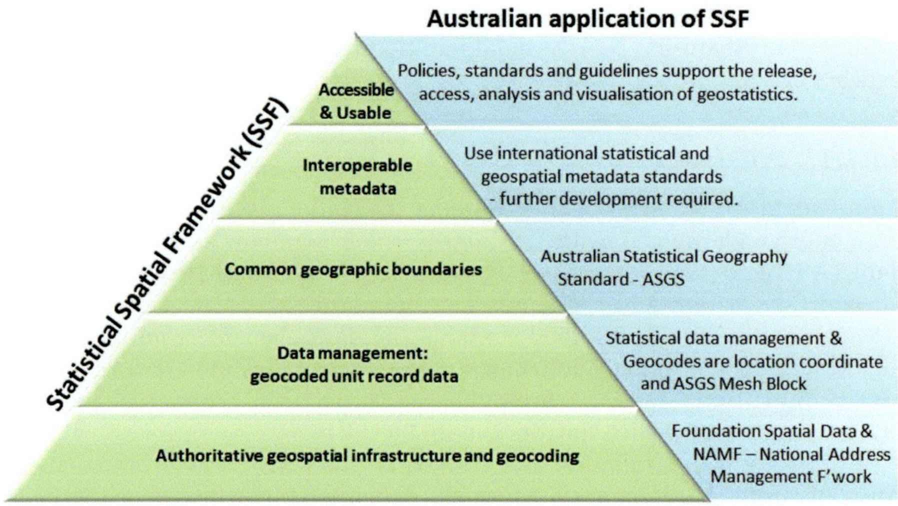

FIGURE 14.2
Australian Bureau of Statistics Statistical Spatial Framework

- The Australian Electoral Commission uses the G-NAF® to deliver an impartial and independent electoral system through active electoral roll management and maintaining equitable electoral boundaries. This underpins Australia's democratic process and stable governance through a transparent representation process.

- Geoscience Australia uses address information to inform the National Exposure Information System (NEXIS), which provides comprehensive and nationally-consistent exposure information that enables users to understand the elements at risk. Resident, Population, Commercial, Agricultural, Infrastructure etc exposure information is produced by sourcing the best publicly available information, statistics, spatial and survey data about buildings, demographics, community infrastructure and agricultural commodities.

This standardised and well governed structure for addresses in Australia has enabled a trial Linked Data API for delivery of the national address file [3]. The hope is that Linked Data will allow new applications of data presentation, access, mining and sharing that improve outcomes for Australians. The API allows users to request G-NAF® data mapped to any number of different 'alternate views', including one modelled on a profile of ISO 19160-1:2015–*Addressing, Part 1: Conceptual model.*

14.4.2 New Zealand

Land Information New Zealand and Statistics New Zealand implemented new systems for managing address data. These two systems, based on ISO 19160-1 and the associated New Zealand profile, manage the addresses used to support New Zealand's electoral system and the collection of official statistics. A standards based approach has made it easier for these two agencies to work together in the collection of up-to-date, authoritative, and accessible address data in New Zealand. ISO 19160 provides a shared understanding of the fundamentals of an address, for both the people involved and also for system interoperability.

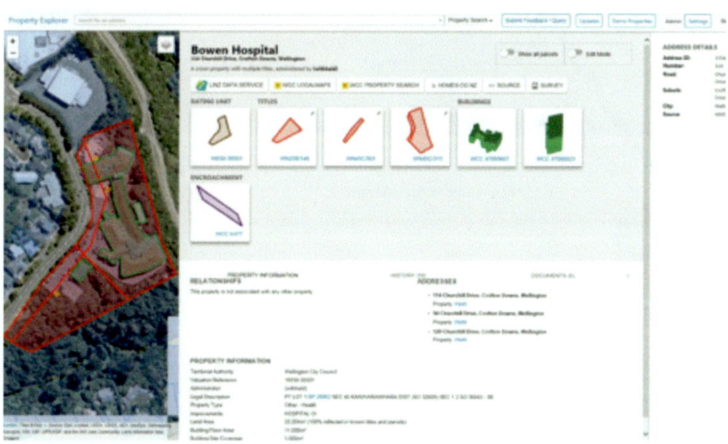

FIGURE 14.3
Example of a property that has multiple addresses

Standardised address data underpins New Zealand's electoral system. For example, electoral boundaries are dependent on accurately locating voters by address. To deliver sustainable development goals, a country relies upon stable governance that is perceived to be fair and representative. Standardised address data also enables critical linking of addresses to other property information, allowing delivery of reliable property-focused data that agencies can use to ensure the right services are delivered to the right citizens (see Figure 14.3). This includes government agencies providing critical social services and others working in fields directly related to the sustainable development goals relevant to address data.

Accurately sampling and capturing statistics requires standardised addresses. These statistics include census, household, economic, and social surveys. Analytics based on these statistics directly contributes to policy that supports citizens' wellbeing and allows outcomes to be measured against sustainable development goals. A standards-based approach to addressing has directly supported the development of New Zealand's Statistical Spatial Framework (see Figure 14.4) and contributions to the UN-GGIM Global Statistical Geospatial Framework.

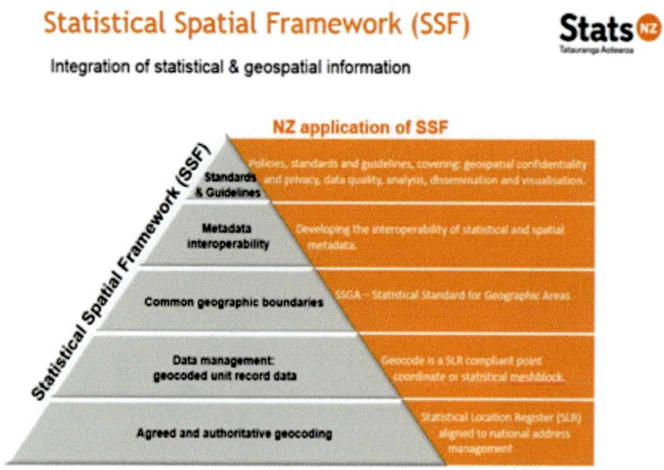

FIGURE 14.4
Statistical Spatial Framework

14.4.3 South Africa

South Africa is a diverse country with many challenges. Thanks to the guide of standards, the subject of addresses now has a solution and firm way forward.

AfriGIS is a private GIS company based in Pretoria, South Africa. 2019 marks our 22nd year of existence. The focus of AfriGIS is to provide actionable answers through location insights to improve decision making. The foundation of the insights lies in the quality of the data and information. At AfriGIS we identified a need for quality information that is accurate and maintained. The spatial data we base our decisions on should be reflective of the world we live in. In 2002, AfriGIS started building and maintaining a host of datasets to assist both private sector and government.

When the United Nations (UN) released the 14 Global Fundamental Geospatial Data Themes, AfriGIS could tick off at least eight of the boxes. We have maintained spatial data layers for South Africa in terms of:

- Geographical Names
- Addresses
- Buildings and Settlements
- Land Parcels
- Transport Networks
- Population Distribution
- Land cover and Land use
- Physical Infrastructure

Addressing is a complex item and challenge in South Africa. The various components that build up an address are divided between different custodians and entities. In order to create a national address dictionary, an estimated 250 different custodians from National and Provincial (9) and Municipal (226) level provide data. AfriGIS rose to the challenge.

SANS 1883-1:2009 was the first Address Standard AfriGIS ever implemented. The data sourced from the different entities are not standardised in any way. Several projects are in progress to ensure that best practices as prescribed in the conceptual model are adhered to.

AfriGIS became involved with ISO/TC211 to guarantee that our data and address offering to our clients remain world standard. Private sector companies have a multinational footprint and have to ensure that data shared across boundaries share the same profile. ISO 19160-1:2015 allowed AfriGIS to map our data to the South African profile as described in SANS 1883-1:2009.

The practical benefits to the company and our clients have been enormous. The following examples would not have been as successful without the foundation of a standard as a beacon in the dark.

Logistics

Ecommerce is one of the greatest emerging markets, not only in South Africa but also in the world. Accurate delivery address is crucial to make the model work. It empowers better delivery time, effective planning and accurate billing. The process works as the point of capture is accurate. The advantage of having a confidence level ensures that drivers have a clear view of the expectation with each delivery.

Emergency Services

In South Africa one of the first private companies to implement SANS 1883-1 is a prominent security firm. They have applied their own business rules to only accept confidence level 1 and 2 addresses. In their business the accuracy, or rather inaccuracy of the address could have life threatening consequences.

Financial Services

Through a host of services an address provides a vast amount of information about an area, a person and the potential of risk. In terms of the financial institutions the risk, the potential and the business profile of an organisation is dependent on the quality of the addresses in their database.

AfriGIS Search

These are just a few small high level examples. How did AfriGIS succeed to bring a Standards compliant address database to clients?

We have developed an API that serves the addresses. AfriGIS Search is an address verification, capture and geocoding tool. It contains intelligence to simplify the complex addressing system of South Africa.

There are six active address types (classes) incorporated into the solution:

- Street Addresses
- Building Addresses
- Farm Addresses
- Site Addresses
- Landmark Addresses
- Intersection Addresses

Informal addresses have been excluded in the product as there are currently no legal custodianship in place to maintain and verify an informal address.

AfriGIS Search gives clients up to 50 million searchable address combinations for South Africa. The reason the number appears to be so high is due to the fact that a single land parcel in South Africa can have multiple addresses; all relevant and legal. The API allows for all components of an address and its history to be maintained and captured.

AfriGIS Search provides a client with the tool to access an address with all the components required to understand the accuracy, maintenance, history and type.

Additional information available contains the land administration information with regards to deeds, ownership and contact information. All the additional information that provide a view of the address can be found within the AfriGIS Search API.

Addresses are a challenge in South Africa. We have multiple address types. We have unconfirmed assigning methodologies. We have numerous role players.

The standard was used as a road map to determine the important pieces of information. To order it within the boundaries of logic and accessibility. It guides us in terms of requirements and possibilities. It made the data sharable and interoperable.

The true test of the benefits of standards were when we took the ISO 19160-1 compliant datasets from different countries. Based on the methodology, the data had to be easily compatible. The results of the combination of the three southern hemisphere address datasets is a story of triumph. The tri-nations may be fiercely competitive on a rugby field. Our addresses work together beautifully. In terms of addressing the conceptual model proved to be successful.

14.5 Case Studies of Relevant Standards for Specific Goals

14.5.1 New Zealand Government Use of WaterML and SOS

Relevant to Goals - 6 Clean Water and Sanitation, Goal 14 Life Under Water

The OGC WaterML 2.0 standard, developed in a working group organised jointly between OGC and the World Meteorological Organisation (WMO). To support requirements for monitoring laid out in the New Zealand Resource Management Act 1991 multiple New Zealand agencies. Monitoring Standards and Technologies for the compilation and reporting of water quality data across New Zealand is underpinned by the WaterML 2.0 and unifies data across regional agencies.

14.5.2 Urban Environment - Multiple Urban Implementations Including UK, Singapore, Germany, Finland, Australia, USA, Canada. Key Standards in Use Include CityGML, SensorWebs, SensorThingsAPI and Others

Relevant to goals: 7 - Affordable and Clean Energy, 11 Sustainable Cities and Communities, 13 Climate Action Two sources of excellent guidance on the

broad range of use cases include the OGC's Future Cities Pilot[3].

The second source is the OGC's Smart Cities Domain Working Group list of use cases. These include topics such as waste management, planning,

[3]www.opengeospatial.org/projects/initiatives/fcp1

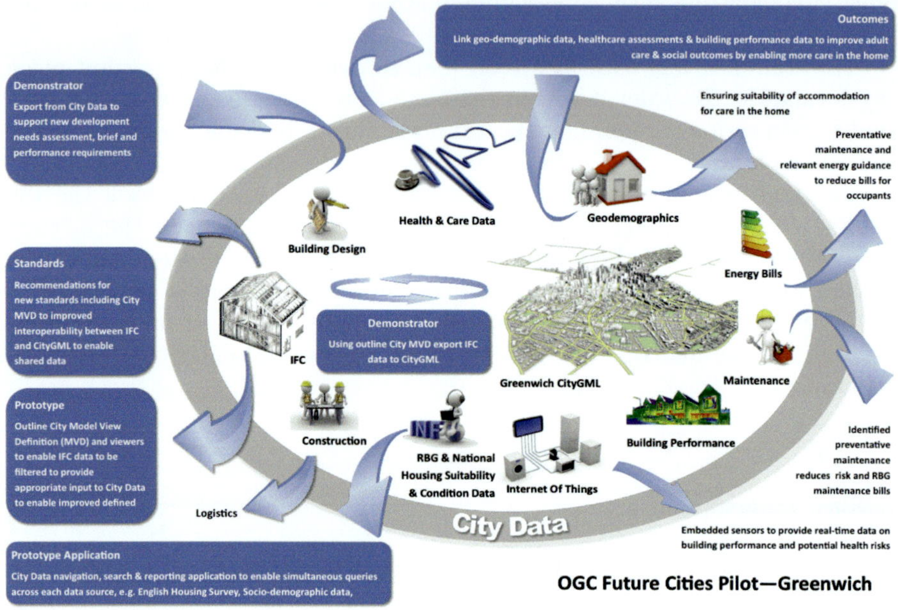

FIGURE 14.5
OGC Future Cities Pilot

disaster management, transport and others. Many of the use cases indicate which SDGs are relevant [4].

14.5.3 Arctic SDP

This project is an international exemplar in the efforts to share data across multiple nations and is perfect example of the vital role that geospatial standards play in striving towards Goal 17 - Partnerships for the goals.

The ArcticSDP proved the capacity of the international standards to achieve interoperability across all the Arctic nations in order to share environmental data vital to understand climate changes and animal migration and behaviour.

Further information on this project can be found at:
http://www.opengeospatial.org/pub/ArcticSDP/index.html

IHO and marine SDGs Goal 14
Standardization forces collaboration - a maritime use case
The maritime sector definitely holds the longest tradition in international

[4] external.opengeospatial.org/twiki_public/SmartCitiesDWG/UseCaseList

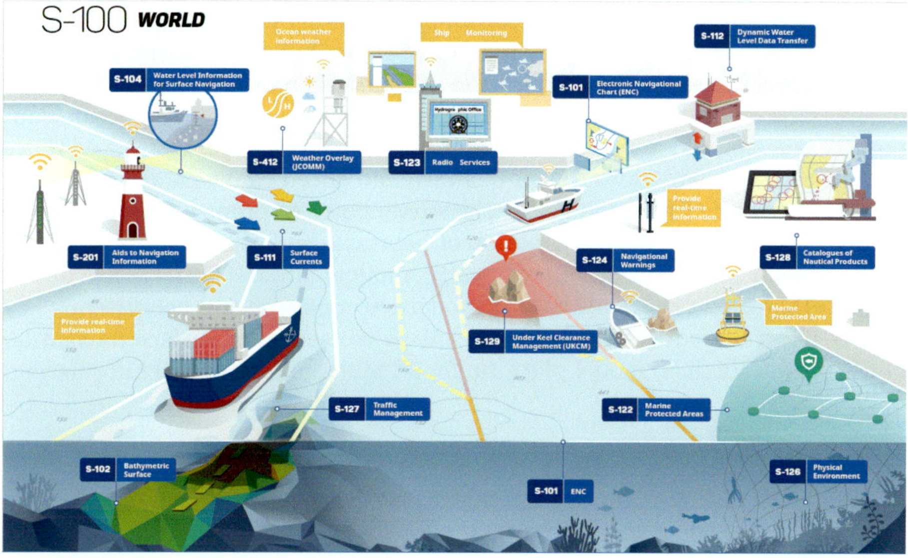

FIGURE 14.6
The S-100 Universal Hydrographic Data Model
The S-100 Standard is a framework document that is intended for the development of digital products and services for hydrographic, maritime and GIS communities. It comprises multiple parts that are based on the geospatial standards developed by the International Organization for Standardization, Technical Committee 211 (ISO/TC211).

standardization of geospatial information. It was the 23rd President of United States, Benjamin Harrison who called for the first International Marine Conference in Washington D.C. in 1889. In order to facility safety of navigation the attending 28 nations agreed – among a good amount of ship's related issues - to improve the regular update of nautical charts and to start the harmonization of the publication of nautical warnings. It was then in 1919 there first International Hydrographic Conference in London agreed to develop firm standards how to technically conduct sea survey and nautical cartography and confirmed the installation of a coordinating intergovernmental body - the International Hydrographic Bureau - later the International Hydrographic Organization (IHO) in Monaco. Since then the IHO has adopted the leading role in global standardization of nautical charting - or in more modern words - to enable the provision of marine geospatial information interpreted and customized for surface navigation in a globally unified manner. Though nautical charts are mainly individually produced by the affected coastal state they all adhere to the same paradigm of information encoding and presentation thanks to the applying technical IHO standards. The late eighties saw the uptake of digital means for navigation and IHO standards for nautical cartography were turned into the digital domain too. Even thanks to Electronic Navigational Charts (ENC) shipping is comparably safe today but there is one significant change of the scenery: The application of nautical information is not limited to the purpose of surface navigation anymore. It likewise has to provide support for efficient and in particular resource saving navigation. Optimised route planning and tracking in terms of distance and speed can generate great savings in fuel burned for propulsion; the surveillance of proper fishery is based on precise charts and the preservation of habitats can be serviced much better if all available information is technically amalgamated. The response of the IHO to this is the installation of a modern standardization ecosystem - the IHO S-100 framework - which is not limited to means for surface navigation but utilizes a common model platform for all maritime geospatial information.

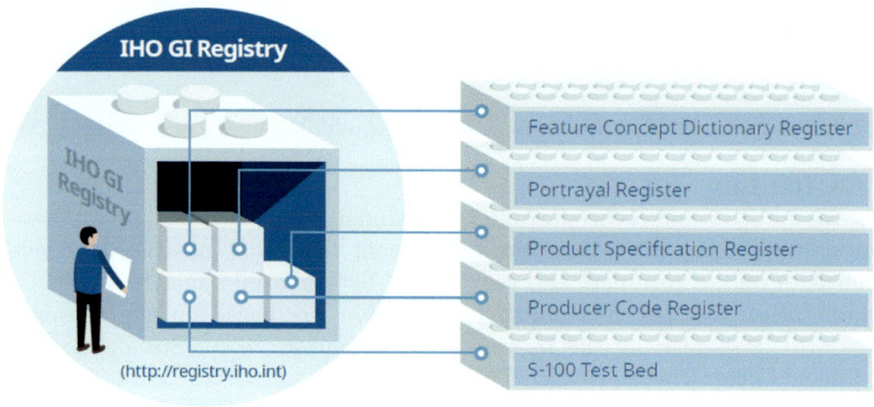

FIGURE 14.7
S-100 Geospatial Information Registry

S-100 basically adopts the fundamental mechanisms of the ISO 19200 series of standards for geospatial information and delivers the most important application of this suite.

The IMO as special UN Organization for the maritime sector has adopted the modelling part of S-100 as its "universal hydrographic data model". In order to achieve interoperability between all data sets relevant for likewise safe and sustainable ship's operation all maritime data providers are called to develop data product specifications based on this model. To facilitate this the IHO runs the web based IHO Geospatial Information Registry which administers objects, attributes and presentation rules [5].

The IHO GI registry interfaces to tools supporting the generation of machine readable catalogues for customized modelling and presentation. The IHO

[5] http://s100.iho.int/S100/

GI registry itself is not limited to the hydrographic domain. Instead altogether nine domains owned and administered by other international and intergovernmental organizations active in geospatial information standardization, e.g. the World Meteorological Organization WMO and the International Association of Lighthouse Authorities IALA are already hosted and the IHO GI Registry is subject for grow further. The ambition is to consolidate comparable standardization activities in the domain of ocean sciences to eventually address all relevant maritime geospatial information in interoperable data product specifications.

Bibliography

[1] ANZLIK. Australian national address management framework, 2019.

[2] Serena Coetzee and Judith Bishop. Address databases for national sdi: Comparing the novel data grid approach to data harvesting and federated databases. *International Journal of Geographical Information Science*, 23(9):1179–1209, 2009.

[3] PSMA Australia Ltd. G-naf ldapi.

[4] PSMA Australia Ltd. G-naf, the foundation geocoded address database for australian businesses and governments.

[5] PSMA Australia Ltd. G-naf view.

[6] Universal Postal Union. Addressing the world–an address for everyone. *White Paper*, 2012.

15

Urban Analytics Data Infrastructure: Critical SDI for Measuring and Monitoring The National and Local Progress of SDGs

Abbas Rajabifard
Centre for SDIs and Land Administration, The University of Melbourne, Australia

Soheil Sabri
Centre for SDIs and Land Administration, The University of Melbourne, Australia

Yiqun Chen
Centre for SDIs and Land Administration, The University of Melbourne, Australia

Muyiwa Agunbiade
Centre for SDIs and Land Administration, The University of Melbourne, Australia

Mohsen Kalantari
Centre for SDIs and Land Administration, The University of Melbourne, Australia

This chapter describes an innovative Spatial Data Infrastructure to support urban analytics and urban research capabilities focused on Australian cities, called Urban Analytics Data Infrastructure (UADI). The UADI provides opportunity for multi-disciplinary, and cross-jurisdictional analytics. The chapter highlights the UADI capabilities to be adopted for deriving the SDG indicators as a response to the UN-GGIM strategic framework 2017 – 2021 technical requirements.

15.1 Introduction

In 2015, the United Nations 2030 Agenda formalised 17 Sustainable Development Goals (SDGs), which consist of 169 targets and 232 indicators. Consequently, the United Nations Committee of Experts on Global Geospatial Information Management (UN-GGIM) aligned their 2017-2021 strategic framework with the 2030 Agenda for Sustainable Development. With the vision of *"Positioning Geospatial Information to Address Global Challenges"* this strategic framework highlighted several key policies and technical points in their operating principles [12]. The technical points span from agreed standards and methods to integration and interoperability of national information systems as well as information sharing and knowledge transfer. These are important to support the evaluation of SDG indicators at a national level. This is with the assumption that these will enable evidence-based policy making and the development of effective implementation strategies towards achieving the goals set by the 2030 Agenda.

The SDGs have established methodologies that allow the generation of comparable indicators worldwide. Therefore, these methodologies, along with the UN-GGIM operating principles, present opportunities to formalise re-usable geospatial tools for producing the indicators. This will allow the UN, and subsequently Member States, to not only compare progress among nations but also to monitor the indicators over time. However, the implementation of the SDG proposed methodologies vary from one jurisdiction and government level to another. This may be due to different terminologies and the subsequent interpretation of their methodologies in various contexts, or the differences in the structural and semantics of the input data used for measuring the indicators. This can result in redundant work for measuring similar indicators and may compromise the comparability of these indicators.

Furthermore, the indicators set forth by governments or other institutions lack transparency of the measurement process and in the case of many existing platforms (e.g. World Council of City Data [WCCD] platform that attempts to present indicators for ISO 37120, urban quality of life), access to such information is limited, affecting the indicator's credibility [14]. These challenges hinder the development of a spatially enabled platform that adopts a set of re-usable geospatial tools for measurement, storage, and effective and transparent communication of the SDG indicators for UN Member States and their respective levels of government. Consequently, customised version of these methodologies are developed to address the subjectivity of indicators and to meet the needs of each jurisdiction. Specifically, this issue can be addressed by a geospatial platform with capability of minimising redundant efforts and encouraging cooperation between different levels of governments, the private sector, academic institutions, and civil society organisations. This is in addition to enhancing evidence-based policy making towards achieving the SDGs by providing a repository of a set of transparent and credible indicators.

At the moment, such a geospatial platform that enables the harmonisation of structurally and semantically heterogeneous datasets is lacking. This makes it difficult to work with an ecosystem of re-usable and shared set of user-generated tools for measuring and communicating the SDG indicators. This chapter aims to introduce a Spatial Data Infrastructure (SDI) developed for urban data analytics in Australia [7] and to highlight the capabilities to be adopted for deriving the SDG indicators as a response to the UN-GGIM strategic framework 2017 – 2021 technical requirements.

The next section explains the global indicator framework developed by the Inter-Agency and Expert Group on SDG Indicators (IAEG-SDGs). It explores data and analytics challenges inherent in the framework highlighted by the latest SDGs report. In section three, the chapter introduces Urban Analytics Data Infrastructure (UADI), the motivation for developing such SDI and its components. The section highlights how the UADI is capable of addressing technical requirements indicated in the UN-GGIM strategic framework 2017 – 2021. Consequently, section four explores the possibilities that UADI can contribute to SDGs. Finally, section five provides an account of ways forward in adopting the UADI for deriving SDG indicators in the global context and its implications in achieving SDGs.

15.2 Global Indicator Framework

The Inter-Agency and Expert group on SDG (IAEG-SDGs) developed the global indicator framework for the SDGs and targets of the 2030 Agenda for Sustainable Development. This framework, including refinements on several indicators, was agreed upon at the 48th session of the United Nations Statistical Commission held in March 2017. Accordingly, the global framework was adopted by the UN General Assembly on 6th July 2017. This framework is part of the Resolution adopted by the General Assembly on work of the Statistical Commission pertaining to the 2030 Agenda for Sustainable Development[1].

The global indicator framework emphasised that *"Sustainable Development Goal indicators should be disaggregated, where relevant, by income, sex, age, race, ethnicity, migratory status, disability and geographic location, or other characteristics, in accordance with the Fundamental Principles of Official Statistics[2]"*. As at April 2018, 232 indicators were listed in this framework. The indicators will be refined annually and will be published in the official website of The United Nations. This will provide information on the develop-

[1]Resolution https://undocs.org/A/RES/71/31371/313
[2]Resolution https://undocs.org/A/RES/68/26168/261

ment and implementation of indicator frameworks to guide the follow up and review of the 2030 Agenda for Sustainable Development[3] .

While these indicators are defined for national level, several national and international organisations have attempted to localise them. For instance, the Association of Flemish Cities and Municipalities (VVSG), an association of 308 Flemish municipalities and cities, attempted to translate the SDGs at the local level [3]. The VVSG developed tools and guidelines to enable local authorities to develop policies to achieve SDGs. These initiatives require measuring, monitoring, and managing SDG's progress at the local levels. As such, the spatial scope of measuring the SDGs indicators defined at the global framework need to be smaller than what is obtainable at the national level.

From a spatial data point of view, some of the indicators are readily presentable in different geographical boundaries (subject to availability of data). For instance, in target 3.c of SDGs, governments in developing countries are required to *"...increase health financing and the recruitment, development, training and retention of the health workforce..."* Indicator 3.c.1 intends to measure *"Health worker density and distribution"*. This indicator potentially can be derived and measured in small geographical boundaries of census blocks or administrative boundaries. For other indicators which are not spatial (e.g. indicator 5.5.2 *""Proportion of women in managerial positions*), they can also be connected to a confined jurisdiction boundary smaller than a state or nation.

It is important to emphasise that the data availability is always a major consideration. As such, a digital platform that is capable of harmonising and standardising the data and analytics tools and then derive SDG indicators is necessary, considering the different data structure and quality. The next section discusses an innovative spatial data infrastructure developed for addressing these challenges.

15.3 The Urban Analytics Data Infrastructure

The Urban Analytics Data Infrastructure (UADI) project is a collaborative effort between a consortium of urban research centres across Australia and is funded by the Australian Research Council [7]. The UADI has been developed to enable multi-disciplinary, cross-jurisdiction, national-level analytics of ISO/DIS 37120 *"Sustainable development and resilience of communities - Indicators for city services and quality of life"*. It provides a digital infrastructure for urban researchers to overcome current challenges related to data access, integration, analysis and sharing.

Since its development, the UADI has improved the state of urban analytics

[3]https://unstats.un.org/sdgs/indicators/indicators-list/

in Australia, and capitalised on previous urban data initiatives, for example the Australian Urban Research Infrastructure Network (AURIN). This has provided opportunities to add more value to the existing initiatives. It also provides the capability to shift the current urban research and planning landscape towards one that is more consistent across jurisdictions. It builds up the requisite intellectual capital to support evidence-based decision-making that transcends traditional disciplinary domains.

In addition, the UADI facilitates analytics tool sharing and provides metadata for both data and tools. These capabilities in the UADI are developed to increase the reliability, trustworthiness, and useability of data, tools, and output information. As such, this Spatial Data Infrastructure (SDI) attempts to address several challenges related to the data and deriving city indicators recently raised by scholars worldwide [4, 14].

As a digital data infrastructure, the UADI enables the integration, harmonisation, connectivity and scalability of multi-source urban datasets. As applied, for an example, in the analysis of urban density, this infrastructure was able to integrate data related to population, building footprints, and land use, which could be used to compare different urban densities (e.g. residential/commercial built-up area per capita, and publicly available open space per capita) in local authorities across Metropolitan Melbourne. The infrastructure developed a new ontological framework [1] and a dictionary to underpin the next generation of data driven modeling and decision-support tools to enable smart, sustainable, productive, and resilient cities.

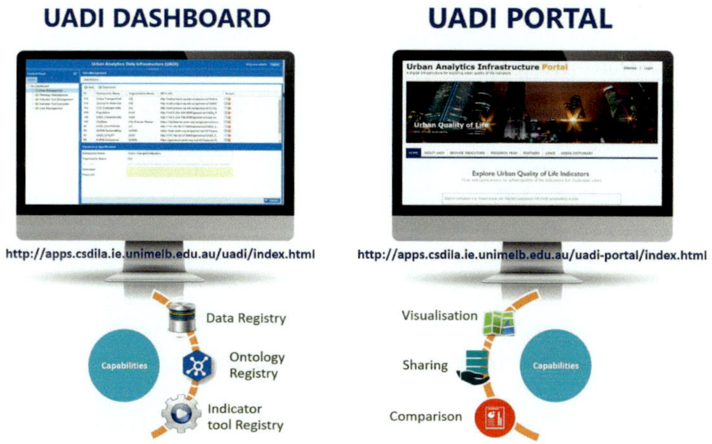

FIGURE 15.1
The UADI components and capabilities

The main objectives in developing UADI are as follows:

1. To provide an underlying framework for harmonisation and integration of urban data by adopting the ISO 37120 and ISO 19115 standards.

2. To develop core system capabilities including data registry, integration and access, as well as analytics tools registry, execution and publishing through web APIs (Application Programming Interfaces) by adopting the OGC (Open Geospatial Consortium) standards.

3. To develop an integrated platform and web-portal to visualise and evaluate the cross-jurisdictional and cross-domain performances.

4. To facilitate open access to those datasets currently accessible through the AURIN Data Hubs and any new open data sets through the development of open access APIs.

The UADI addresses challenges such as those associated with data access, data integration, and the use of varying terminology between disciplines. This infrastructure is comprised of two main components (Figure 15.1):

1. UADI Dashboard: The aim of this component is to allow users to access, integrate, and semantically enrich the data as well as manage and execute their own or others' analytics tools. Users can also preview and publish the results of the analytics in the dashboard.

2. UADI Portal: The aim of this portal is to enable public users to discover, access, explore, and compare urban quality of life indicators calculated by a variety of contributors from different sectors such as research and development, government, and the private sector.

The core capability of the UADI is using ontology that consists of one (or more) upper-level and domain ontologies, describing the generic (e.g. space and time) or specialised concepts pertinent to one or more domains of knowledge [2]. As an example, one of the upper-level ontologies developed in UADI is for Austrian Statistical Geography Standard (ASGS), which determines the relationships of statistical and administrative geographical boundaries across Australia (Figure 15.2). Furthermore, using METHONTOLOGY [1], a number of application ontologies for urban density and urban accessibility measurements were developed.

In addition to the definition of concepts, their relationships are also defined by ontology in the UADI. Therefore, the mapping between any data and its attribute in a dataset to one or many concepts within the ontology can be used to describe the dataset for discovery, and also data integration purposes [1].

The data in the UADI model refers to the datasets that are available from the data providers through standard web services and over the Internet. The data can be spatial or non-spatial and can be structured and non-structured. This data should be exposed via the data custodians to UADI through standardised OGC protocols (e.g. Web Feature Services) [11]. These services can then be registered using their metadata and via the data catalogue component in the UADI.

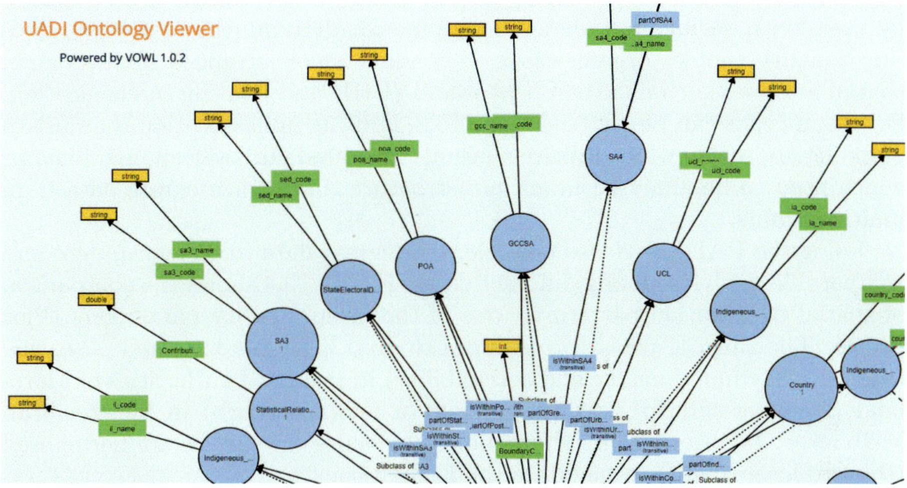

FIGURE 15.2
An example of using VOWL (v1.0.2) to show the concepts (blue circles), properties (green blocks) and relations (blue blocks) defined in the ASGS Ontology. The subclass of a concept is represented in dashed arrows.

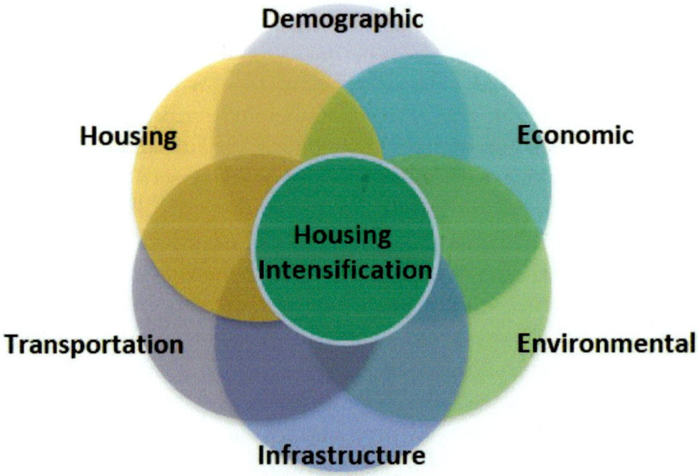

FIGURE 15.3
Housing intensification; a multi-domain urban planning approach.

This infrastructure is a smart technology that can be used to generate indicators for decision makers, such as urban density [1] and accessibility [10]. For this purpose, the UADI encapsulates and registers each indicator tool as a Web Processing Service (WPS) [11] endpoint, which can then be executed by users with various parameters as inputs for deriving the indicators. This functionality enables organisations to conduct more advanced multi-domain spatial analyses such as urban heat island (UHI) and housing intensification. For instance, as can be seen in Figure 15.3, housing intensification is an urban planning approach, which involves several integrated and overlapping domains which need to be analysed in an infrastructure that can integrate data from multi-domains.

Using the UADI, it is also possible to integrate data to do transparent and comparable analysis among different cities. Figure 15.4 shows the comparison of spatial distribution pattern for one of the urban density parameters (Plot Factor: The ratio of private land areas to land available for public use, distinguished by functionality and accessibility) in two local authorities of Moreland (an inner-suburb) and Whitehorse (a middle-suburb) in Metropolitan Melbourne. Figure 15.4 also shows how the results of analyses are normalised (the two lower maps) to facilitate the comparisons.

15.4 UADI's Contribution to SDGs

The literature on city indicators has highlighted several challenges about transparency, reliability, and usefulness of the indicators [4]. As explained in the first section, the lack of data provenance and uncertainty in indicator measurement process in the case of existing platforms limits the credibility of indicators. These challenges apply to SDG indicators as well as there are specific local conditions for one region which may have not been the case in others. So how can countries register the process based on which indicators are derived?

In 2016, when the transforming plan from Millennium Development Goals (MDGs) to SDGs was prepared by the member states, several limitations about data and decision making processes were highlighted. One of the bottlenecks was the lack of quality data to enable regular monitoring and support evidence-based decision making. Accordingly, international agencies including UNDP, World Bank Group, and UN-Habitat suggested using real-time data, adopting geospatial data, strengthening statistical capacity, utilising new technologies, changing the methods of data collection and dissemination, developing global standards for integrated statistical systems, and promoting open data [13, 5].

As such, in order to measure, monitor and compare the SDG's progress by the UN member states, a spatially enabled information decision making platform is critical. Such a platform should enable the member states to derive

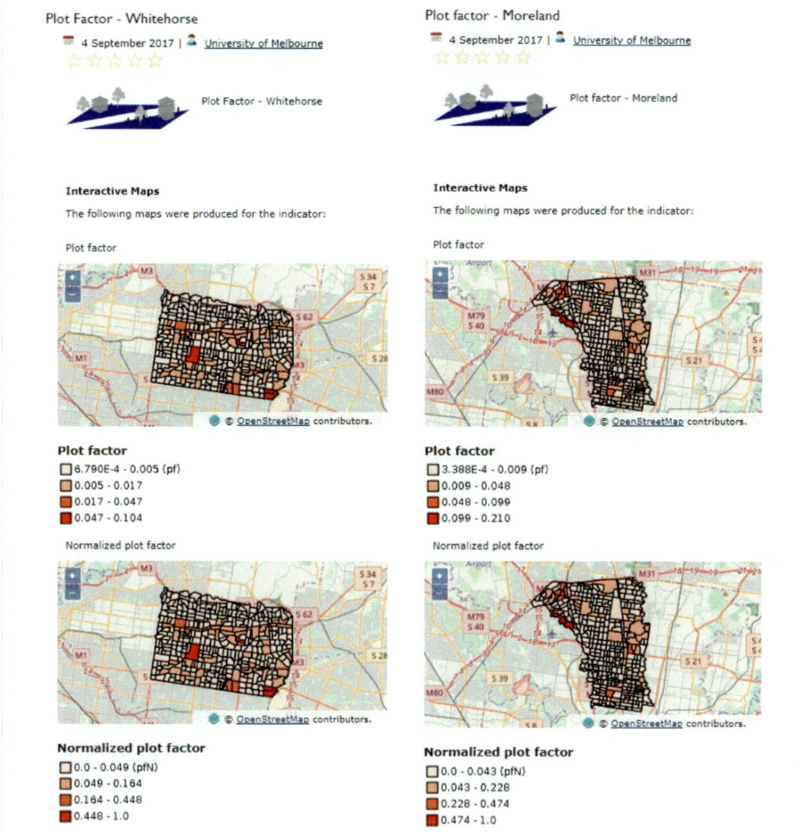

FIGURE 15.4
Spatial comparison of plot factor in two local authorities in metropolitan Melbourne.

their indicators through data access and integration facility using reliable and replicable tools in order to visualise and share the outcomes of the SDG indicators (Figure 15.5). The UADI, as explained in the previous section, is a digital infrastructure with the capability to meet the SDG's progress management requirements in conceptual framework presented in Figure 15.5. Furthermore, the UADI is potentially an enabler to progress the UN-GGIM's strategic framework. It is also designed to operate the principles that allow the formalisation of re-usable geospatial tools in capacity building, thereby enabling the UN and member states to compare, monitor and manage SDG progress.

The UADI's capability in registering spatial and non-spatial data enables deriving SDG indicators that are non-spatial as well. In some cases, it is possible to connect the non-spatial indicators to a certain geographical boundary.

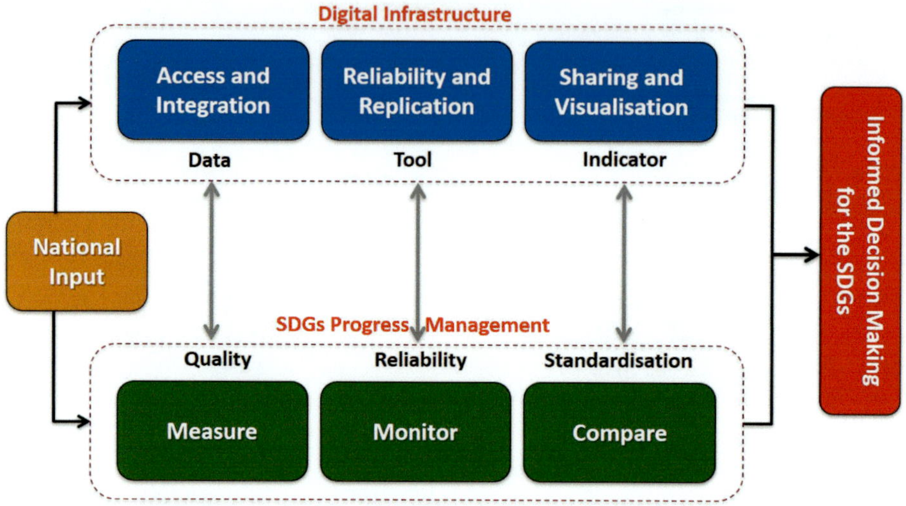

FIGURE 15.5
Conceptual framework for a digital infrastructure enabling the SDG's progress management.

As an example, as part of target 1.4 of the SDGs, governments are required to *"...ensure all men and women have equal rights to economic resources, and access to basic services, and control over land and other forms of property..."*Indicator 1.4.2 intends to measure *"proportion of total adult population with secure tenure rights to land, with legally recognised documentation and who perceive their rights to land as secure..."*. Subject to the availability of data, this SDG indicator can be attributed to a particular census or administrative boundary. As a result, regarding the 1.4.2 indicator, UN Member States can provide a better understanding of the rate of progress in a spatially and temporally visualised fashion, which identifies the deficiencies that require further improvement at local, state, and national government levels.

In addition, the UADI enables the evaluation for potential future scenarios. This capability can additionally facilitate strategic planning and informed decision making for optimal solutions amongst various alternative options. This capability of UADI enables defining use cases such as land development, housing affordability, emergency and energy efficiency, and inclusive infrastructure development (e.g. transport, telecommunications, and other utilities). The use cases along with indicators will enable different government levels to localise the implications of SDGs (Figure 15.6). It is also important to improve the UADI by using the live data (e.g. sensor network), and big data for advanced analytics.

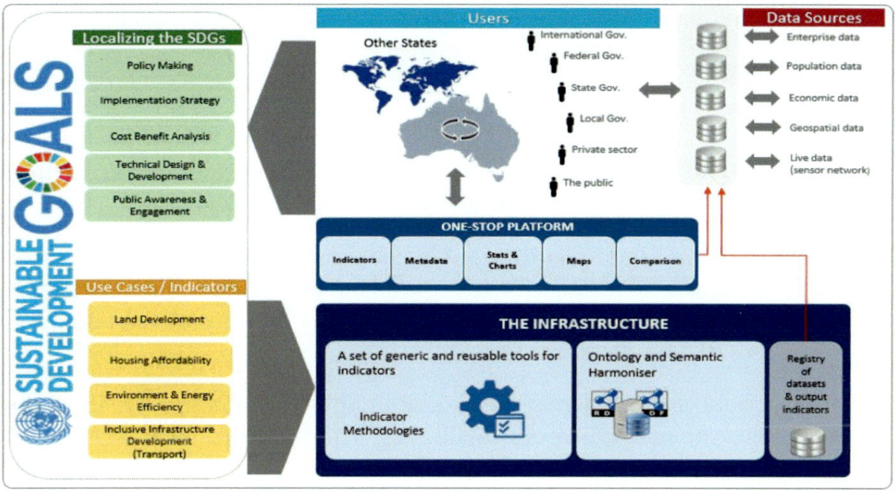

FIGURE 15.6
Potential contribution to SDGs and localising the SDG indicators.

15.5 Discussion and Conclusion

This chapter discussed how the development of national indicators for the SDGs and localising the 2030 Agenda for Sustainable Development require a digital infrastructure that can facilitate the registration and sharing of indicators locally and globally. The chapter then introduced a spatial data infrastructure (SDI) that enables the harmonisation of structurally and semantically heterogeneous datasets to work with an ecosystem of re-usable and shared set of user-generated tools for measuring and communicating the SDGs indicators.

The chapter has also discussed the capabilities that UADI platform offers for deriving the SDG indicators as a response to the UN-GGIM strategic framework 2017 – 2021 technical requirements. The UN-GGIM can facilitate the adoption of such a digital infrastructure for member states to register new indicators related to land administration, disaster risk reduction, food security, and the implementation of standards in order to measure and monitor the inclusive progress of the SDGs.

One of the major advantages of the UADI is its capability to spatially visualise the indicators, which help to benchmark and compare different jurisdictions. Each local, state, and national government can also compare their own progress in SDG management by investigating the changes of SDG indicators through time.

The chapter has also shown that the UADI system offers several innovations, which highlight future research directions for new developments in areas such as:

- Ontology and its role in national and international linked data projects. The UADI system has already addressed several challenges in the Australian National Linked Data project, which put Australia at the forefront of open data and semantic web development.
- Formalising the Open Geospatial Consortium (OGC) smart city framework [6] and a new agenda for incorporating sensor data, crowd-sourcing volunteered geographic information data (VGI) [9] for real time analytics in deriving SDG indicators.
- Adding more use cases related to sustainability, including liveability [8] and quality of life indicators facilitating the localisation of SDGs. The use cased will help in local policy making, refining strategies and public awareness and engagement for implementing SDGs.

Acknowledgment

The UADI project is supported by an Australian Research Council Linkage Infrastructure Equipment Facilities (ARC LIEF) grant (LE160100174). The authors would like to acknowledge the contribution of Maryam Rabiee in developing Figures 5 and 6 and Sam Amirebrahimi in developing Figure 6.

Bibliography

[1] Yiqun Chen, Soheil Sabri, Abbas Rajabifard, and Muyiwa Elijah Agunbiade. An ontology-based spatial data harmonisation for urban analytics. *Computers, Environment and Urban Systems*, 72:177–190, 2018.

[2] Gilles Falquet, Claudine Métral, Jacques Teller, and Christopher Tweed. *Ontologies in urban development projects.* Springer Science & Business Media, 2011.

[3] Bart Van Herck, Valentijn Vanoeteren, and Katrin Janssen. Local indicators for the 2030 agenda (sustainable development goals), 2019. Brussels, Belgium.

[4] Rob Kitchin and Gavin McArdle. Urban data and city dashboards: Six key issues. 2016.

[5] United Nations. *The Millennium Development Goals Report 2015.* 2015.

[6] George Percivall, C Ronsdorf, S Liang, D McKenzie, and L McKee. Ogc smart cities spatial information framework. *OGC Internal reference*, (14-115), 2015.

[7] Abbas Rajabifard, Serene Ho, and Soheil Sabri. Urban analytics data infrastructure: Critical sdi for urban management in australia. *SMART WORLD*, page 95, 2016.

[8] S Sabri, S Ho, and A Rajabifard. Implementing the '20-minute neighbourhood': Leveraging new spatial data and technologies to support planning for a sustainable metropolitan melbourne. *Healthy Future Cities; Sun, SH, Wenqi, L., Eds*, pages 502–517, 2016.

[9] Soheil Sabri, Abbas Rajabifard, Serene Ho, Sam Amirebrahimi, and Ian Bishop. Leveraging vgi integrated with 3d spatial technology to support urban intensification in melbourne, australia. *Urban Planning*, 1(2):32–48, 2016.

[10] Yuchao Sun, Gary McCarney, Soheil Sabri, Sam Amirebrahimi, Muyiwa Agunbiade, Yiqun Chen, Azadeh Keshtiarast, Mohsen Kalantari, Callan Bright, Doina Olaru, et al. A flexible accessibility analysis tool for enhanced urban analytics. In *Proceedings of the Australasian Transport Research Forum, Auckland, New Zealand*, pages 27–29, 2017.

[11] Chet Bing Tan, David A McMeekin, Geoff West, and Simon Moncrieff. Ciao-wps-utilizing semantic web (web 3.0) techniques to assist in the automatic orchestration of geospatial processes and datasets. In *International Conference on Geographical Information Systems Theory, Applications and Management*, pages 32–48. Springer, 2016.

[12] UN-GGIM. Strengthening geospatial information management. Technical report, 2017.

[13] UNDP and World Bank Group. Transitioning from the mdgs to the sdgs, 2016.

[14] Yetian Wang and Mark S Fox. Consistency analysis of city indicator data. In *International Conference on Computers in Urban Planning and Urban Management*, pages 355–369. Springer, 2017.

16

New Technical Enabling Tools for Data Acquisition and Maintenance of Topographic Data of Urban and High Mountain Areas to Support SDGs

Gottfried Konecny

Leibniz University Hannover, Germany

This chapter discusses the need of geospatial data and the new technical tools for data acquisition and maintenance of (topographic) data to support SDGs. The chapter provides an example of data acquisition for urban and high mountain areas.

16.1 Introduction

The definition of Sustainable Development Goals depends on the availability of data. Most data required for this are geospatial. The data are dependent on resolution and their object definition at a specific resolution.

Traditionally geospatial information was displayed in the form of maps at different scales. In order to be able to display the information for a certain object location and object size an appropriate scale is required. For data concerning the environment small scales of 1:250000 or 1:500000 may be sufficient, for urban data large scales ranging from 1:1000 to 1:10000 are more appropriate.

The larger the scale, the more effort in data acquisition with respect to time and cost is required. This is the reason why historically the global coverages in map scales have increased from 1:1 million in 1950 to 1:50000 in 2000.

These changes have been made possible due to improvements in technology. While in 1900 only terrestrial surveys by many surveyors were possible, by 1950 the accepted data acquisition technology was aerial photogrammetry based on aerial photographs. The developments from analogue photogrammetry to analytical photogrammetry and then to digital photogrammetry between 1950 and 2000 have made significant improvements in mapping technology.

16.2 Global Progress in Mapping From 1900 to 2000

In 1900 terrestrial surveys were only permitted to generate maps in Europe, parts of India and parts of the United States of America (see Figure 16.1).

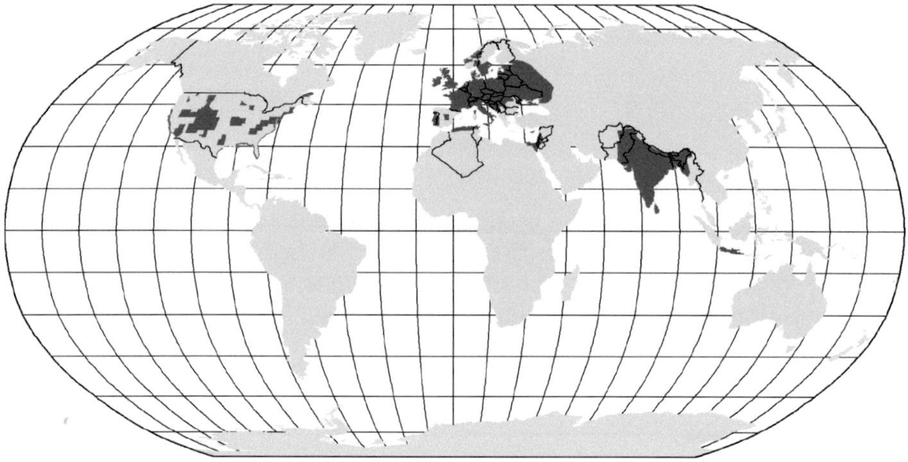

FIGURE 16.1
Global Topographic Mapping Coverage in 1900 [9]

Between World War I and World War II the mapping by photogrammetry was promoted by the International Society for Photogrammetry, founded in Vienna in 1910 [8]. Progress came about by the design and production of stereoplotting instruments in Germany, France, Italy, Switzerland and the USA and Britain. Russia and the soviet block had a rather independent development from the West developing its own instruments. This instrumentation was used greatly during the war-faring nations in World War II to produce maps for the areas affected by the military war operations in Western and Eastern Europe as well as in great portions of Asia and the Pacific.

After the war ended in 1945 the International Society for Photogrammetry met in 1948 in the Hague, Netherlands. Its President was the first post war Prime Minister of the Netherlands, William Schermerhorn. He convinced the Dutch government of the mapping needs of the third world countries. In order to cope with the problem the Dutch government made a donation to the United Nations by establishing an International Training Centre for Photogrammetry (the ITC) in Delft in 1950. This school in addition to the instrument industry became responsible for spreading the photogrammetric mapping technology to most countries of the globe with a total of 1900 students from 170 countries. When Schermerhorn retired as director of the ITC the institution was relocated at Enschede in the 1960s and still contributes to the further development of mapping technology.

In Russia a similar effort to make photogrammetry known in the Soviet controlled areas took place by the educational institutions MIIGAiK in Moscow and by NIIGAiK in Novosibirsk.

The United Nations Sectretariat in New York began to show an interest in the spread of photogrammetric mapping technology in the third world countries as early as 1955 when it organized the first United Nations Cartographic Conference for Asia and the Pacific and in 1976 for the Americas. These UNCC Conferences were the root for the establishment of the United Nations Global Geospatial Information Management (UNGGIM) Secretatiat in New York since 2011, now holding annual conferences.

Both UNCC as well as UNGGIM showed an interest in the global progress of mapping. The results of the progress of mapping document the global coverage for 4 scale ranges: range IV: 1:250000, range III: 1:100 000, range II: 1:50000 and range I: 1:25000 with the following figures for the global data coverage [7, 6] (see Table 16.1.).

TABLE 16.1
Global Topographic Mapping Coverage: Progress Between 1968 and 2012

Year	Range IV (1:250000)	Range III (1:100000)	Range II (1:50000)	Range I (1:25000)
1968	80.0%	38.2%	24.3%	7.7%
1974	80.0%	40.5%	35.0%	11.6%
1980	80.0%	42.2%	42.0%	13.3%
1986	87.4%	46.4%	49.3%	17.9%
2012	98.4%	67.5%	81.4%	33.5%

The second issue is the current state of these maps. The update rates have been determined for the years 1968 and 2012 as shown in Table 16.2:

TABLE 16.2
Global Update Rates of Topographic Maps in 1968 and in 2012

Year	Range IV (1:250000)	Range III (1:100000)	Range II (1:50000)	Range I (1:25000)
1968	27.7 years	37 years	55 years	31.2 years
2012	37 years	31.2 years	26.3 years	22.4 years

The regional update rates are shown in Table 16.3 for the year 2012:

TABLE 16.3
Regional Update Rates for 2012

Region	Range IV (1:250000)	Range III (1:100000)	Range II (1:50000)	Range I (1:25000)
Africa	43.7 years	36.3 years	35.1 years	24.2 years
Asia	38.0 years	35.3 years	27.2 years	22.8 years
Australia and Oceania	30.4 years	29.0 years	16.5 years	22.4 years
Europe	21.8 years	21.1 years	17.1 years	13.8 years
N America	44.0 years	29.2 years	26.3 years	35.4 years
S America	31.5 years	30.1 years	34.7 years	9.8 years

Global base data for SDGs are therefore not available in updated form at the listed scale ranges in the form of geospatial base data. At the age of high resolution space imaging there are, however multiple coverages of global geospatial image data, which can be geocoded and spatially referenced. These images often permit to update areas of the existing global dataset and extract the missing information at these scale ranges at a cost.

16.3 Large Scale Mapping of Urban Areas

No global information is available for the urban scales 1:1000 to 1:10000. If mapping at these scales exists, it generally originates from local administrations at unclear and often unacceptable specifications for a transfer to global datasets.

But the problems for large scale data coverage and map age are similar to what exists for medium scale datasets originating from national mapping agencies shown in Tables 1 to 3.

It is possible to obtain these data from countries, which have included the generation and the maintenance of large scale datasets as a task for the national map agencies, as is the case for the Kingdom of Saudi Arabia, in which the main concentration of development is in rapidly changing urban areas.

The relevant large scale mapping efforts made by the Ministry of Municipal and Rural Affairs (MOMRA) concentrate on the mapping of urban areas for the country of 2.15 million km^2 with a population of 33 million, of which the urban population is 83%.

The dense urban area of the country covers 24450 km^2 and the suburban area including the dense urban area covers 318278 km^2 [1].

The topographic map data coverage of MOMRA is as shown in Table 16.4:

TABLE 16.4
Topographic Map Data Coverage of Urban and Semiurban Areas in Saudi Arabia

Scale	(1:1000)	(1:2500)	(1:10000)	(1:20000)
Coverage in km^2	24450	39838	338278	1743032
Type of area	urban core	semi urban	rural	open country

In addition the topographic database contains orthoimagery coverage and DTM grid data for the mapped areas (Table 16.5):

TABLE 16.5
Orthoimagery and DTM Coverage of Urban and Semiurban Areas

Ortho Imagery Coverage in km^2 GSD		DTM Coverage in km^2 Grid Size	
10 cm	39838	1m	39838
20 cm	17000	2m	17000
40 cm	1781500	5m	1781500

These vector maps are administered in an ArcGIS geodatabase.

Updating using new aerial imagery by photogrammetric line mapping is currently being done. However, due to budget restrictions, it is not possible to schedule re-flights faster than every 5 to 7 years.

From these flights digital orthophotos are generated from that imagery within one year. The digital orthophotos may be overlaid for the mapped areas from 5 to 7 years ago. These overlaps will be able to assess the need for new map updating contracts within a 2 year period. Depending on priorities an update of the mapped areas is possible in 5 to 10 years. For advanced countries, typical of what exists in Europe, the base data situation is listed for Great Britain and for Germany:

16.4 Large Scale Mapping in Europe

Great Britain with a territory of 219931 km^2 and a population of 60 million is covered by the Ordnance Survey Master Map at scales 1:1250 in urban areas and 1:2500 in rural areas. The Ordnance Survey with a staff of 1190 and an operating annual budget of 83 million British Pounds keeps the topographic data system based on an object structure, up to date with help from outsourcing contracts within 6 months by terrestrial surveys with total stations based on GNSS CORS networks.

Germany with a territory of 357386 km^2 and a population of 82 million is covered by cadastral geometric records equivalent to a scale 1:1000 (ALKIS) including property boundaries and buildings. The topographic data for objects other than buildings are resurveyed for the topographic database (ATKIS) equivalent to a scale 1:5000. The cadastral records including the parcel boundaries are maintained by 226 survey authorities with a total staff of about 10 000 professionals in the 16 States of the Federal Republic on a transaction basis. These updates in ALKIS for parcel boundaries and buildings are available in the system within 1 to 2 months. The updates in ATKIS for other topographic data are available every 1 to 2 years using photogrammetry or GNSS based terrestrial surveys.

Both Britain and Germany possess a costly professional infrastructure, which does not exist in most countries around the globe, with the exception of some countries (in Europe, East Asia, Canada and Australia), and which cannot be built up in less than a generation.

For countries not having such a traditional professional infrastructure involving the legal, educational and cultural prerequisites the challenge must be in using digital automation technology to overcome the handicaps of traditional approaches.

16.5 Future Alternatives by New Technology

The alternatives for improving the situation are shown in Table 16.6:

TABLE 16.6
The Alternatives for Improvement by New Technology

Source	Frequency	Acquisition	Detail	Feature Accuracy	Problem Areas
high resolution satellite imagery	1 year	easy	moderate	0.5 to 1m	limited to 1:5000
mobile mapping	as per request	easy	high	0.2 to 0.5m	hidden areas
object oriented 3D oblique imaging via models	1 year	easy	high	0.2 to 0.5m	automation technology for object generation

These improvements by application of new technology need some further remarks:

16.5.1 High Resolution Satellite Imagery

The biggest provider of high resolution satellite imagery is Digital Globe.

Since 1999 Ikonos provides imagery with a GSD 0.8m. In 2007 the resolution was improved on World View 1 to 0.5m GSD. In 2008 a GSD of 0.41m became available on GeoEye 1 and since 2017 World View 4 images with GSD 0.3m are available on World View 4.

Similar imagery is also available from ESA for Pleiades and Sentinel as well as for Kompsat from South Korea.

After the images have been ordered, delivered and rectified on form of an ortho-image, multiple stereo coverages permit to derive DSMs and true ortho-images, which can serve to identify changes of topography, especially of buildings in urban areas. Investigations gave shown, that stereo images of satellites with 0.5m GSD can derive changes for monitoring data bases for buildings at the scale 1:5000. [2].

16.5.2 Mobile Mapping

The competitors for updating roads are Google, Here and Tomtom. Each update traffic routes by mobile vans. Some companies, such as Cyclomedia, use mobile vans with optical cameras to update building facades at high accuracy. Yet others, such as Tesla and the car manufacturing industry prefer high accuracy surveys by service companies in preparation for automatic driving.

The application of mobile mapping for map updating has nevertheless the handicap to overcome hidden areas by obstructed views.

16.5.3 3D Oblique Imaging via 3D City Models With Automated Object Creation of Buildings

A new more automated possibility exists by creating urban city models from overlapping oblique imagery. This imagery source may be combined with other airborne or satellite imagery, including UAV imaging and existing vector data and lidar surveys. The dataset is subjected to a sequence of algorithms using point cloud processing, image processing, computer vision tools, adjustments and deep learning for sparsity driven DTM extraction for optimized partial automation to generate DSMs and DTMs for later change detection, to extract LOD2 building models according to CityGML standards.

The semiautomatic deep learning based object extraction helps to generate building roof prints and to extract building facades for automated texture mapping.

The software system developed by the Turkish- Saudi Arabian company Geotech (headed by Kamil Eren) has been introduced internationally by the name 'CitiGenius'. It was used in a pilot project for Istanbul, in projects of Saudi Arabia, in the City of Hannover in Germany and in U.S. Cities. The point cloud extraction succeeded to extract DSMs and DTMs, true ortho-images and 3D building models according to City GML standards including object generation for buildings and their administrative use. An object-oriented database consisting of some 5900 buildings was obtained within one week. This database could be linked with relational data for property registration, which were generated and updated on a transactional basis.

CitiGenius follows the tradition of CityGML used for the urban city model of the city of Berlin. It covers within the city area of Berlin of 890 km^2 together with 590 000 buildings.

Urban change detection by CitiGenius promises to become an efficient tool for the acquisition and the maintenance of urban data, even though a number of steps are still required to customize it for use in a particular city or country, with a view to create an updatable urban information system.

16.6 The Use of New Stereo Satellite High Resolution Satellites by China for the Mapping of High Mountain Areas

A special application of the use of simultaneous stereo satellite high resolution images has been made possible by the launch of the Chinese satellite Ziyuan-3 in 2012, which permits near simultaneous stereo imaging with 2.1m GSD images. The imagery is particularly useful to map the neighbouring countries of China with high mountain areas at the scale 1:50 000. The simultaneous stereo views are imaged at the same illumination conditions. Combined with the utilization of accurate Beidou GNSS data it is said to be possible to survey mountain peaks within the accuracy of the Beidou positioning capability, estimated at 10 to 15m. As a result, an international cooperative project has been initiated to map the 14 highest mountain peaks of the world with an elevation of over 8000m.

Again this project involving partners in China, Germany, Austria and the USA can compare the stereo satellite capability with earlier attempts to produce 1:50 000 maps of the Nanga Parbat [5] and Mount Everest [10, 3, 4] with the stereo capability of new high resolution stereo satellites, such as Ziyuan-3.

Bibliography

[1] M. N. Alrajhi and G. Konecny. The updating of geospatial base data. *ISPRS Annals of Photogrammetry, Remote Sensing and Spatial Information Sciences*, IV-3:3–5, 2018.

[2] Muhamad Nasir A. ALRAJHI. *A Semi-Automated Procedure for Orthophto Generation from High Resolution Satellite Imagery*. November 2013.

[3] W Altherr and A Grün. Mt. everest karte 1: 50 000–faszination eines kartenprojektes. *Schweizerische Zeitschrift für Vermessung, Kulturtechnik und Photogrammetrie, Febr*, pages 90–97, 1990.

[4] Andreas Roditakis Armin Gruen. Visualization and animation of mount everest. International Archives of the Photogrammetry, Remote Sensing and Spatial Information Sciences, Vol. XXXIV-5/W10, 1990.

[5] Richard Finsterwalder. Deutscher alpenverein, map of nanga parbat expedition 1:50 000, reprint 1936, 1934.

[6] A. Konecny, G.; Breitkopf.U.; Radtke. The status of topographic mapping in the world- a unggim-isprs project 2012-2015. www.isprs.org, 2015.

[7] G. Konecny. E/conf78/bp8 1986. World Cartography XIX, (1990), 1990.

[8] G Konecny. 100 years of the society. Internatonal Society for Photogrammetry and Remote Sensing, 2010.

[9] Mark S Monmonier. *The History of Cartography: cartography in the twentieth century*. University of Chicago Press, 2015.

[10] Bradford Washburn. Mount everest: surveying the third pole. *National geographic*, 174(5):653–659, 1988.

17

Night-Light Remote Sensing: Data, Processing and Applications

Xi Li

Wuhan University of Science and Technology, China

Deren Li

Wuhan University of Science and Technology, China

Huayi Wu

Wuhan University of Science and Technology, China

Night-light remote sensing is a tool which can be used when working towards achieving the UN Sustainable Development Goals. It has attracted a lot of attention and brought many new research directions as an emerging subject.

17.1 Introduction

Night-light remote sensing has attracted a lot of attention and brought many new research directions as an emerging subject. Satellites acquire night-light images by detecting visible light sources such as city lights, fishing boat lights and fire spots under cloudless conditions at night. Unlike daytime remote sensing, night-time remote sensing has unique capabilities for reflecting human activities, which can be used to study in the following fields: regional development [9, 3], conflict evaluation [17], light pollution [1], and fishery [5]. At present, night-light data is mainly obtained through sensors in the visible and near-infrared bands, mainly including Defense Meteorological Satellite Program/Operational Linescan System (DMSP/OLS) night-light, Visible Infrared Imaging Radiometer Suite (VIIRS) Day/Night Band DNB (DNB), Earth Remote Observation System-B (EROS-B), Luojia 1-01, and Jilin1-03B night-light images. Besides, photographs taken by astronauts from the International Space Station (ISS) can also reflect night light on the earth's surface, and other research used aircrafts to take photos to study night light for cities. Figures 17.1 to 17.4 show the night-light images from different sensors.

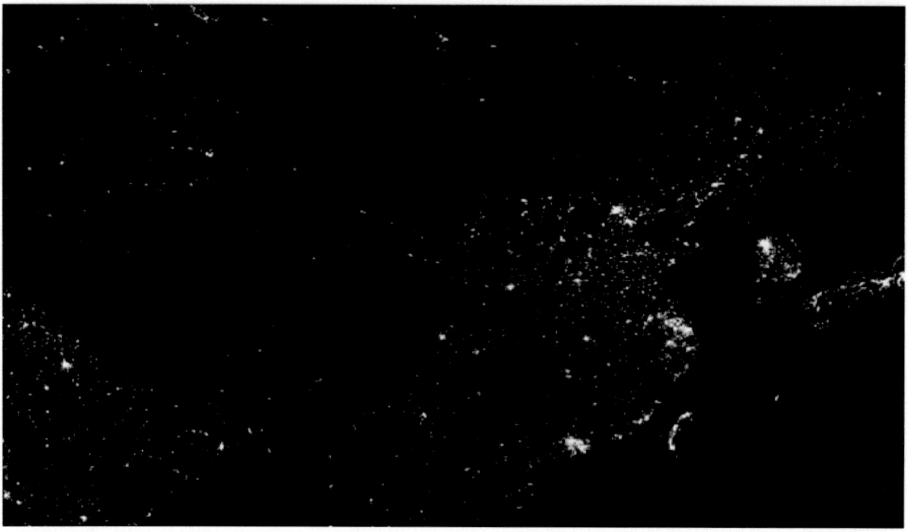

FIGURE 17.1
The DMSP/OLS night-time light image of East Asia in 2012.

17.2 DMSP/OLS Night-light Data

As the first night-time remote sensing satellite, DMSP has gradually been used to conduct research on night-time remote sensing. The DMSP/OLS was originally designed to detect night clouds, but scientists found that it also has the ability to detect night-time light on the earth's surface[14]. National Oceanic and Atmospheric Administration (NOAA) has released the DMSP/OLS night-light annual global composites (1992-2013) from six satellites (F10, F12, F14, F15, F16, F18) [1].The spatial resolution of the image is 0.0083 degree, and the images have only digital values and no radiation units. Because the DMSP/OLS image is pressed as 6 bits digital number (DN) values, saturation commonly exists in urban cores. Therefore, DMSP/OLS images are mostly used for macroscopic research.

[1] https://ngdc.noaa.gov/eog/dmsp/downloadV4composites.html

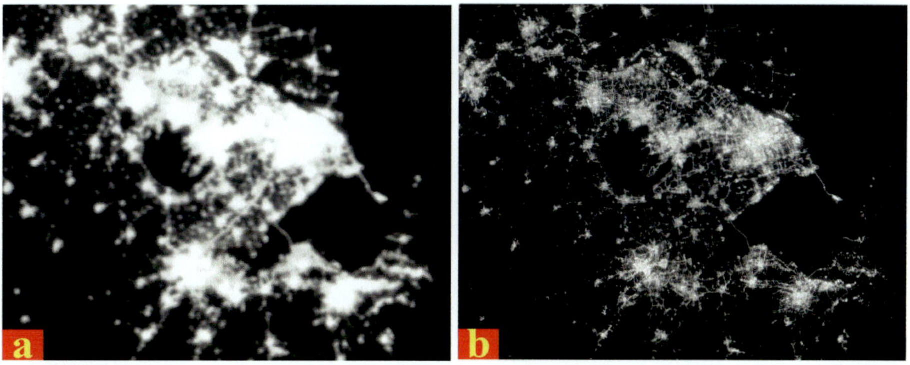

FIGURE 17.2
The (a)DMSP/OLS and (b)VIIRS DNB night-time light images of Yangtze River Delta in 2012.

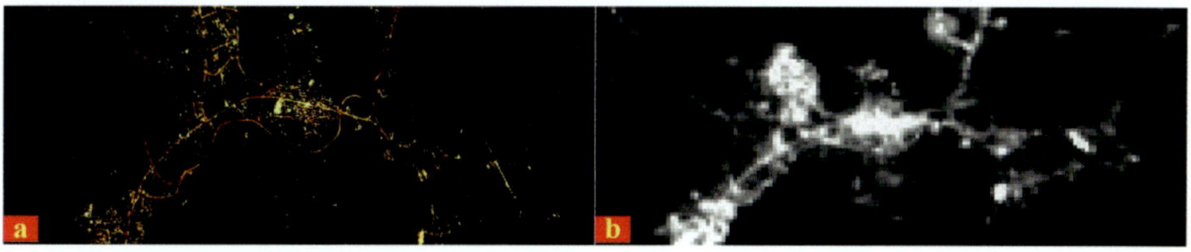

FIGURE 17.3
The (a) Jilin1-03B (June 27,2017) and (b) Luojia 1-01 (January 31, 2019) night-time light images of Hong Kong.

17.2.1 VIIRS DNB Night-light Data

The VIIRS was carried by Suomi National Polar-Orbiting Partnership (Suomi NPP) satellite which was launched on October 2011. VIIRS DNB has the ability to detect surface light and has higher spatial and radiation resolution than DMSP/OLS data [21]. NOAA has released three VIIRS DNB products, including daily products, monthly products and annual products (2012 - present). Its spatial resolution is 742 m and it can revisit most ground points in 12 hours. VIIRS DNB images can be used for quantitative research after radiation correction processing. It can be used not only for macroscopic research, but also for microscopic research on the development structure of the city [4].

17.2.2 VIIRS DNB Night-light Data

The EROS-B satellite was launched on April 25, 2006 from Svobodny Launch Complex in eastern Siberia. It has a panchromatic band ranging from 0.5 to $0.9\mu m$, and the scan width of the imagery was about 8.3 km. EROS-B images have a high spatial resolution which is 0.7 m. The data is described as 16 bits

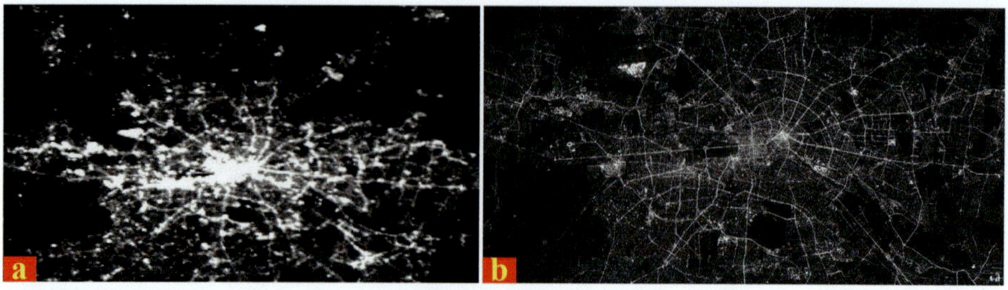

FIGURE 17.4
The (a) Luojia 1-01 (August 27, 2018) and (b) aerial survey using CCD (September 11, 2010) night-time light images of Berlin.

DN values (0 - 65535) to represent brightness values. There is no information on how to calibrate its DN values into radiance values [15].

17.2.3 Jilin1-03B Night-light Data

Chinese commercial satellite - Jilin1-03B was launched on January 9, 2017 by the Chinese commercial satellite company—Chang Guang Satellite Technology Co., Ltd. It has multi-spectral bands, including 430-512 nm (blue), 489-585 nm (green) and 580-720 nm (red), and has a high spatial resolution of 0.92 m. It is the first multi-spectral bands satellite which can detect low light at 7E-7 W/cm^2/sr [28]. Jilin1-03B data can be used to study light pollution and light distribution inside cities.

17.2.4 Luojia 1-01 Night-light Data

Luojia 1-01 satellite was launched on June 2, 2018. It has been equipped with a 4-megapixel scientific CMOS Image Sensor [25]. It is the first remote sensing satellite used for night light study in China and also the first low-orbit satellite with earth observation and satellite navigation enhancement functions. The image of Luojia 1-01 has a high spatial resolution (129 m), and a wide band range from 460 nm to 980 nm. It can revisit a point in 15 days. The data can be used in socio-economic parameter estimation, eco-environmental disaster monitoring, major event assessment, public health, etc. Table 17.1 shows the comparison of different night-light remote sensing satellites.

TABLE 17.1
The comparison of different satellites.

Variable	DMSP / OLS	VIIRS DNB	EROS-B	Jilin 1-03B	Luojia 1-01
Available	1992 - 2013	2012 - present	2006 - present	2017 - present	2018 - present
Country	The U.S.	The U.S.	Israel	China	China
Swath	3000km	3000km	8.3 km	11.6 km	250km
Spatial Resolution	2.7 km	742 m	0.7 m	0.92 m	129 m
Quantization	6 bits	14 bits	16 bits	8 bits	14 bits
Saturation	Saturated	Not Saturated	Not Saturated	Not Saturated	Not Saturated
Radiance Calibration	No	Yes	No	Yes	Yes
Bands (μm)	0.50-0.90	0.50-0.90	0.50-0.90	Blue: 0.43-0.51; Green: 0.49-0.59; Red: 0.58-0.72	0.46-0.98

17.3 Data Processing

17.3.1 DMSP/OLS Intercalibration

DMSP/OLS sensors have collected a long time series night-time light (NTL) data, which is useful in studying changes of human activities including but not limited to energy usage, population distribution and urbanization dynamic. However, due to the systematic differences in satellite orbits and sensor degradation, the NTL data collected by different sensors are very different, and as a result, its potential is not fully utilized. In order to eliminate the inconsistency of NTL data and facilitate scientific research, it is necessary to perform intercalibration to NTL data.

According to Zhang et al.[26] , the existing strategies in calibrating NTL time series can be mainly grouped into two categories: 1) Manually selecting the invariant area as reference area based on experience [12, 10, 23, 2, 20, 24] and 2) Automatic identification of stable pixels as pseudo-invariant features [16]. Although there are many differences among all calibration methods, the two assumptions are the most critical: 1) there is very little ground change at the calibration sites, and 2) sensor deviations in different regions are consistent. For example, Elvidge et al.[10] calibrated global DMSP/OLS NTL time series using the Sicily of Italy as the calibration site, Li et al.[16] assumed that land change is an outlier and used an algorithm to iterative the outlier and discard them, and Zhang et al.[26] automatically identified the data points along the ridgeline and used those points to derive calibration.

We can tell from the graph below, intercalibration significantly improved the consistency of TSOL (total sum of lights).

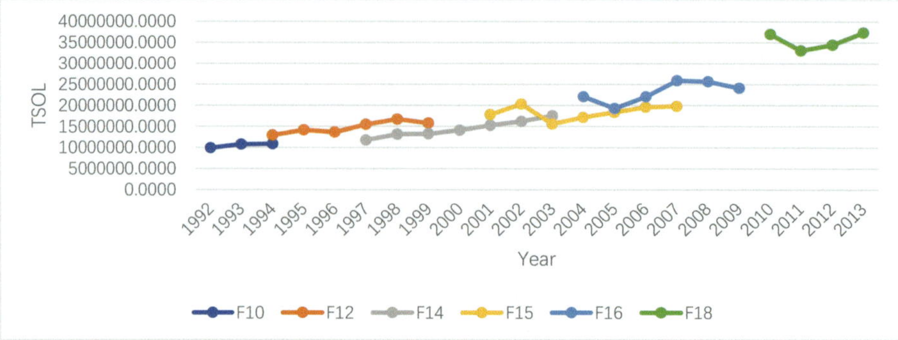

FIGURE 17.5
Sum of lights time series from raw NTL (China).

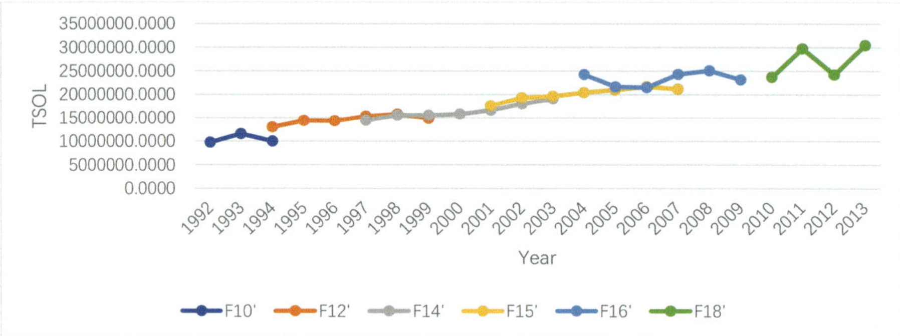

FIGURE 17.6
Sum of lights time series from intercalibrated NTL (China).

17.3.2 Improvement of VIIRS DNB Daily Data

The radiance of VIIRS DNB night light data is affected by moonlight, atmosphere, snow, etc. after radiation calibration, and further correction is necessary to more accurately describe the radiance of the artificial light on the surface. Moreover, since the acquisition time of some images is in the day and night, the refraction of sunlight through the atmosphere will affect the night light data of the night-time region. Therefore, we need to eliminate those affects and follow the method of Román and Stokes to get radiance L_{NTL} [22].

$$L_{NTL} = \frac{L_{DNB} - L_{path}}{T_{\uparrow}} - L_m T_{\downarrow} \quad (1)$$

where L_{DNB} represents the radiance of the night light of the VIIRS DNB band at the top of the atmosphere, L_{path} represents the total radiance of the path radiation received by the satellite observation direction, L_m represents the lunar radiance, $T_{\uparrow}$ represents the total path transmittance of the surface to the sensor, and $T_{\downarrow}$ is the total transmission rate of the moon to the surface.

17.3.3 Saturation Correction

When it comes to NTL data application, saturation is a very serious problem, which leads to the lack of details in urban cores in NTL imagery. As DMSP/OLS data has only 8-bit quantization and low dynamic range, saturation of data values in core urban areas limits the utility of NTL data for many applications.

In general, the correction methods of saturation fall into two categories: those that utilize only NTL data, such as linear regression method [13], and those that use other satellite data to correct the saturation of NTL data, like Vegetation Adjusted NTL Urban Index (VSNUI)[27] and Human Settlement Index (HSI)[19].

For the former category that utilize only NTL data, the main idea is based on the assumption that the tendency of DN change in the saturated area is similar to that of the non-saturated area. So we can extract the NTL of non-saturated area, and then develop a regression model, either linearly or none-linearly, to adjust the saturated area. And for the latter category, the main idea is that NTL has a strong relationship with human activities. A typical example is urbanization results in land cover change, specifically the loss of agricultural, natural vegetation, or abandoned lands to construct the built environment in most industrialized countries and emerging economies. Therefore, any data source about vegetation abundance and environment change can be used to generate an index to adjust the saturation of NTL data.

17.4 Applications

17.4.1 The Applications of Night-light Data

Socio-economic data often contains measurement errors and lacks spatial continuity in many developing and underdeveloped regions of the world. Night-time light, a special tool has a positive correlation with human activities,

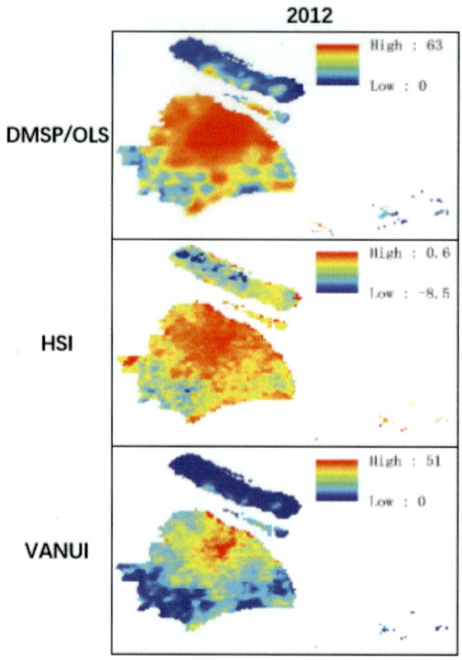

FIGURE 17.7
The spatial pattern of DMSP/OLS data, HSI and VANUI in Shanghai in 2012.

which provides potential possibilities to quantify these socio-economic data. Elvidge et al.[11] utilized the DMSP/ OLS NTL imagery of 21 countries in the America to do regression analysis, and found out that the correlation coefficient between total NTL and GDP in light-emitting area is 0.97. Since then, scientists carried on similar researches on different regions like European Union[7], China[18], etc.

All of those researches show a strong relationship between NTL and GDP. Now it is widely believed that the NTL is an efficient economic indicator in economic application, such as urbanization research, economic growth and decline analysis, and regional policy assessment. Recently, more studies reveal that NTL data is also an objective tool that can be used in the following fields.

Light pollution: Artificial night-time lighting has diverse and problematic environmental impacts. These include effects on the physiology, behavior and phenology of organisms. Sánchez de Miguel, A. et al.[6] first presented a method, color-color diagrams, to classify outdoor lighting types from ISS(International Space Station) imagery, then determined the relations between the spectral information that can be obtained from the imagery and some key environmental indices, including photopic vision, the Melatonin Suppression Index, the Star Light Index, the Induced Photosynthesis Index, production of $NO_2 - NO$ radicals, energy efficiency and CO_2 emissions, and Correlated Color Temperature.

Demography: Combination of NTL data with other auxiliary data can help to disaggregate demographic data more accurately. By combining NTL imagery, vegetation index products, and population statistics, Zhuo et al.[29] simulated the population density distribution at 1-km resolution grids by establishing different population spatial distribution models.

Natural disaster loss assessment: Elliott, R. J. R., et al.[8] examined the impact of typhoons on local economic activity in coastal China by combining historical typhoon track data and damage proxy with satellite derived night-light intensity data to construct a panel data set, which allows researchers to estimate the impact of typhoons at a spatially highly disaggregated level. They found that typhoons have a negative, significant, but short-term impact on local activity.

Humanitarian support: Since reliable witness reports are hard to gather in a war zone, satellite images, as one of the few sources of objective information, are potentially of great importance. Li, X. and D. Li[17] used 38 monthly DMSP/OLS composites covering the period between January 2008 and February 2014 as a NTL time series, and found that NTL and lit areas in Syria during the crisis declined by about 74% and 73% respectively. In 12 of 14 provinces, the NTL declined by >60%. These findings lend support to the hypothesis that NTL can be a useful source for monitoring humanitarian crises such as that unfolding in Syria.

17.4.2 Case: The Study of Syrian Crisis

At the beginning of 2011, the Syrian civil war broke out. Many Syrians were forced to flee their homes and lost their lives due to lack of food and medical resources. At present, the number of deaths in the Syrian civil war has exceeded 400,000, causing an almost devastating blow to the entire country. However, how to assess the impact of the war timely and accurately is a very important and difficult problem. Witness reports are currently the main source of war assessment, but during violent periods, the comprehensiveness and neutrality of witness reports are difficult to guarantee. Consequently, remote sensing satellite data has been used as a supplement in war assessment. The remote sensing satellite image is an objective and accurate data source, and the night-time light remote sensing image can be regarded as a representation of human activities and social economy. When the social and economic system of a certain place changes significantly due to disasters, its night light will also fluctuate greatly. Therefore, the NTL data has been widely used for assessing the impact of war.

Li et al.[17] investigated the impact of the Syrian crisis using DMSP/OLS NTL remote sensing data. Firstly, the DMSP/OLS time series images were registered, denoised and calibrated using invariant region-based intercalibration method. Then visual comparison, analysis by administrative regions and spatial analysis were applied to demonstrate the impact of the Syrian civil war objectively. By comparing the DMSP/OLS images in two periods, it can be seen that the night light in Syria has been sharply reduced since the outbreak of the civil war, and the lights in some towns have even disappeared completely. The change of NTL were calculated as Equation (2):

$$SNL_{change} = \frac{SNL_m - SNL_n}{SNL_n} \tag{2}$$

where $\text{SNL}_{\text{change}}$ represents the change of NTL, SNL_{m} and SNL_{n} represent the sum of night lights at two different periods. According to the calculation results, the sum of night light in Syria decreased by about 74% from March 2011 to February 2014. Among the 14 provinces in Syria, 12 provinces have reduced their lights by more than 60%, and the areas with the most serious night light reduction are also the most intense places in the civil war. While areas such as Damascus, controlled by Syrian government forces, have seen relatively little reduction in night light. In addition, the linear regression method was used to analyse the relationship between the reduction of night light and the number of refugees in different provinces. The study found there is a significant positive correlation, confirming that refugee migration is one of the important factors causing the reduction of night light.

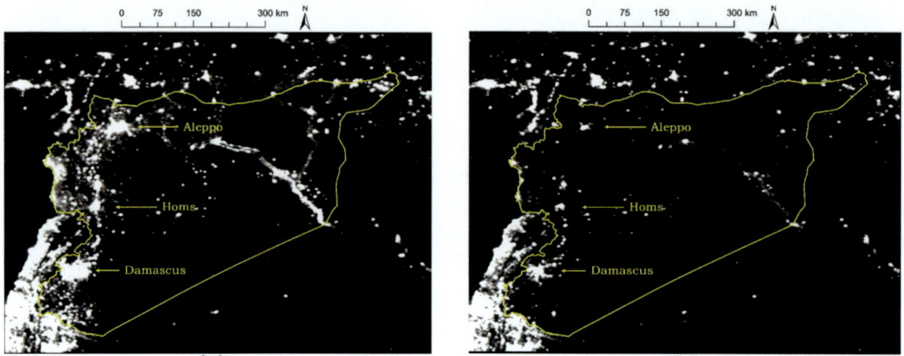

(a) (b)

FIGURE 17.8
The DMSP/OLS NTL monthly composites in (a)March 2011 and (b)February 2014.

The results of the research on Syria were cited in a report of the United Nations Security Council. On March 11, 2015, the results were released in a telebriefing entitled ““WithSyria 4th Anniversary Campaign: Turn the Lights Back On””. On this briefing, the paper author presented the assessment images of the Syrian civil war and clarified that since the war, Syria has lost 83% of its night lights. Most areas of Syria have been plunged into darkness, with

Aleppo Province being the most serious and the loss of night lights reaching 97%. This shows that Syria is falling into a serious humanitarian disaster. This result has been reported by more than 600 media outlets and NGOs around the world, including the 7416th meeting of United Nations Security Council, CNN, BBC, AOL, CBC, Fox News, Al Jazeera, AP, AFP, Reuters, France 24, New York Times, Guardian, Le Figaro, Jerusalem Post, etc. It can be seen from these that the night-light remote sensing can play a very important role in the assessment of humanitarian disasters.

Bibliography

[1] Sarah E. Bauer, Sara E. Wagner, Jim Burch, Rana Bayakly, and John E. Vena. A case-referent study: light at night and breast cancer risk in georgia. *International Journal of Health Geographics*, 12(1):23, Apr 2013.

[2] Jonathan Bennie, Thomas W Davies, James P Duffy, Richard Inger, and Kevin J Gaston. Contrasting trends in light pollution across europe based on satellite observed night time lights. *Scientific reports*, 4:3789, 2014.

[3] Xin Cao, Jin Chen, Hidefumi Imura, and Osamu Higashi. A svm-based method to extract urban areas from dmsp-ols and spot vgt data. *Remote Sensing of Environment*, 113:2205–2209, 10 2009.

[4] Z. Chen, B. Yu, W. Song, H. Liu, Q. Wu, K. Shi, and J. Wu. A new approach for detecting urban centers and their spatial structure with nighttime light remote sensing. *IEEE Transactions on Geoscience and Remote Sensing*, 55(11):6305–6319, Nov 2017.

[5] Kwangho Choi, Chung Il Lee, Kwangseok Hwang, Sang-Woo Kim, Jong-Hwa Park, and Yeong Gong. Distribution and migration of japanese common squid, todarodes pacificus, in the southwestern part of the east (japan) sea. *Fisheries Research - FISH RES*, 91:281–290, 06 2008.

[6] Alejandro Sánchez de Miguel, Christopher CM Kyba, Martin Aubé, Jaime Zamorano, Nicolas Cardiel, Carlos Tapia, Jon Bennie, and Kevin J Gaston. Colour remote sensing of the impact of artificial light at night (i): The potential of the international space station and other dslr-based platforms. *Remote Sensing of Environment*, 224:92–103, 2019.

[7] Christopher NH Doll, Jan-Peter Muller, and Jeremy G Morley. Mapping regional economic activity from night-time light satellite imagery. *Ecological Economics*, 57(1):75–92, 2006.

[8] Robert JR Elliott, Eric Strobl, and Puyang Sun. The local impact of typhoons on economic activity in china: A view from outer space. *Journal of Urban Economics*, 88:50–66, 2015.

[9] C. D. Elvidge, K. E. Baugh, S. J. Anderson, P. C. Sutton, and T. Ghosh. The night light development index (nldi): a spatially explicit measure of human development from satellite data. *Social Geography*, 7(1):23–35, 2012.

[10] Christopher Elvidge, Daniel Ziskin, Kimberly Baugh, Benjamin Tuttle, Tilottama Ghosh, Dee Pack, Edward Erwin, and Mikhail Zhizhin. A fifteen year record of global natural gas flaring derived from satellite data. *Energies*, 2(3):595–622, 2009.

[11] Christopher D Elvidge, Kimberley E Baugh, Eric A Kihn, Herbert W Kroehl, Ethan R Davis, and Chris W Davis. Relation between satellite observed visible-near infrared emissions, population, economic activity and electric power consumption. *International Journal of Remote Sensing*, 18(6):1373–1379, 1997.

[12] Christopher D Elvidge, Feng-Chi Hsu, Kimberly E Baugh, and Tilottama Ghosh. National trends in satellite-observed lighting. *Global urban monitoring and assessment through earth observation*, 23:97–118, 2014.

[13] Masanao Hara, Shuhei Okada, HIROSI YAGI, Takashi Moriyama, Koji Shigehara, and Yasuhiro Sugimori. Progress for stable artificial lights distribution extrvction accuracy and estimation of electric] power consumption by means of dmsp/ols nighttime imagery. *International Journal of Remote Sensing and Earth Sciences (IJReSES)*, 1(1), 2010.

[14] Marc L. Imhoff, William T. Lawrence, Christopher D. Elvidge, Tera Paul, Elissa Levine, Maria V. Privalsky, and Virginia Brown. Using nighttime dmsp/ols images of city lights to estimate the impact of urban land use on soil resources in the united states. *Remote Sensing of Environment*, 59(1):105 – 117, 1997.

[15] Yali Katz and Noam Levin. Quantifying urban light pollution — a comparison between field measurements and eros-b imagery. *Remote Sensing of Environment*, 177:65 – 77, 2016.

[16] Xi Li, Xiaoling Chen, Yousong Zhao, Jia Xu, Fengrui Chen, and Hui Li. Automatic intercalibration of night-time light imagery using robust regression. *Remote sensing letters*, 4(1):45–54, 2013.

[17] Xi Li and Deren Li. Can night-time light images play a role in evaluating the syrian crisis? *International Journal of Remote Sensing*, 35(18):6648–6661, 2014.

[18] Xi Li, Huimin Xu, Xiaoling Chen, and Chang Li. Potential of npp-viirs nighttime light imagery for modeling the regional economy of china. *Remote Sensing*, 5(6):3057–3081, 2013.

[19] Dengsheng Lu, Hanqin Tian, Guomo Zhou, and Hongli Ge. Regional mapping of human settlements in southeastern china with multisensor remotely sensed data. *Remote Sensing of Environment*, 112(9):3668–3679, 2008.

[20] Bhartendu Pandey, PK Joshi, and Karen C Seto. Monitoring urbanization dynamics in india using dmsp/ols night time lights and spot-vgt data. *International Journal of Applied Earth Observation and Geoinformation*, 23:49–61, 2013.

[21] Proceedings of the Asia-Pacific Advanced Network. *Why viirs data are superior to dmsp for mapping nighttime lights*, volume 35, 2013.

[22] Miguel O Román and Eleanor C Stokes. Holidays in lights: Tracking cultural patterns in demand for energy services. *Earth's future*, 3(6):182–205, 2015.

[23] Ye Wei, Hongxing Liu, Wei Song, Bailang Yu, and Chunliang Xiu. Normalization of time series dmsp-ols nighttime light images for urban growth analysis with pseudo invariant features. *Landscape and Urban Planning*, 128:1–13, 2014.

[24] Jiansheng Wu, Shengbin He, Jian Peng, Weifeng Li, and Xiaohong Zhong. Intercalibration of dmsp-ols night-time light data by the invariant region method. *International journal of remote sensing*, 34(20):7356–7368, 2013.

[25] Guo Zhang, Litao Li, Jiang Yong-hua, Xin Shen, and D Li. On-orbit relative radiometric calibration of the night-time sensor of the luojia1-01 satellite. *Sensors*, 18, 12 2018.

[26] Qingling Zhang, Bhartendu Pandey, and Karen C. Seto. A robust method to generate a consistent time series from dmsp/ols nighttime light data. *IEEE Transactions on Geoscience and Remote Sensing*, 54, 05 2016.

[27] Qingling Zhang, Crystal Schaaf, and Karen C Seto. The vegetation adjusted ntl urban index: A new approach to reduce saturation and increase variation in nighttime luminosity. *Remote Sensing of Environment*, 129:32–41, 2013.

[28] Qiming Zheng, Qihao Weng, Lingyan Huang, Ke Wang, Jinsong Deng, Ruowei Jiang, Ziran Ye, and Muye Gan. A new source of multi-spectral high spatial resolution night-time light imagery—jl1-3b. *Remote Sensing of Environment*, 215, 09 2018.

[29] Li Zhuo, Jin Chen, Peijun Shi, Zhihui Gu, Yida Fan, and Toshiaki Ichinose. Modeling population density of china in 1998 based on dmsp/ols nighttime light image. *Acta Geographica Sinica*, 60(2):266–276, 2005.

Part V

SDGs Perspectives: Current Practices and Case Studies

18

Why and How Informal Development Should Be Formalized Quickly, Inclusively and Affordably- Experience From UNECE Region

Chryssy Potsiou

National Technical University of Athens & Former FIG President, Greece

In order to address a selected SDGs goals and indicators, and the fact that there are many informal developments happening worldwide, it is very important to improve the level of land records. In this context, this chapter is discussing about why and how informal development should be formalized quickly, inclusively, and in affordable manner, particularly discussing the experiences from UNECE region.

18.1 Introduction

As a result of a 20-year research by the author, partially in cooperation with FIG, the World Bank, UNECE and government agencies and local authorities in the field of formalization of informal development in the European region, and mainly in Albania, Cyprus, Georgia, Greece, Italy, Kosovo, Kyrgyzstan, Montenegro, North Macedonia and Turkey we have learned to identify the problem of current informality in real estate within the South-Eastern European and Caucasus regions as to its size and definition as well as its causes [15, 12].

"Informal building" is defined as an unauthorized property unit which may be lacking planning and building permits and in many cases may be lacking property titles as well, or real estate built in excess of legally granted permits. In most cases it is residential real estate. Such properties are considered to be illegal, therefore are out of the economic circle. They have been characterized as illegal, are not accepted to be registered and cannot be transferred, rented, inherited or mortgaged.

We have also learned that there are a variety of types of informal buildings in each one of the various countries, varying from good and even multi-story constructions, as shown at Figure 18.1, up to settlements of poor but permanent construction which constitute a considerable investment of labor and revenue and are worth rescuing as shown at Figure 18.2. There are many similarities, as well. We have estimated that about 50 million people live in informal, self-made buildings in UNECE today [15].

FIGURE 18.1
Various types of multi-story informal constructions

Within the UNECE region there are slums, too, as shown at Figure 18.3, but the core research interest so far is not directly focused on how to formalize slums, as this is not the major problem in the region [11]. Past experience from Greece has shown that Roma slums may as well be formalized in a similar manner as other informal buildings. There are many examples in which slums have been formalized by recognizing informal tenure on occupied state or municipal land and providing ownership rights on the land to the occupants, or by providing ownership titles through a judicial procedure based on the adverse possession principle when privately-owned land was occupied illegally by Roma slum dwellers. Planning and structural improvements and integration of the land into a city plan was then provided according to the general practice in Greece [14]. Recent examples of good practice of re-settlement

FIGURE 18.2
Informal settlements of poor quality but permanent construction

projects for slum dwellers have been identified in Kosovo region by providing ownership rights as well as job opportunities to the dwellers[10].

Major political changes coupled with rapid urbanization, poverty, massive internal migration, conflicts, marginalization, natural disasters, cumbersome authorization processes (planning and building permitting) and corruption may be listed as some of the causes. However, corruption should not be considered as a major reason for informality; one cannot claim that 50 million citizens, as well as the authorities who tolerated the phenomenon, were corrupt. Why is informal development not a major issue in western European countries? Human beings are similar, but there the infrastructure makes it easier and more attractive to be and remain legal, while in the regions under study there are weaknesses in the infrastructure and authorization process.

But the list of causes is even longer, including the absence of policies by the states and their failure to adopt pro-growth planning as well as affordable housing policies; serious weaknesses of the private sector and lack of professional regulations; the lack of knowledge and political will to develop land policies which would facilitate the recognition of existing tenure and provision of private property rights and would aid the transition from centrally planned to market economies; and the failure or reluctance of state agencies to implement measures to support structural reforms to facilitate the digital economy and the UN Sustainable Development Agenda 2030.

In brief, so far this research has identified the obvious: that when neither

FIGURE 18.3
Roma informal settlements are common in many UNECE countries (top); example of a formalized Roma settlement (by provision of ownership titles) (bottom).

the state nor the private sector provide the supply of appropriate real estate types and quantities to satisfy the current demand, people build informally. It is also important here to remember that demand in property markets is defined both by the need but also by the desire and the purchasing power of the consumers. In most of the informal cases in this region the state's housing policy is inadequate to meet the demand and people have built informally either because of their need for housing they could afford or because the private sector was not providing such a product (in many places the private sector is interested to serve the high income rather than the low-middle and middle income classes), while at the same time the state had not provided planning and permitting tools for affordable or social housing. In some cases there has been a demand for second housing, or a desire of people to "move up" to larger or better housing. The industry had not planned for an efficient mechanism to provide for such products and services.

Informality however, in the real estate sector and inevitably in real estate markets, is directly related to a general informal culture, a characteristic of development in the so called "frontier markets"; it may also exist in developed economies where its extent is less significant. Much of the building labor is self-provided by the occupants but there is also a great amount of construction material consumed, and services provided, informally. Informality is usually accompanied with fraud and lack of transparency; it affects public revenue, productivity and job opportunity; it creates non-productive capital, dead capital, and construction completion is indefinite.

18.2 Informality Is Considered a Social, Economic and Environmental Challenge

The most important social challenge of the existing informal settlements is *weak ownership rights.* Unclear ownership rights on a property unit are created when people either have built on land: (a) to which they have no ownership titles (e.g., occupied land that belongs to the state, or the municipality, or to a social enterprise, or to a third party); or (b) that they only have the right to use (recognized tenure), but usually the state or the municipality delays or refuses to provide ownership titles; or (c) has been illegally subdivided though they may own legally as a whole but due to zoning regulations parcel subdivision is forbidden or not regulated, and therefore the newly created parcels or property units in general cannot be legally registered; or (d) that they own legally but they have built without obtaining a planning and/or building permit, or they had obtained a planning and/or building permit but they have built beyond the scope of the permit and the newly created property unit cannot be registered.

In all the above cases people finally have weak or limited property rights and the property cannot be registered in the cadastre, transferred legally, taxed or mortgaged. Dwellers in informal settlements, informal land and real estate rarely have ready access to capital. The common practice applied so far in tenure regularization for informal settlement dwellers in other regions is of no value in the regions under research; this process often starts with the delivery of an administrative permit to occupy the land that can be conditionally upgraded to a leasehold and, at a later stage, to a long-term registered freehold. In general, improving tenure security incrementally by recognizing the occupation and providing dwellers with legally recognized tenure reaching from occupancy certificates to full property rights is a long, bureaucratic procedure that cannot directly provide for full exploitation of property assets and cannot help to achieve wealth for the poor. It simply delays the implementation of many SDGs of the UN Sustainable Development Agenda 2030 and therefore it is not recommended for any region when people have already invested a relatively significant part of their labor income to build a house to provide for their housing needs.

Registration improves security of tenure, establishes property rights over investment, minimizes lending risks and provides access to credit and funding mechanisms; it also improves legal protection, as well as legal empowerment of occupants/owners.

According to [2] the great economic divide in the world today is between the 2.5 billion people who can register property rights and the 5 billion who are impoverished, in part because they have no ability to registered their property rights. Private rights provide people the assurance they need in order to invest and protect their properties from abuse. Security of property rights is one of the drivers of economic growth and freedom.

When informal development was identified in large numbers ((e.g., as indicated at [13], in Greece, in 2005, it was estimated that 1M buildings were informal, in Albania about 500,000, in Cyprus about 80% of condominiums and 40% of single family houses, in North Macedonia about 350,000, in Kosovo 450,000 and in Montenegro about 130,000)) the phenomenon demonstrated a systemic failure which shows that a great proportion of the population had no access to clear property rights, property units could not be registered and were kept outside the economic circle. Such a system needs improvement.

Allowing such large numbers of informal buildings as dead capital - that is, property units that cannot become productive for the people who have informal tenure - is contrary to the UN Sustainable Development Agenda 2030. Goal #1 (*end of poverty in all its forms everywhere*), and especially target *1.4.* that "*by 2030, countries should ensure that all men and women, in particular the poor and the vulnerable, have equal rights to economic resources, as well as access to basic services, ownership and control over land and other forms of property, inheritance, natural resources, appropriate new technology and financial services, including microfinance*".

Similarly, SDG 11 (*make cities and human settlements inclusive, safe, resilient and sustainable*) and in particular SDG 11.1 expresses that *by 2030, states should ensure access for all to adequate, safe and affordable housing and basic services, and upgrade slums.*

Tenure security is a major step toward provision of adequate housing. The subject of adequate housing, though, is closely linked to a country's general land policy, economic development and to the provision of urban infrastructure. A country's housing policy is connected to its basic infrastructure development policy, such as provision of land for urban development and provision of utility services. It is broadly recognized that almost every country of the world will never have enough public funds to efficiently address the adequate housing issue for all, without the private sector participation, meaning that all countries are borrowing money to lend to their citizens. In order for states to be able to provide credit at low interest there is an urgent need to reduce lending risks by providing clear property titles for mortgaged-backed bonds. Therefore, providing clear ownership rights is a major priority if governments are to facilitate credit at low interest or other affordable housing tools for those in need.

In addition, land is referred to in SDG #2 (*ending hunger*) target 2.3: *By*

2030, double the agricultural productivity and incomes of small-scale food producers, in particular women, indigenous peoples, family farmers, pastoralists and fishers, including through secure and equal access to land, other productive resources and inputs, knowledge, financial services, markets and opportunities for value addition and non-farm employment, which requires access to credit and funding mechanisms in order to improve agriculture and business, and maximize quality and quantity of products.

In general, the right to adequate housing, good management of land and security of property rights is referred in the New Urban Agenda, and in many other SDGs such as Goal #5 on *gender equality*, Goal #13 on *climate action*, Goal #15 on life *on land* and Goal #16 *on peace, justice and strong institutions*. As it is stated at [3] these goals and targets will never be achieved without good land governance and well-functioning nationwide administration systems in place. In order for society to be able to meet these SDGs within the expected time limit major land reforms are required; to make such reforms successful unity is important. People should understand, trust and be willing to support the necessary activities and should voluntarily participate to provide information and to enhance procedures. In order to ensure unity and fairness and to eliminate conflict among society and the local communities, existing tenure and weak property rights on land, both formal and informal, should be recognized and registered. As mentioned in the [5] Voluntary Guidelines on the Responsible Governance of Tenure of Land, Fisheries and Forests in the Context of National Food Security (Voluntary Guidelines on Tenure) VGGTs, "States should facilitate the operations of efficient and transparent markets to promote participation under equal conditions and opportunities for mutually beneficial transfers of tenure rights which lessen conflict and instability ... states should take measures to prevent undesirable impacts on local communities."

Once property rights are issued and registered the property units should be in the economic cycle. Past policies applied in various countries that required planning and construction improvements prior to formalization are not applicable today because they are time-and-cost consuming and they delay the resulting formalization and expected economic, social and environmental benefits [4]. Improvements of neighborhoods and basic services provision must be made available for social and environmental reasons but also to make such properties more attractive and improve their value for the benefit of the owners and the national economies immediately following formalization. There should not be a general rule and guide on how such planning and construction improvements may be achieved, however, as conditions may vary from place to place and/or from property to property. Such improvements may be initiated and funded according to local needs either by the dwellers (once they have access to credit) in cooperation with the private sector through development of small or medium-sized development and planning projects that will also include consolidation of parcels (as soon as relevant tools and regulations are in place), or by local authorities through urban planning and land readjustment projects; the latter is more complicated, expensive and time consuming, however.

In general, a large number of informal buildings are rather self-made constructions (of 1 or max 3 to 4-story buildings), that may not comply with local planning and construction standards and they are usually built to provide for the housing needs for the next 30-40 years; such buildings once formalized should gradually be improved - if feasible - or replaced by better constructions as the financial situation of their owners and the state policies will improve. Planning and building regulations should be modified accordingly to provide incentives and facilitate such reforms as well as to enable the gradual increase of urban density in such areas and the sustainable development of such self-made cities in the 3rd dimension in order to avoid further urban sprawl; this maybe also valid for urban areas build formally but following such a pattern of construction.

In general, large demolitions and forced evictions are considered to be a serious violation of the international norms for adequate housing and, where demolition is necessary, decent resettlement is required.

In an effort to discourage informal development governments frequently deny basic utility services to informal settlements despite the many years of their existence. Informal settlements' dwellers are then led to proceed to further illegalities such as illegal connection to electricity that increases the risk of disasters in the settlements, or illegal drilling for water, and sewage disposal which has a negative environmental impact in the management of underground waters. Thus, such state policies should be considered to be among the most severe violations of the right to dignity, security, health and life. People living in informally developed areas are constantly at higher risk for fires, flooding and other disasters, and services must be provided to minimize such risks. Depriving informal settlement dwellers of fundamental services is a violation of the international objective of human rights for all. Adopting strategies for enabling the improvement of living conditions in informal settlements should be one of a government's priorities to ensure the fundamental human right to life, health and safety of dwellers.

Informality is an economic challenge, also, because it affects public revenues, productivity and job opportunities. The assets invested in informal real estate represent non-productive, dead capital. As a result of informality the tax base is limited while higher taxes must be levied on a subset of "legal" real estate and related business. Moreover, competition within real estate markets with a great level of informality is distorted due to a lack of transparency; in such markets personal, off-record negotiations matter more than rules-based transactions.

Through this research it became broadly recognized that indeed, informality remains substantial in countries where overregulation and bureaucracy, taxation, fees, penalties and related costs give significant incentives to build and work "under the radar".

Informality in real estate, its construction and operation as well as its demolition, is also considered a significant environmental challenge especially when it is spread over highly protected lands, or when, for instance, construction methods do not meet current standards for energy consumption.

When demolition is needed, not only is it expensive but it causes environmental impact and should by all means be followed by a special treatment of the debris. Where demolition is indicated it is important to communicate properly and to provide information in a timely manner to the occupants; the occupants should be involved as well as all affected other community members for the agreement for a meaningful and fair solution. Adequate compensation and alternative housing must be provided. It is also important that affected occupants and communities have access to affordable legal assistance.

18.3 Fit-for-Purpose Formalization Policies

As a result of this research and the derived knowledge we have systematically encouraged countries to initiate - where possible - formalization projects, and through cooperation with them we have discovered and assessed the several policies adopted in order to address this problem.

We have together identified policies that require improvements to informal properties in order to be in compliance with regulations prior to formalization. We have considered the high fees imposed upon owners or occupants of informal properties when buying land with missing property titles, and the penalties imposed for having illegally occupied or built thereon. Such policies are seldom affordable and do not provide for quick and inclusive settlement. If it was do-able and affordable to build legally according to existing rules and regulations the majority of people would not have chosen to do it illegally in the first place.

There are also policies that would provide planning amnesty, though accompanied by high penalties. However, in cases where property titles are missing this process seems politically more difficult. Again, such policies are not inclusive and affordable as they fail to solve the most important issue: weak ownership status.

In addition, there are policies that even when well intended, measures become bogged down due to administrative bottlenecks, or there are changes in government or government policy.

As a result, formalization is a lengthy, bureaucratic and expensive process in most places.In the meantime, through this research we have also managed to identify and quantify the annual GDP loss caused by delaying the formalization of the informal real estate sector, and have proved that it can be significant and worthy of serious consideration. The formalization of informal development is the most important step towards formalization in the real property market.

Today we have reached a stage at which we have decided to proceed with the compilation of an FIG/UNECE WPLA guide for the formalization process of informal buildings to assist policy makers, managers and staff of government agencies, as well as private sector specialists and members of civil society organizations. This guide is anticipated to become a mindset changer to provide understanding, inspiration and knowledge and to contribute to the global and national efforts towards the eradication of hunger and poverty by achieving the SDGs within the framework of the UN Sustainable Development Agenda 2030.

The guide spells out the accumulated knowledge and experience in a compact way, and is aligned with broadly recognized WB, UN Habitat, GLTN, and FIG publications such as the FAO VGGT in the Context of National Food Security and the FIG Fit-for-Purpose Land Administration, to mention a few.

A "fit-for-purpose" formalization procedure is giving priority to clearing out weaknesses in property titles and to providing for registration of property titles in an inclusive, affordable and timely manner. Informalities related to zoning, planning and/or construction regulations should never be connected to ownership rights or block property registration and transactions. Such informalities should not have an impact on the provision and registration of property titles, or affect the transaction or mortgaging of a property. It should be possible for a parcel with a damaged or half-finished construction that does not comply with regulations to be in the market for sale or inheritance. Construction efficiency and stability issues should not block the marketing of the parcel and/or building, however this important issue should be considered prior to issuing an occupation permit. Controls must necessarily be imposed upon comparatively large formalized constructions that will accumulate large numbers of people (e.g., for issuing an occupancy permit for multi-story, multi-family residential buildings, or operational permits for commercial real estate such as hotels, restaurants, schools, offices, cinemas). If the construction compliance of such buildings can be guaranteed by the private sector (maybe already involved in the construction of such buildings, as it is the case in some countries) then formalization should be quick, affordable and inclusive.

In the formalization process governments need to go beyond the established policies and practices in order to successfully deal with the property market challenges, the funding challenges, the structural stability challenges, the environmental challenges and the difficult ethical challenges and any hostile reactions to a formalization project. They need to argue for the contradictory concept to "legalize the illegal". The above described rationale provides all the required justification for developing a strategy and a communication policy with all involved stakeholders and society.

A legal framework and administrative and regulatory process should be prepared and tested in pilot studies. As well, ways to raise awareness, when differing priorities among the parties exist; the socio-economic realities must be clearly demonstrated, with the relevant significant data [4].

The clear economic and social benefits of formalization of informal structures must be advertised via public awareness programs demonstrating, also, the economic cost of informality while allowed to continue.

Modern technology and its products (e.g., mobile services, apps, UAVs, satellite images, orthophotos, VGI and crowdsourcing) will help to overcome the lack of transparency and provide the base map and the methodology for data collection and cadastral mapping of informal and unregistered settlements (e.g., identification of constructions, adjudication of occupants, formalization of titles and registration of properties and property rights). [8, 9, 7, 1, 6]. The use of any available cadastral information or development plans is highly advisable in order to identify and register informal real estate and occupants/owners. This may be accomplished by engaging the involved professionals including notaries, constructors, civil engineers, developers and real estate agents as well as the local authorities that may have records, the various state agencies such as the tax office; occupants and society should also be engaged in the recording process.

For the formalization of the informal to be successful it is important that technical advice be employed on how to build inclusively, affordably and in a timely manner an efficient framework. It is vitally important to revise the planning and permitting systems to discourage and eliminate a continuation of informality in order to support market needs and growth while in parallel to define the lands and real estate that should be protected.

In considering formalization of all informal real estate by 2030 it is necessary to determine the duration and the costs of the project in each case and to prioritize needs. This will influence decisions related to title provision, planning amnesty, the requirement for any controls, planning for future improvements, optional future stability controls for issuing operational permits to commercial real estate, inspections, monitoring, demolitions, and resettlement.

Important legal decisions are required on how to deal with occupied private, state and municipal lands as well as land that belongs to social enterprises, and to ensure gender and ethnic equity when possible. Governments should prioritize the provision of good title, when possible, in order to be aligned with SDGs, VGGT, FFP LA and other objectives.

Administrative aspects such as the determination of the responsible authority for formalization and the required fees, the registration of informal constructions into the cadaster using modern and low-cost technology, should be considered, as well. Once registration is accomplished transactions and mortgaging should be facilitated.

Urban regeneration methods and planning improvements - if necessary - for informally developed areas should adopt simplified norms and standards. Similarly, structural stability controls may be classified according to the size and the operational use of buildings and should be adjusted to the limits and abilities of local knowledge and practice.

Information on how to implement a formalization framework based on best practice through country-specific approaches including technical tools and methods, the role of professionals, state agencies and citizens, should be taken into account.

During and following formalization there must also be monitoring of the progress and of the situation especially in environmentally or socially sensitive areas to avoid future informal development. Automated monitoring of protected lands is highly recommended to avoid the need for on-site inspections and costs, while limiting the opportunity for corruption and bribery.

Bibliography

[1] K Apostolopoulos, M Geli, P Petrelli, C Potsiou, and C Ioannidis. A new model for cadastral surveying using crowdsourcing. *Survey Review*, 50(359):122–133, 2018.

[2] Hernando de Soto. *The Mystery of Capital: Why Capitalism Triumphs in the West and Fails Everywhere Else.* Basic Books, 2000.

[3] S. Enemark. Supporting the 2030 global agenda, April 2017.

[4] Potsiou et al. The cost of a lack in real property formalization on the greek national economy, 2016.

[5] FAO. Voluntary guidelines on the responsible governance of tenure of land, fisheries and forests in the context of national food security, 2012.

[6] Maria Gkeli, Chryssy Potsiou, and Charalabos Ioannidis. Crowdsourced 3d cadastral surveys: looking towards the next 10 years. *Journal of Geographical Systems*, pages 1–27, 2019.

[7] Lemmen C. Molendijk M. Gorton K. Jones, B. Fit-for-purpose and fit-for-future technology for cadastral systems. World Bank, March 2017.

[8] Mathilde Molendijk, JM Morales Guarin, and Christiaan Lemmen. Light mobile collection tools for land administration: proof of concept from colombia. *GIM International*, 27(1):20–23, 2015.

[9] G Mourafetis, K Apostolopoulos, C Potsiou, and C Ioannidis. Enhancing cadastral surveys by facilitating the participation of owners. *Survey review*, 47(344):316–324, 2015.

[10] C Potsiou. Study, recommendations and related training concerning informal buildings in kosovo unpublished manuscript. 2016.

[11] Chrysi Potsiou. Study on illegally built objects and illegal development in montenegro, January 2012.

[12] Chryssy Potsiou. Informal urban development in europe: Experiences from albania and greece. *United Nations Human Settlements Programme (UN-HABITAT)*, 2010.

[13] Chryssy Potsiou. Formalizing the informal: Challenges and opportunities of informal settlements in south-east europe., 2015.

[14] Dimopoulou E. Potsiou, C. Access to land and housing of the greek roma. *Surveying and Land Information Sciences (SaLIS)*, 72(1):37–49, 2012.

[15] Potsiou C. Badyina A. Tsenkova, S. Self-made cities – in search of sustainable solutions for informal settlements in the united nations economic commissions for europe region, 2009.

19

SDGs and Geospatial Information Perspective From Nigeria-Africa

Muyiwa Agunbiade

University of Lagos, Nigeria

Olajide Oluwafemi

University of Lagos, Nigeria

Oluyemi Akindeju

University of Lagos, Nigeria

This chapter discusses SDGs connectivity by exploring the nature of interlinkages between the SDGs from the lens of geospatial information and geospatial data infrastructure. It also focuses on evolving an integrated framework towards achieving SDGs in developing economies.

19.1 Introduction

To develop further on the achievements of the MDGs and to complete what has not been achieved, the United Nations adopted, in 2015, the 2030 Agenda for sustainable development, setting 17 development goals and 169 targets. The world has witnessed tremendous progress in the living condition of many people, notwithstanding the challenges associated with the implementation of the MDG goals. This progress, however, was uneven. It was anticipated that Africa progress towards sustainable development will have implications on the global achievement. Paradoxically, while many countries in developed world achieved many of the goals and were able to monitor progress, many African countries, including Nigeria, did not make appreciable progress. Nigeria's record in achieving the MDGs has been abysmally low [5].

Like the preceding MDG goals, implementing the *indivisible* SDG agenda by policy actors is faced with challenges [3] especially regarding the nature of interlinkages between the SDGs. The complex nature associated with the indivisible characteristic of the 2030 Agenda also requires an unbundling. In other words, it is important to understand the possible trade-offs as well as synergistic relations between the different SDGs in a way that is enough to achieve long-lasting sustainable development results. Most importantly, in an era when communities and their various challenges are interconnected across different domains and jurisdictions. The implementation of SDGs is challenged by the required science-based analysis that is anchored on robust fit-for-purpose geospatial information. One of the long-standing issues, in this regard, is the lack of reliable data in appropriate format to aid decision-making and monitor progress. Where data are available, they are fragmented among different institutions and agencies (private and public), and as such not integrated and accessible. In addition, the available data are not disaggregated in a usable format. Specifically, it is established that accurate and reliable geospatial data, integrated across sectors, are central to the implementation and monitoring of progress towards the attainment of SDGs [14]. It is also established that geospatial data and urban processes interact in complex and integrated ways to foster sustainable development. While the application of geospatial technologies and Spatial Data Infrastructure in development processes is increasingly emerging globally, the regional spread has not been proportionate, with most developing countries lagging behind.

It has been recognised that the 169 targets, currently being further defined by the measurable indicators, are designed to help evaluate and monitor the implementation of SDGs and determine if the 17 goals are achieved. Indicators are important tools to support decision making process and measure what matters. Indicators are particularly necessary for the monitoring and evaluation of SDGs for quality, consistency and comparability of data over time and space and across sectors and regions [1]. However, Africa, in particular, faces a unique challenge of measuring the attainment of these goals due to the paucity of adequate and appropriate geospatial information that is in the right format and that is fit-for-purpose. This is further compounded by lack of interagency collaboration and uncoordinated policies. This lack of synergy between geospatial data infrastructure and SDGs and clear interactions within the SDGs is evident as there is increasing need for reliable baseline data to monitor and evaluate progress. This identified gap is a huge obstacle to many African countries in their quest for sustainable development [15]. Couple with this challenge, is the lack of clear road map for integration of geospatial data into national development and policy making process.

Most often, the baseline data are "set on the available information, which in many cases, is scanty, unreliable and dubious" ([19], p. 5). Considering this precarious situation, it is apparent that Geospatial information is a requisite prior to the monitoring phase of SDGS. This provides the necessary background to establish the prevailing condition [19]. Currently, there is a general lack of adequate geospatial information policy and governance due to the ineffective and inefficient institutional framework and tools. Africa is also challenged by issues around capacity and knowledge transfer. Yet, there is an increasing array of global challenges, including interregional issues such as peace, security, natural disasters and climate change that needs to be measured, which no nation or region can solve independently. Thus, the justification for calls regarding global coordination that is firmly anchored on good Geospatial Information and geospatial data infrastructure.

While geospatial information is essential, the production of geospatial data infrastructure domain across Africa is subject to particular institutional and knowledge constraints. There seems to be an enduring knowledge gaps between the supply and demand sides of data as demonstrated by the lack of institutional capacity to coordinate. As such, both supply and demand sides are problematic, and they are huge obstacles to development. In this regard, as illustrated by Jerven (2013), one of the challenges is the unstructured nature of interactions between producers and consumers [10]. There seems to be a lack of understanding on the nature of data and how such data should be collected, analysed, disaggregated and integrated for effective policy making and national development. Before now, Riddel (1990) observed that maybe the major challenge with the existing Africa data is that they are generally known to be erroneous, but the extent of inaccuracy cannot be easily ascertained [13]. Even today, nothing seems to have significantly changed in many of the African countries despite the steady improvement in geospatial data technologies globally and despite the urgent need for integrated data platform to support the complex interrelationship of the sustainable development agenda. As noted by International Council for Science (2017), the "SDGs interact with one another as an integrated set of global priorities and objectives that are fundamentally interdependent" [3]. Strategies to understand the range of positive and negative interactions among SDGs is critical to revealing their full potential at any scale.

However, there are different dimensions to the issues of connectivity between, geospatial information and SDGs. The central arguments are that: i) while the 17 SDG goals are broad, they are interdependent, and the success of the 2030 Agenda lies in demystifying the interlinkages between policy areas as presented in the sustainable development goals (SDGs). Meaning that to achieve all the 169 targets it will require, specifically, the development of Geospatial Information infrastructure, to help facilitate the connectivity. ii) It could also be argued that it is important to localize the SDGs by highlighting the role of local institutions and local actors. These are considered as key in achieving and contextualising separate list of targets for respective goals, thus, the perspective from Africa as a region is critical for fit-for-purpose analyses and solutions. In this regard, indicators must be consciously formulated to meet regional needs, aspirations and priorities. iii) the "*leaving no one behind*" principle of the SDGs that is intended to ensure, on the assumption that, a sustainable and smart future will develop a better future for all communities appears high-level and difficult to achieve in Africa. iv) the unavailability of effective and efficient land administration, that is, the lack of matured land information (LIS) and formal land registration systems, can be argued to be the root of most problems in African countries. Thereby incubating insecure land rights and social exclusion, economic regression, and environmental degradation.

This chapter starts by providing insights into the past events at implementing the MGDs, in Africa especially in Nigeria. This is based on the consideration that Nigeria's record in achieving the MDGs has been significantly low and the assessment of current SDGs' level of success or achievements will continually depend on the harmonization and utilization of a whole lot of tremendous data both spatial and non-spatial. As such, it requires the development of a platform like the geospatial data infrastructure. The next section reviews the existing knowledge and discusses the interconnection between the SDGs and geospatial information, taking into consideration: geospatial information, urban and rural resilience; approaches to integrating geospatial information and technologies in the implementation of the SDGs. Considering that all developments have spatial dimension, that is location-based, the key information required is geospatial information ([19], p. 5).

19.2 Existing Knowledge About Interconnection Within SDGs and Between Geospatial Information

This section provides a review of interdisciplinary analysis and multisectoral expertise on the usefulness and application of geospatial information. It reviews the interconnection between the SDGs and geospatial information while drawing inferences as regards the implications for urban and rural resilience. As noted by International Council of Science (2017), the underlying principle is that SDG goals interact with one another [3]. However, the "approaches for how to more systematically identify, characterize and address interactions between all sustainable development policy issues remains a challenge" ([9], p. 1499; [8]).

19.2.1 The Interconnection Between the SDGs, Geospatial Information, Urban and Rural Resilience

As noted by Nilsson, *et al.* (2018), understanding both the negative and positive interactions between the SDGs is essential for decision-making that promotes the implementation of sustainable development[8]. In another words, the interactions across the SDGs should be thought through systematically. Nilsson, *et al.* (2018) provide insights into the mapping and assessing of SDGs interactions, using a defined typology and characterization approach, and summed that negative interactions are outnumbered by positive ones. This suggests that a more integrated policy making has the potential to provide more effective development outcomes. Nilsson, *et al.* (2018) further observed the challenges in identifying and assessing all the key interactions comprehensively at the global scale. They, therefore, argued for context-specific understandings. This argument is consistent with the overarching *premise of the 2030 Agenda that underscores a nationally adapted interpretations and action on the SDGs* ([8], p. 1499).

An essential component of determining integration is the consideration for the contextual meaning of the SDGs targets. There are nuance interpretations of the expectations as to the progress expected from each target. This simply means that, before assessing interactions, one needs to articulate what should be looked out for to articulate progress on a target, especially from the (subnational or national) context of implementation in terms of actual, observable outcomes ([8], p. 1499).

While it is essential to determine the linkages between different targets set for the SDGs, it is also important to consider that understanding the interdependent existence of people and the space they inhabit, is largely determined by the availability of geospatial data. One could infer, in this regard, that the level of connectivity at local, national, and global levels shapes future urban forms. Especially now that there is a noticeable intensification of challenges in our cities due largely to: increase in population growth and human mobility with majority of the world population now living in cities [4]; significant disasters that are weather related traceable to climate impact; limited resources to cope with the unintended consequences of population growth and related disasters. Significant part of this is linked to the lack of a comprehensive spatial data infrastructure, that has been argued to impede the process of strengthening community and infrastructure resilience, thus, preventing the protection of social and environmental sustainability, and narrowing the development gap, especially in the developing countries like Africa.

However, one of the key findings is that there is clearly no one-size-fits-all approach to understanding target interactions and infrastructure resilience. Noting that building on the existing knowledge will require a commitment to continuous iteration and improvement.

19.2.2 Geospatial Information to Support Inclusive Urbanisation, Resilient Development, and the SDGs

The logical question to ask is that how does geospatial information supports and informs inclusive and even urbanisation, resilient development, and the SDGs? Earlier, Feeney *et al.* (2001) linked the increasing need to organize data across sectors and institutions, through the development of SDIs, to the growing need of addressing the complex and multiple challenges associated with sustainable development [2]. The primary aim for developing SDIs is to achieve better outcomes from spatially related development decision-making across economic, social and environmental spheres.

Despite the widespread adoption of digital technology and a high internet penetration rate (highest in Africa), existing databases are isolated, uncoordinated and lack harmony for a holistic policy framework for decision making. The isolation of digital information gathered by various agencies and department on individuals, spatial components and events provides a veritable start point to implement geospatial information framework for a localized connectivity of the global SDG agenda. The domestication of the global agenda is, therefore, necessary to set the stage for effective implementation and monitoring.

As noted by Scott & Rajabifard (2017), the challenges of achieving sustainable development is not only about some sets of significant social, economic and environmental issues that are almost entirely geographic in nature [15]. Rather, it could also be noted that, geospatial information can provide a set of science and time-based monitoring solutions to these challenges, especially those that are driven by spatially enabled data.

Pesch (2014), posited that irrespective of rational interactions and connections over a long period of time, the reality is that there has been a limited connection and integration between sustainable development and geography, geospatial information and associated enabling infrastructure such as National Spatial Data Infrastructures (NSDIs) at both the technical and political levels [11]. This is not peculiar to developing nations but also highly data-rich and technology-driven nations. Wu Hongbo, UN Under-Secretary-General for Economic and Social Affairs, emphasized the role of geo-statistical data in improving governments' ability to 'examine, monitor, manage, propose and predict development and growth options for a sustainable future'. Wu also stressed the importance of geospatial information in decision making, policy formulation, measuring and monitoring development, including the post-2015 agenda.

As an all-encompassing, comprehensive global blueprint, the applicability of the 2030 Agenda in all countries, in all contexts, and at all times earns it the *universality* attribute as one of the core underpinning principles. Other principles include *Leaving no one behind, Interconnectedness and Indivisibility, Inclusiveness and Multi-Stakeholder Partnerships.* The geospatial community is strategically positioned to integrate geospatial information into the global development agenda, specifically in a way that will support measuring and monitoring the targets and indicators of the SDGs, with the core principles at the heart of it.

19.2.3 Approaches to Integrating Geospatial Information and Technologies in the Implementation of the SDGs

This section assesses the existing framework for a more holistic approach to integrating geospatial information and technologies in the implementation of the SDGs, by first reviewing the challenges for achieving the SDGs. Scott and Rajabifard (2017), out of serious concern for the attainment of SDGs, raised a fundamental question: How can geospatial information be implemented and integrated into national information systems, at a policy level, in order to contribute more holistically to measuring and monitoring the targets and indicators of the SDGs at a technical level? ([15], p. 60). This is, perhaps, against the realisation that while the evolution of sustainable development and the development of geospatial information have progressed in parallel period, frameworks for their integration have remained largely undeveloped even in developed countries [15]. In this regard, Pesch (2014) earlier notes that for a very long time, even in a highly data-rich and technology-driven global countries, there has been very little connection between sustainable development and geospatial information at either the political or the technical arena [11]. For example, while global leaders have committed much effort in developing targets and indicators to benchmark progress, there has been little understanding about the strategic direction for the integration of geospatial data information for efficient monitoring the implementation and achievements of the goals [15].

It is even paradoxical that while the United Nations - the proponent of SDGs - report on the '*Future We Want*' acknowledged the value of reliable geospatial data for sustainable development [6], the report failed to clearly demonstrate strategies to mainstream geospatial data infrastructure into sustainable development [15]. For an effective integration, a clear integrated strategic direction, which takes into consideration national realities and regional peculiarities, is required. The challenges of developing such an integrated approach - a fit-for-purpose geospatial data framework - is proving difficult for many decision makers around the globe. This is, however, acute in developing countries, especially the African countries.

In recent times, however, with much international advocacy and dialogues coupled with the need for baseline data to monitor and evaluate progress towards sustainable development goals, the effort for integration is gaining momentum. The research efforts of the global geospatial community have provided useful frameworks, at both the policy and practical levels, to facilitate the integration of geospatial information and technologies in the implementation of the SDGs.

Scott and Rajabifard (2017); UN-Habitat (2016) works on sustainable development and geospatial information offer a useful strategic framework for integration which could serve as building blocks for implementation within the national geospatial strategic framework [15, 4].

19.3 Framework and Methods

This section adopts conceptual and empirical approaches for understanding contextual interactions between the SDGs targets, drawing on SDGs interactions framework as developed by Nilsson et al. (2018) [8]. The interactions depend on the meaning and the transparency of the assumptions associated with the interactions. One of the major benefits of this approach is the ability to critically and systematically navigate the several dimensions of the 2030 agenda, with particular focus on the contextual meaning of the targets. This is important to be able to establish the interactions between the targets. It should also be noted that policy and/or regulatory mandates also have the capability to affect the nature of interactions.

The central consideration for the framework revolves around the typology and scoring of interactions on a 7-point scale to identify causal and functional relations as it relates to the achievement of the sustainable development goals and targets [4]. In addition, key contextual determinants that impact on the interaction are governance and geographical contexts, implementation technologies, policies and time-horizon. This chapter adopts this interaction framework and emphasizes on: governance context, geographical context, spatial data infrastructure as embedded in implementing technology, and the time dimension.

With regards to the governance context, the assessment of the SDGs and targets are critically dependent on good governance. Inappropriate governance measure can potentially impact interactions to the extent that positive interactions can be reversed and turned into negative one. Regarding the geographical context, interactions is reinforced depending on where such interactions take place. Especially where cross-scale and cross-geographical interactions occurs.

Equally important is the '*implementing technology*' that focuses on geospatial data /information. This framework has necessitated a need to set up and monitor policy level mechanism, leading to a harmonious integration of Geospatial and Statistical information for sustainable development in Africa [19]. Interaction is also impacted within the consideration of the time frame for the assessment. Therefore, articulating the trade-offs, synergies and spin-offs between the goals of the SDGs is important to unlocking their full potential. Thus, whatever the scale, it is an important consideration that progress made in some areas is not made at the expense of progress in others.

To expand the spatial data infrastructure as embedded in the '*implementing technology*', it will be important to consider the 5ps Model. This involves the categorisation of SDG goals based on the five critical dimensions of: *people, prosperity, planet, partnership* and *peace.* As such, the geospatial information efforts shall, on the basis of defined spatial limitations or extent (LGAs, States, Geopolitical zones or Geographic regions) be gathered on the basis of the People's demographic characteristics, their economic and industrial prosperity, the state and manner of planet resource utilization, fostered partnership among stakeholders and entrenched peaceful co-existence in the society. The categorisation of the goals based on the 5ps model include: People (goals 1,2,4); Prosperity (goals 3,6,7,8,9, 11); Planet (goals 13, 14, 15); Partnership (goals 17, 12); Peace (goals 6, 10, 16). The next section is structured through this framework and model.

19.4 Findings and Analysis

19.4.1 The Interactions Between Various Elements of GIM and the SDGs in African Context

Going by the mission statement of GI4SD, it is anticipated that "*Africa produces and uses authoritative and evidence-based Geospatial Information for the attainment of its sustainable development goals and agenda 2063 objectives*". The statement went further to provide insight into what is meant by authoritative and evidence-based GI as referring "*to rigorously controlled best quality and "official" - consensus-based - GI, and its attribute of objective, logically-led and uncertainty-free or reduced source of decision making*" ([19], p. 8). However, the reality is different from this expectation. There are few challenges preventing this from happening as anticipated.

The challenge is not only limited to the availability of reliable sources of GI, but also on 'access, quality, completeness, currency, availability of standardized metadata, interoperability of GI datasets, traceability of GI products, rights of data producers, liability of GI service providers, GI products and services pricing' ([8], p. 1490). In addition, it is also about systematically focusing on the means of implementation considering issues like finance, technology, capacity building, trade, policy coherence, partnerships, data, monitoring and accountability.

As documented by UNGGIM (2016), "*efforts to build capacity in GIM in Africa over the past 20 years have been supply driven and have typically reflected the mandates of mostly external actors. Local, national and regional applications of GIM have continued to expand in scope and relevance, but without a strong demand-driven agenda for building capacity in GIM. The outcomes of such efforts will continue to fall short of their true potential*" ([19], p. 37).

19.4.2 Geospatial Information: Strengthening Community, Infrastructure, and Institutional Resilience

The role of geospatial information is twofold, linking the 'where' component of SDGs and make challenges in various locations more visible and assisting with spatially tracking progress. The areas of monitoring and review, focusing on high quality, timely, reliable, and disaggregated data, including earth observation and geospatial information, was captured in the UN 2015 General Assembly text:

> *We will support developing countries, particularly African countries, LDCs (least developed countries), SIDS (small island developing States) and LLDCs (land-locked developing countries), in strengthening the capacity of national statistical offices and data systems to*

> *ensure access to high quality, timely, reliable and disaggregated data. We will promote transparent and accountable scaling-up of appropriate public-private cooperation to exploit the contribution to be made by a wide range of data, including earth observations and geospatial information, while ensuring national ownership in supporting and tracking progress* [7].

It is very essential to integrate information systems at a national level that flow up into a regional and global level. The framework is a national bottom-up approach. In developing countries, the use of data construct framework depends on institutional and architectural arrangements. As the world increasingly moves to rich data paradigm turning data into valuable information to support decision making, regarding development challenges, also requires change. Achieving SDGs requires the use of geospatial information to overcome challenges such as land rights, food production, disaster risk reduction, safe human settlements, and other social, economic, and environmental issues at local, national, and global levels.

As Nigeria reaches for the top global positions in the urbanization ranking in a couple of decades, the myriad of environmental, social, economic, socio-cultural and infrastructural challenges consequent upon this growth trend widens the risk and vulnerability factors and stretches the resilient limits of the cities, thereby posing enormous threats to the actualization of the SDGs. The geographical complexities and the demographic dynamics/socio-cultural diversities of Nigeria portends a huge challenge for fostering implementation of the goals, and therefore requires a more unifying, inclusive and localized approach.

19.4.3 The Role of Geospatial Data Infrastructures and Services in Achieving the SDGs in African Context

Many of the challenges associated with sustainable development can be analyzed, modeled, and mapped within a geographic context [17]. However, while many of the challenges have spatial dimension, at the development policy making level, not much is understood concerning the role of spatial attributes in sustainable development processes [12, 15]. In this regard, one of the most important questions in development community today, which requires evidence, is: "how can geospatial data infrastructures and services enhance the attainment of SDGs?"

In response to this question, the global geospatial community has focused discussion on the role and value of geospatial data for governance and development [12]. This research effort received a major boost, in recent years, with the global adoption of the Sustainable Development Goals and, coincidentally, considerable advancement in the level of awareness of geospatial technologies [15].

Sustainable Development Goals, in its conceptualisation, recognise the complex, interdependent, integrated and indivisible nature of the physical, economic and social systems, and the diverse associated challenges. Its 17 goals and 169 targets, though global, also recognise the need for a diverse range of quality, accessible, timely and reliable disaggregated data for monitoring and evaluating the implementation and achievements at national and regional levels.

The interconnected nature of SDGs and its multisectoral and multilevel implementation approach, for developments that leave nobody or situation behind, call for a holistic approach that integrates spatially enabled data platform into the national development policy framework. Geospatial Data Infrastructure has, therefore, emerged as a valuable platform that enhances access and sharing of geospatial information across sectors and inter-agencies as well as integration for evidence-based decision making and sustainable policy formulation [15]. In essence, geospatial data infrastructure can provide enabling and coherent capability and the needed unifying platform for multisectoral and inter-regional collaboration, consensus and evidence-based decision-making [15]. As argued by Feeney *et al.* (2001), the primary objectives of geospatial data infrastructure is to provide a platform for data coordination across disciplines and institutions for a better development outcome across space and time [2]. Scott and Rajabifard (2017) expressed similar view that the need for geospatial data infrastructure is essentially for achieving 'better outcomes from spatially related economic, social and environmental decision-making' ([15], p. 64).

Geospatial Data Infrastructure is, therefore, increasingly being embraced globally, though with regional variations, as the world gradually coming to terms with the need for an integrative framework for evaluation and monitoring development progress. For instance, in Europe, while policy actors are conscious of the need to standardize geospatial infrastructure in order to enhance data quality, many of the African countries are facing data fragmentation [12]. While it may not be totally accurate to argue that developing countries are data poor, lack of platform to bringing together the existing fragmented data might be a more possible argument. Geospatial data infrastructure offers such a unique opportunity to overcome this challenge, as it provides holistic and sustainable platform to bridge gaps between data, data providers and data users, as well as time and space, thereby enhancing the visibility of data to support policy and development decisions.

Geospatial data infrastructure progressively became one of the valuable components of the infrastructure required for socioeconomic prosperity, ecological management and liveability across levels of human settlements. The United Nations report on the establishment of UN-GGIM stressed the role of geospatial data infrastructure in shaping the formation and implementation of sustainable development programmes and polices [16]. The report of the United Nations Committee of Experts on Global Geospatial Information Management, at its second annual session, aptly captured the many roles geospatial data infrastructure can play in monitoring and evaluating the implementation and outcomes of sustainable development across sectors. It stated that:

> *Perhaps most importantly, there is a strong belief that geography provides the integrative framework necessary to support the requirements of multiple information communities in a timely and effective manner––providing the right data at the right time to the right place. The same geospatial content, repurposed, can support applications ranging from agricultural management, to emergency planning and response, to scientific collaboration on climate change, to transportation planning. All of these applications have implications for sustainable development and liveability* [17].

The absence of National Geospatial Data Infrastructure platform means that governments, at all levels, will rely on unstandardized and fragmented data for decision making. This is the situation in the developing countries, especially African countries, where governments are continually challenged with lack of timely data compounded with poor data quality and a general lack of interoperability between different sources of data [15, 18]. One of the immediate consequences of this is that development decisions are based on inaccurate data. Data have the power to mislead or inform development and policy making. In any context, the capability to implement either national standards or globally agreed goals, such as SDGs, largely depends on the quality of available resources and data. For many reasons, the current data use for development and policy making in many African countries provide little guide for effective development. In general, institutional and structural characteristics of many Africa nations pose huge barriers in collecting and evaluating data to implement and monitor both local and global development goals.

Data essentially influence what is known about the state of development and subsequently shape decision making process. The United Nations report on the achievements of MDGs provided some reflections on the challenges and lessons learned from the implementation of MDGs. It recognises the global achievements of MDGs. It however acknowledges that reliable and timely data in appropriate format, an essential component of any development programme, to effectively prioritise policy agenda and monitor progress were inadequate in the implementation of MDGs. This is particularly acute in Africa. Commenting on these reflections, Scott and Rajabifard (2017) concluded that the implication was that MDGs were largely implemented in many developing regions, especially Africa, without reliable data or a sustainable data platform to aid consistent measure and monitor of implementation progress towards sustainable development [15].

Essential for global agenda, such as SDGs, that seeks to address complex and multiple challenges, is the need to have an adequate understanding of the interrelationship of the challenges in relation to space, time and people.

Equally important is the ability to monitor trends of events, provide timely information, particularly to the population at risk, and prioritise responses and actions. Geospatial data infrastructure can help in these regards, as asserts by United Nations (2015) [7]:

> *Knowing where people and things are and their relationship to each other is essential for informed decision-making. Comprehensive location-based information is helping governments to develop strategic priorities, make decisions, and measure and monitor outcomes. Once the geospatial data are created, they can be used many times to support a multiplicity of applications.*

One of such areas of applications is the possibility to link ecological and socioeconomic data in a way that clearly presents interconnections across the spheres of sustainable development - environment, economic and social - and how they influence one another. Geospatial data infrastructure provides such a unique opportunity to integrate geospatial data into national development framework in a more holistic and sustainable way. Considering the peculiarity of each geographical region and location in terms of environmental configuration and level of development, geospatial data, with standardised indicators as the object of measurement, provide necessary transparency and accountability for development governance and evidenced for policy making. Meeting the numerous goals of sustainable development requires the integration of geospatial data infrastructure platform.

Geospatial data infrastructure is essential for enhancing the political and social engagement of hitherto marginalised people and to shape policy and development outcomes through evidence. Though well-intended, it is proving difficult to be integrated into development framework in many developing countries. Geospatial Data Infrastructure, as an evolving platform, is much more than just data. Achieving SDGs requires conscious and evidence based spatial and socioeconomic decision making. SDI will play a pivotal role in enhancing the efficiency and adequacy of such decision. As argued earlier, the challenge is not limited to the availability of reliable sources of GI, but also on unrestricted access, completeness, currency, quality, availability of standardized metadata, provenance, interoperability of GI datasets, GI products and services pricing.

19.5 Inferences, Future Direction and Conclusion

As noted by UN-GGIM (2016), to achieve the UN-SDGs and AU Agenda 2063 targets, at national, sub-regional and regional levels in Africa, will require good governance and sound policies in Geospatial Information Management

([19], p. 24). As discussed in the preceding sections, these will guide the way African countries, will get organised and operate in order to achieve maximum benefits of GIM efforts. In addition, leveraging the already initiatives such as UN-GGIM, UN-Expert Group on Land Administration and Management, UN-GGIM Private Sector Network and UN-GGIM Geospatial Societies will facilitate the harmonisation and standardisation of data and integration of multi-domain analytics.

Griggs *et al.* (2017), assert that "*the process of systematically identifying and scoring interactions across the 17 SDGS using a common terminology is very valuable*" ([3], p. 8). Consistent with this position, one could argue in support of Griggs *et al.* (2017) that the process allows broad multi-disciplinary and multisectoral conversations [3]. It also allows synthesis and scoping of knowledge needs while providing rational and concrete clustering of targets that need to be addressed together to allow integrated approach for implementation and monitoring. The major limitations, however, is the challenge relating to selecting the important interactions from all the possible alternatives, especially considering the different expert's characterisation of interaction. That is, the contextual meaning of each of the possible interactions.

In conclusion, geospatial information policy is required to effectively manage geospatial information for sustainable development. It should be seen as a compulsory requirement that is anchored on legal and coherent institutional environment, that is set to achieving the most cost-effective and fulfilling impact. Therefore, GI applications should not only be encouraged, but should be layered on political and institutional structure designs to strengthen transparency in governance. A strong political will, built on strong GI policy has capacity to produce good governance that respect objective, fair and equity-driven decision making.

Bibliography

[1] Daniel Lage Chang, Jamile Sabatini-Marques, Eduardo Moreira Da Costa, Paulo Mauricio Selig, and Tan Yigitcanlar. Knowledge-based, smart and sustainable cities: a provocation for a conceptual framework. *Journal of Open Innovation: Technology, Market, and Complexity*, 4(1):5, 2018.

[2] M Feeney, Abbas Rajabifard, and Ian P Williamson. Spatial data infrastructure frameworks to support decision-making for sustainable development. 2001.

[3] DJ Griggs, Mans Nilsson, A Stevance, David McCollum, et al. *A guide to SDG interactions: from science to implementation*. International Council for Science, Paris, 2017.

[4] UN Habitat. Urbanization and development: emerging futures. *World cities report*, 3(4):4–51, 2016.

[5] Nigeria MDGs. Millennium development goals end point report, 2015, 2015.

[6] United Nations. The future we want. *United Nations General Assembly. Resolution 66/288*, 2012.

[7] United Nations. Transforming our world: the 2030 agenda for sustainable development. *United Nations General Assembly. Resolution 70/1*, 2015.

[8] Måns Nilsson, Elinor Chisholm, David Griggs, Philippa Howden-Chapman, David McCollum, Peter Messerli, Barbara Neumann, Anne-Sophie Stevance, Martin Visbeck, and Mark Stafford-Smith. Mapping interactions between the sustainable development goals: lessons learned and ways forward. *Sustainability science*, 13(6):1489–1503, 2018.

[9] Måns Nilsson, Dave Griggs, and Martin Visbeck. Policy: map the interactions between sustainable development goals. *Nature News*, 534(7607):320, 2016.

[10] Poor Numbers. How we are misled by african development statistics and what to do about it, 2013.

[11] Udo Pesch. Sustainable development and institutional boundaries. *Journal of Integrative Environmental Sciences*, 11(1):39–54, 2014.

[12] Bhanu Rekha. Geospatial data: Key to achieve sdgs, 2016. Geospatial World.

[13] Roger Riddell. *Manufacturing Africa: performance & prospects of seven countries in sub-Saharan Africa*. James Currey Publishers, 1990.

[14] Michael Rigby and Serryn Eagleson. Spatial data infrastructure challenges for addressing the sustainable development goals in australia. 09 2018.

[15] Greg Scott and Abbas Rajabifard. Sustainable development and geospatial information: a strategic framework for integrating a global policy agenda into national geospatial capabilities. *Geo-spatial Information Science*, 20(2):59–76, 2017.

[16] UN-GGIM. Report of the secretary general on global geospatial information management. *United Nations General Assembly E/2011/89*, 2011.

[17] UN-GGIM. Monitoring sustainable development: Contribution of geospatial information to the rio+20 processes, 2012.

[18] UN-GGIM. Statement of shared guiding principles for geospatial information management, 2015.

[19] UN-GGIM. Geospatial information for sustainable development in africa (gi4sd) african action plan on global geospatial information management (aap-ggim) 2016-2030. *United Nations Economic and Social Council, E/ECA/GGIM/16/1*, 2016.

20

Openness and Community Geospatial Science for Monitoring SDGs – An Example From Tanzania

Maria Antonia Brovelli

Politecnico di Milano, Italy

Maria Ilie Codrina

Technical University of Civil Engineering of Bucharest, Romania

Serena Coetzee

University of Pretoria, South Africa

This chapter focuses on two aspects that enable the monitoring of the United Nations (UN) sustainable development goals (SDGs): openness and community geospatial science.

20.1 Introduction

Openness typically refers to transparency, to free and unrestricted access to information, and to inclusive consensus-based decision-making. Community science is a branch of citizen science that involves a commitment from citizens, not only to collecting data, but also to designing and planning project activities in a more egalitarian (if not bottom-up) approach between (professional) scientists and citizen scientists. When the focus of the research is geospatial, we are dealing with community geospatial science. Examples of this approach are some of the projects related to OpenStreetMap (OSM), where citizens are significantly more than just "active sensors", playing an instrumental role in the definition and shaping of the campaign [5].

Based on the two pillars of openness and community geospatial science, a relevant example for monitoring SDGs was developed for the OSGeo UN

Committee Educational Challenge [22]. In 2018, the Committee sent out a call for the development of educational material, comprising three challenges. The first two challenges were closely related to UN operations. The third challenge was aimed at addressing the current lack of training material for using open source software together with freely available high-resolution global geospatial datasets for environmental, social and economic analysis in support of UN SDGs. The hands-on training material was conceived in such a way to be available and replicable anywhere in the world. These characteristics are especially relevant in developing countries where data is often scarce and resources for buying software are limited. Winners were guided by mentors to ensure that the material met the requirements of the target audience. In this chapter, the material developed for the third challenge is described.

The training material refers to SDG 9 (Build resilient infrastructure, promote inclusive and sustainable industrialization and foster innovation), and more specifically to the computation of indicator 9.1.1 (Proportion of the rural population who live within 2km of an all-season road) for a rural area of Tanzania. The methodology and technical tools (data and software) in this use case are presented in detail, so that anybody can replicate the method for other parts of the world. Since it is based on open data and open software, the method is cost effective and completely sustainable. Moreover, the possible lack of data can be overcome by actively involving people and communities in mapping their specific region of interest with freely available tools.

In the remainder of the chapter we present concepts and examples of open data and open software, followed by the concept of community geospatial science with reference to citizen science and volunteered geographic information (VGI). Next, we describe the use case from the training material and then conclude.

20.2 Open Data and Open Software

The principles of openness and transparency are widely advocated these days - as in open data, open software, open knowledge and open government - but what do they really mean?

The architects of the twenty-first century digital age proclaim that openness is their foundational value. The technological foundations that sustain this vision of openness are digital: the internet, mobile telephony and distributed systems. According to Russell (2014), 'openness' is a "marriage of technology and ideology and a fusion of technology, democracy, and entrepreneurial capitalism" [19]. The work described in this chapter exemplifies open principles: for the use case, open data is collected and analysed with open source software, and the training material is made available as an open educational resource.

Open source software has its origins in the early days of computing when programming problems were solved through scientific collaboration. Software was shared and each programmer added a new aspect to existing knowledge [4]. It evolved into a software development and licensing approach that ensures transparency through access to the source code and collaboration through a set of rights that protect the copyright to the source code. Through the free redistribution of the software and works derived from it, it is possible to create software products based on each other's work [18].

The interesting point, especially from the point of view of the poorest countries with limited resources for technologies, is that there is at least one mature sophisticated open source product for every geo-technology area and geospatial information need and application - from data collection in the field, crowdsourcing, desktop applications, spatial extensions to database management systems and software stacks. Together, they can be used to create sophisticated free and open Web and cloud-based systems [12, 21]. In developing countries, economic motivations rank high when choosing to use free and open source geospatial software [3].

OSGeo (Open Source Geospatial Foundation) is a not-for-profit organization aimed at fostering the global adoption of open geospatial technology. According to OSGeo, open source starts off with a license that provides royalty free (re)use of software. Next, open source guarantees access to the source code for audit and modification and the ability to redistribute the software with no additional costs[1]. A wide range of open licenses are in use. Creative Commons licenses [2], a set of well-defined licenses that each describe a different permitted use of copyrighted materials by the public at large are widely used for content. The training material described in this chapter is made available under one of these Creative Commons licenses. For source code, OSI-approved licenses, such as the GNU General Public License (GPL), are frequently used [3].

Besides software, open data or knowledge is based on the principle that some information should be shared and available to anyone, without any restrictions to rights of access or use. According to Open Knowledge International, open data and content can be freely used, modified, and shared by anyone for any purpose [17]. Generally, transparency and collaboration are well aligned with the principles that democratic governments stand for and with the principles embodied in the Charter of the United Nations. Organizations, such as Open Knowledge International, promote the use of open data and knowledge, e.g. to support citizens in taking action on social problems. Monitoring SDGs with open data makes it possible for citizens to track the status of SDGs in their countries or cities. It empowers them with an understanding of the challenges at hand so that they can work on addressing these, for example, by supporting or lobbying for appropriate initiatives or by holding governments and other actors to account. OpenStreetMap, started in

[1] www.osgeo.org
[2] https://creativecommons.org/licenses/
[3] https://opensource.org/licenses

the UK in 2004, is the most widely known example of global open geospatial data. The success of its open characteristics is further described in the next section.

Through open education, barriers to education are removed by making educational resources freely available for anyone to study and use or by eliminating admission requirements. Through GeoForAll, OSGeo promotes open education among its members based on the belief that knowledge is a public good and that open principles in education provide opportunities for everyone. Teachers and educators provide open access to their educational resources for teaching related to geo-technologies and geospatial data. Access to education is a challenge in the poorest countries, confirmed by the UN SDG 4 on Education. Open and freely available educational resources contribute to addressing this challenge.

The UN OSGeo Committee works towards identifying and developing open source geospatial software and services that meet the requirements of UN operations, taking full advantage of the expertise of mission partners, including partner nations, technology developed by contributing countries, international organizations, academia, NGOs, and the private sector [4]. The 2018 Challenges were aimed at supporting these goals.

20.3 Community Geospatial Science

Public participation in scientific achievements has a long history but the last few decades have seen more attention and an impressive increase in the number of people involved. Citizen science, i.e. scientific research conducted, in whole or in part, by amateur (or non-professional) scientists, the term used for denoting such an approach, is a diverse practice, encompassing various forms, depths, and aims of collaboration between scientists and citizen researchers and a broad range of scientific disciplines [8].

Different classifications of citizen science projects exist based on the degree of influence and the extent of the contributions by citizens. Haklay *et al.* (2018) distinguish three kinds of citizen science projects [9]:

1. Long-running citizen science, the traditional projects, similar to those run in the past [11, 1].
2. Citizen cyberscience, strictly connected with the use of technologies [6], and which can be subclassified into:
   (a) volunteer computing, where citizens offer the unused computing resources of their computers;

[4] https://wiki.osgeo.org/wiki/UnitedNations_Committee

   (b) volunteer thinking, where citizens offer their cognitive abilities for performing tasks difficult for machines; and
   (c) passive sensing, where citizens use the sensors integrated into mobile computing devices to carry out automatic sensing tasks.

3. Community science, involving a more significant commitment from citizens, also in designing and planning the project activities in a more egalitarian (if not bottom-up) approach between scientists and citizen scientists [10, 13, 2]. Community science is further classified into:
   (a) participatory sensing, where citizens use the sensors integrated into mobile computing devices to carry out sensing tasks;
   (b) Do-It-Yourself (DIY) science, which implies that participants create scientific tools and methodologies for carrying out their research; and
   (c) civic science, which is "explicitly linked to community goals and questions the state of things" [9].

Because of the bottom-up approach, community science is the most interesting when it gets to activities and programs that are closely related to the life and well-being of people. If a geospatial dimension is involved, i.e. location plays a fundamental role in interpreting the phenomena under study, we can refer to this as community *geospatial* science.

The best example of community geospatial science is OpenStreetMap (OSM) [5]. Many people consider it to be an object (i.e. a map or its modern version, a geodatabase): "a free, editable map of the whole world that is being built by volunteers largely from scratch and released with an open-content license" [5]. It is also commonly referred to as a geo-platform or project where as many as 5 million users contribute, edit, download and assess the data that is shared. OpenStreetMap is most of all a community of communities [20], in the sense that the OpenStreetMap community is diverse and incorporates the motivations of many different groups, depending on how they approach their volunteer activity.

Examples of communities are the community "dedicated to humanitarian action and community development through open mapping", the Humanitarian OpenStreetMap Team (HOT) [6]; the community that is "putting the world's vulnerable people on the map", Missing Maps [7]; the community that works to close the gender gap in OSM, GeoChicas[8]; the community of university students, YouthMappers [9]; the community that helps end female genital mutilation and aids community development in rural Tanzania, Crowd2Map[10].

[5] https://www.openstreetmap.org
[6] https://www.hotosm.org/
[7] https://www.missingmaps.org/
[8] https://wiki.openstreetmap.org/wiki/GeoChicas
[9] https://www.youthmappers.org/
[10] https://crowd2map.wordpress.com/

Starting from the community level, OSM has been able to attract the attention of institutional entities, ranging from small local ones (such as the local civil protection agencies) to national mapping agencies and international organisations, like the World Bank and the UN (see for example the Open Cities Africa project[11]) [7]. Universities are often involved in community geospatial science, even if their role is that of co-creators of solutions together with the other involved actors.

Apart from the social ecosystem that has been established in this global community geospatial science experience, the technological ecosystem that has been established is worth mentioning. One of the main reasons for the success of OSM is that the technology behind the project allows everybody to contribute, independently of their level of expertise. The tools and systems, developed by different actors in the social ecosystem of OSM (volunteers; small and medium companies; universities; local, national or international agencies) are generally characterized by being free and open source, i.e. they can be passed on for further development by other people in the community; and by the different applications very often accessible simply through the personal account on the OSM platform.

Apart from facilitating contributions by individuals, the OSM ecosystem is designed to elicit and simplify collaboration. One fundamental tool (the Tasking Manager) for instance allows the subdivision of large areas to be mapped into a set of smaller ones, each of a size that can be mapped by an individual. This facilitates collaboration among mappers and avoids problems of overlap and confusion. Moreover, this tool allows the validation of the mapped data, so that a quality assessment of the mapped area is possible.

Citizen science and community geospatial science represent a new step in the history of science and these examples, like OSM and its communities, are relevant cases of what can be done within the new paradigm of collaboration and openness.

20.4 The Use Case and Training Material

The UN 2030 Agenda defines the challenges to be overcome in order to achieve prosperity for all in a sustainable manner for the entire planet up to the year 2030 [14]. These challenges are embodied in the 17 SDGs that are broadly interdependent and address all aspects, from poverty to peace and justice, from environmental protection to human health, from food security to gender equality. To track progress towards achieving the 2030 Agenda, it is essential to understand the attainment level for each of the 17 SDGs. Progress can be

[11] https://opencitiesproject.org/

quantified by measuring, gathering data and calculating the indicators that define each goal using a consolidated methodology [15].

As the 2030 Agenda fulfilment has a worldwide scope, the monitoring and reporting activities must be done globally. Yet, the quantity, quality, precision and consistency of the necessary data vary significantly across the globe and so do the resources to gather it. However, in the last two decades, we have witnessed a significant increase in the availability of open data, from open public data to citizen science data, and in the case of geoscience, satellite imagery and related products [12] .In this context and within the framework of the third OSGeo UN Committee Educational Challenge related to open geospatial data and software for UN SDGs, we developed a replicable use case to demonstrate that open geospatial data, which very often are contributed by citizens and communities, are available globally; that free and open source solutions for geospatial have sufficiently developed to conduct a global geospatial analysis at small and intermediate scales; and that these data and software can be used to monitor a geospatial SDG indicator.

The selection of the indicator has been done following these guidelines:

1. to have a spatial dimension;

2. to avoid indicators that are already addressed through an advanced initiative, such as the GEO Wetlands Initiative[13], WHO Interactive Air Pollution Maps[14], GEO AquaWatch[15] or ESA CoastColour[16].

Indicator 9.1.1, Proportion of the rural population who live within 2 km of an all-season road (C0901010), was selected, which supports the target of developing quality, reliable, sustainable and resilient infrastructure, including regional and transborder infrastructure, to support economic development and human well-being, with a focus on affordable and equitable access for all. In December 2018, the UN published an updated but still in progress work on the metadata for indicator 9.1.1. that includes clear definitions for the terminology, methodology, data sources and their availability [16]. A significant limitation is highlighted: "*The Indicator relies substantially on data collected by road agencies and national statistics offices for their operational work. As such, its update is dependent on the frequency of update of the road condition surveys and national census.*" ([16], subchapter Methodology, paragraph Comments and limitations). This is exactly the type of limitation we aim to address with our use case. In comparison to data collected by a multitude of national agencies, globally produced datasets have the advantage of worldwide coverage and a coherent and consistent technical and logical structure. Yet, both kinds of dataset can have a similar scope. Even though the open dataset is the result

[12] https://wiki.osgeo.org/wiki/Training_Material_for_UN_Open_GIS_OpenData
[13] Retrieved from https://www.earthobservations.org/activity.php?id=122 on 3.02.2019
[14] Retrieved from http://maps.who.int/airpollution/ on 3.02.2019
[15] Retrieved from https://www.geoaquawatch.org/ on 3.02.2019
[16] Retrieved from http://www.coastcolour.org/ on 3.02.2019

of a collaborative community effort, its structure and semantics are described and allow a clear understanding of it. Thus, reliable cleaning of the data can be done, if necessary.

For the present use case, we selected Tabora Region, one of the 31st administrative regions in the central-western part of Tanzania. However, the same use case is replicable for any other region in the world, provided the data are available. This is possible because we based the computation only on open global datasets. In Table 20.1, we show the geospatial information required to calculate the Rural Access Index (RAI) for a region, together with the open global datasets in our use case.

The most challenging geospatial information to obtain is the road network and the road condition. At the moment, the only global dataset that could provide such information is OpenStreetMap. The amount and quality of OpenStreetMap data for various regions around the world can vary significantly. As our region of interest is located in Africa, specific OpenStreetMap developments for Africa must be taken into consideration, namely the Highway Tag Africa Typology of Road Network in African countries[17], a roads classification designed for the context of African countries, and the East Africa Tagging Guidelines[18], which provide guidance for tagging roads in Kenya, Tanzania, Uganda, Rwanda, Burundi, and South Sudan. Even though the road classification is specific to Africa, the clear and consistent definition of each element[19] and tag[20] in OpenStreetMap makes this use case reproducible in any other part of the world. With this in mind, apart from describing the processing steps for the Tabora Region RAI calculation in detail, we present the conceptual workflow for calculating the RAI in Figure 20.1.

The workflow consists of three main stages: (1) preparation of the geospatial data; (2) calculation of the RAI; and (3) presentation of the results. The first stage is the most time consuming as preparing the data implies a thorough analysis, ranging from the structure (i.e. format) to aspects related to consistency, precision, scale etc., and subsequent cleaning of the dataset. The outcome of the first stage is a dataset with only the information necessary for the analysis, any redundant information is removed.

The use case and the training material are based on a well-known free and open source GIS package, QGIS3[21]. For the RAI calculation, a topologically correct road network is desirable, but not necessary. Because OpenStreetMap is an open collaborative mapping project, there may be inconsistencies in the data. After close analysis of the Tabora region road network, a series of in-

[17]Retrieved from https://wiki.openstreetmap.org/wiki/Highway_Tag_Africa on 3.02.2019

[18]Retrieved from https://wiki.openstreetmap.org/wiki/East_Africa_Tagging_Guidelines on 3.02.2019

[19]An element is the basic component of the OpenStreetMap conceptual data model of the physical world

[20]A tag describes the element to which is attached and it is defined by a key and a value that are conventions agreed upon by the OSM community and openly published on the OSM wiki.

[21]Retrieved from https://qgis.org/en/site/ on 3.02.2019

TABLE 20.1
Geospatial information and corresponding open datasets used in the RAI calculation for the Tabora Region. Data from different sources: 1)*gadm.org*, 2)*worldpop.org.uk*, 3)*sedac.ciesin.columbia.edu*, and 4)*openstreetmap.org*

Theme	Geospatial Information	Dataset Used	RAI Layer	Producer / Collector of Data
Administrative Units	Administrative Units	Database of Global Administrative Areas[1]	Administrative Units	University of California, Berkeley, Museum of Vertebrate Zoology, and the International Rice Research Institute (Global Administrative Areas 2009)
World Population	Estimates of numbers of people per grid square	WorldPop[2]	Population Numbers	GeoData Institute, University of Southampton
World Population	Polygon representation of urban areas with city or agglomeration name and time series	Global Rural-Urban Mapping project (GRUMP), v1[3]	Urban Geometries	Socioeconomic Data and Applications Center (SEDAC)
General Geospatial Data	OpenStreetMap is built by a community of mappers that contribute and maintain data about roads, trails, cafés, railway stations, and much more, all over the world	OpenStreetMap[4]	Road Network, Road Condition	OpenStreetMap contributors

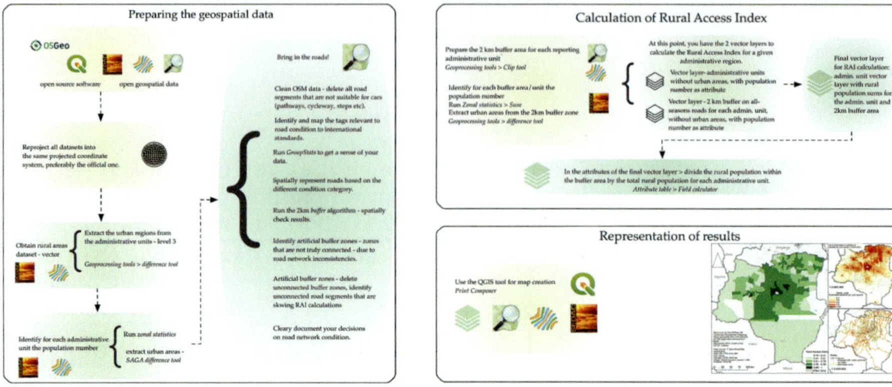

FIGURE 20.1
Conceptual workflow for calculating the RAI

consistencies were identified and corrected to eliminate as much as possible any artificial results in the RAI. Even though the use case was developed for a particular region, the types of inconsistencies addressed include the most common situations, thus making the study replicable. The types of inconsistencies and solutions for addressing each one of them are presented in detail in the training material.

The geospatial layer of the RAI is obtained by dividing the rural population within the 2km buffer area around the all-season roads in a specific administrative unit by the total rural population for that administrative unit. The last step of the workflow provides suggestions for presenting the results of the RAI analysis (Figure 2) so that the indicator can be used as the basis for future management and development of the region in question. The complete training material, for those who want to repeat the calculation of the RAI for Tabora or do something similar in other regions of the world is available at the website[22].

20.5 Conclusion

The Tabora region use case for calculating an SDG geospatial indicator proves that through the exclusive use of open global datasets, some of which contributed by citizens, and free and open source software, complex geospatial analyses can be conducted to better understand, manage and protect our environment. The use case was deliberately developed using only global datasets so that it can be replicated for any other region in the world.

[22]https://wiki.osgeo.org/wiki/Training_Material_for_UN_Open_GIS_OpenData

Undoubtedly, the issue of data quality remains when considering an open collaborative environment such as OpenStreetMap. However, this kind of initiative can and should work as a driving force towards improving the open datasets, either by defining new significant attributes - referred to as tags, in OSM – or by cleaning and maintaining the attributes in the dataset.Working with citizens and communities like those surrounding OpenStreetMap, following a community geospatial science paradigm, "will ensure that the challenges are addressed for all populations in different locations, leaving no one behind" (Rajabifard, 2019).

Bibliography

[1] Rick Bonney, Heidi Ballard, Rebecca Jordan, Ellen McCallie, Tina Phillips, Jennifer Shirk, and Candie C Wilderman. Public participation in scientific research: Defining the field and assessing its potential for informal science education. a caise inquiry group report. *Online Submission*, 2009.

[2] Jessica Breen, Shannon Dosemagen, Jeffrey Warren, and Mathew Lippincott. Mapping grassroots: Geodata and the structure of community-led open environmental science. *ACME: An International E-Journal for Critical Geographies*, 14(3), 2015.

[3] Junyoung Choi, Myung-Hwa Hwang, Hyungtae Kim, and Jaeseong Ahn. What drives developing countries to select free open source software for national spatial data infrastructure? *Spatial Information Research*, 24(5):545–553, 2016.

[4] Arnulf Christl. Free software and open source business models. In *Open source approaches in spatial data handling*, pages 21–48. Springer, 2008.

[5] OpenStreetMap Wiki contributors. About openstreetmap, 2017.

[6] François Grey. Viewpoint: The age of citizen cyberscience. *Cern Courier*, 29, 2009.

[7] The World Bank Group. Measuring rural access: using new technologies. Technical report, Washington, D.C 20433, 2016.

[8] Trisha Gura. Citizen science: amateur experts. *Nature*, 496(7444):259–261, 2013.

[9] Mordechai Muki Haklay, Suvodeep Mazumdar, and Jessica Wardlaw. Citizen science for observing and understanding the earth. In *Earth Observation Open Science and Innovation*, pages 69–88. Springer, Cham, 2018.

[10] Paul Jepson and Richard J Ladle. Nature apps: Waiting for the revolution. *Ambio*, 44(8):827–832, 2015.

[11] Hiromi Kobori, Janis L Dickinson, Izumi Washitani, Ryo Sakurai, Tatsuya Amano, Naoya Komatsu, Wataru Kitamura, Shinichi Takagawa, Kazuo Koyama, Takao Ogawara, et al. Citizen science: a new approach to advance ecology, education, and conservation. *Ecological research*, 31(1):1–19, 2016.

[12] Rafael Moreno-Sanchez. Free and open source software for geospatial applications (foss4g): A mature alternative in the geospatial technologies arena. *Transactions in GIS*, 16(2):81–88, 2012.

[13] S Nascimento, Â Guimarães Pereira, and A Ghezzi. From citizen science to do it yourself science; 2014.

[14] United Nations. Transforming our world: the 2030 agenda for sustainable development. *United Nations General Assembly. Resolution 70/1*, 2015.

[15] United Nations. Work of the statistical commission pertaining to the 2030 agenda for sustainable development. *United Nations A/RES/71/313*, 2017.

[16] United Nations. Sdg indicators metadata repository, metadata for indicator: 9.1.1. proportion of the rural population who live within 2 km of an all-season road. *United Nations, Department of Economic and Social Affairs, Statistics Division*, 2018.

[17] Open Knowledge Foundation (OKF). The open definition, 2018.

[18] OSD. The open source definition, 2018.

[19] Andrew L Russell. *Open standards and the digital age.* Cambridge University Press, 2014.

[20] Patricia Sols. Building mappers not just maps: challenges and opportunities from youthmappers on scaling up the crowd in crowd-sourced open mapping for development, April 2017. AAG Mapathon; AAG Annual Meeting.

[21] Stefan Steiniger and Andrew JS Hunter. The 2012 free and open source gis software map–a guide to facilitate research, development, and adoption. *Computers, environment and urban systems*, 39:136–150, 2013.

[22] OSGeo Website. Osgeo un committee challenge, 2018.

21

Modernizing Land Administration Systems to Support Sustainable Development Goals - Case Study of Victoria, Australia

Hamed Olfat

The University of Melbourne, Australia; Land Use Victoria, Australia

Davood Shojaei

The University of Melbourne, Australia; Land Use Victoria, Australia

This chapter discusses the concepts of both land administration system (LAS) and sustainable development goals (SDGs) as well as some previous works that have linked these two concepts together. It then recommends the requirements of a LAS that can support SDGs. As a case study, the LAS modernization journey in the State of Victoria, Australia to support SDGs is reviewed next. Finally, the chapter concludes with some recommendations for future enhancements of the Victorian LAS.

21.1 Land Administration Systems

To achieve sustainable development goals (SDGs), countries require access to an effective, efficient and modern land administration system (LAS) based on a cadastre engine that contains spatially accurate land parcels and corresponding rights, restrictions and responsibilities (RRRs).

LAS is a simple tool for managing natural resources, environmental monitoring and protection, disaster management, physical and economic planning [8]. In order to support a land market, LAS is a must for all nation states to support and protect ownership rights. This requires having proper data sets prepared by surveyors to support trading land in the market and having a system to provide access to the ownership information. For many countries, cadastre is the engine of land administration which provides integrity and security to land market.

Cadastre also provides a base map for various other purposes such as utility services, urban planning, and disaster management which facilitates spatial enablement government and the wider society [18].

Many developed countries use land parcel layer as a main component of Digital Cadastre Database (DCDB) and attach required attributes to this layer for managing cadastral information. This layer requires a survey network as a base for maintaining the integrity and accuracy. In addition, a proper process is also necessary to keep this important data set updated. This adds value to DCDB by adding other layers and connecting the processes to it. However, while the concept of cadastre is simple, implementation of that is very complex and changing an established cadastre takes long as it has lots of connections to various business processes and regulations.

Some initiatives started to highlight the role of cadastre and proposed some changes to make it compatible with the current and future needs of land administration. For example, Cadastre 2014 proposed by FIG provided a vision for future cadastre [17] and the United Nations-FIG Bathurst Declaration on Land Administration for Sustainable Development is another example of required changes in the cadastre [3]. In addition, the Intergovernmental Committee on Surveying and Mapping (ICSM) in Australia has developed Cadastre 2034 Strategy and proposed the future cadastre in Australia [7]. Based on Goal 1 of the Cadastre 2034 Strategy, the cadastral system should sustainably manage land ownership. This keeps the integrity and societal benefits of the cadastral systems.

The following section briefly introduces the SDGs developed by the United Nation (UN).

21.2 Selected LAS Works

Land administration theory requires the implementation of the land management paradigm to drive systems dealing with land rights, restrictions and responsibilities towards supporting sustainable development. The land management paradigm, where land tenure, value, use and development are considered holistically as essential and omnipresent functions performed by organized societies, is the cornerstone of modern land administration theory [19].

The land management paradigm makes a national cadastre the engine of the entire LAS, underpinning the country's capacity to deliver sustainable development (Figure 21.1). The cadaster should assist the functions of land tenure, land value, land use, and land development. In this way, within the LAS, the cadastre or cadastral system becomes the core technical engine delivering the capacity to control and manage land through the four LAS functions.

Cadastres are regarded as the foundation for sustainable social, economic and environmental development of societies [13].

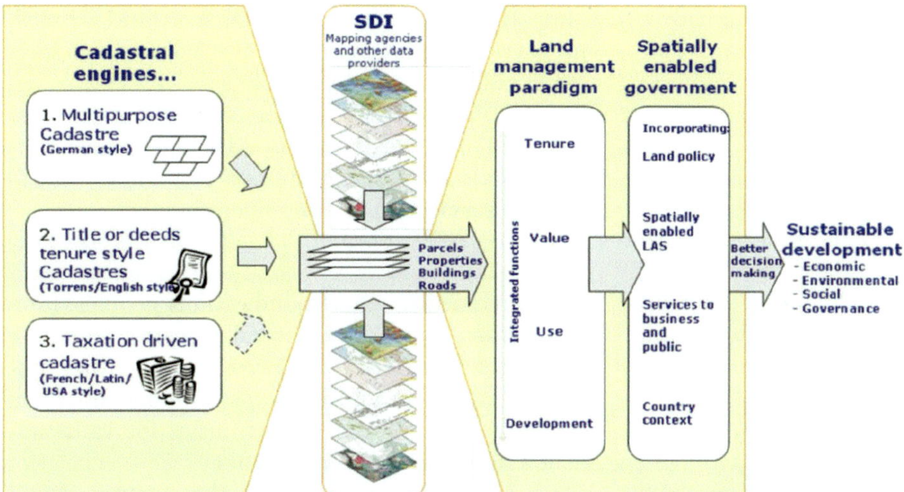

FIGURE 21.1
The cadastre as an engine of LAS - the "butterfly" diagram [19]

The relationship between 17 SDGs and LAS was also reviewed by Dawidowicz and Zrobek (2017) in Poland to build a LAS to support the SDGs [2]. They identified the key challenges that a LAS should address to support sustainable development. In Poland, the Integrated Real Estate Information System (IREIS), is being implemented based on sustainable development.

In the next section, the SDGs with direct and indirect relationship with LAS are identified and the requirements of a LAS to support SDGs are explored.

21.3 Land Administration Systems Related Requirements to Support Sustainable Development Goals

SDGs require access to LAS. However, the relationship between goals and LAS can be direct or indirect. Direct relationship means that a specific goal cannot be achieved at all without a LAS. Whereas, an indirect relationship means that a specific goal might not be efficiently achieved without a LAS. Table 21.1 shows the goals and targets that have a direct link to LAS.

The rest of SDGs and targets have an indirect link to LAS. As an example, Goal 4, "Ensure inclusive and equitable quality education and promote lifelong

TABLE 21.1
SDGs and targets that have a direct link to LAS

Goal	Target
Goal 1. End poverty in all its forms everywhere	(1.4) By 2030, ensure that all men and women, in particular the poor and the vulnerable, have equal rights to economic resources, as well as access to basic services, ownership and control over land and other forms of property, inheritance, natural resources, appropriate new technology and financial services, including microfinance
Goal 2. End hunger, achieve food security and improved nutrition and promote sustainable agriculture	(2.3) By 2030, double the agricultural productivity and incomes of small-scale food producers, in particular women, indigenous peoples, family farmers, pastoralists and fishers, including through secure and equal access to land, other productive resources and inputs, knowledge, financial services, markets and opportunities for value addition and non-farm employment
Goal 5. Achieve gender equality and empower all women and girls	(5.a) Undertake reforms to give women equal rights to economic resources, as well as access to ownership and control over land and other forms of property, financial services, inheritance and natural resources, in accordance with national laws (5.b) Enhance the use of enabling technology, in particular information and communications technology, to promote the empowerment of women
Goal 11. Make cities and human settlements inclusive, safe, resilient and sustainable	(11.1) By 2030, ensure access for all to adequate, safe and affordable housing and basic services and upgrade slums (11.3) By 2030, enhance inclusive and sustainable urbanization and capacity for participatory, integrated and sustainable human settlement planning and management in all countries (11.6) By 2030, reduce the adverse per capita environmental impact of cities, including by paying special attention to air quality and municipal and other waste management (11.7) By 2030, provide universal access to safe, inclusive and accessible, green and public spaces, in particular for women and children, older persons and persons with disabilities (11.a) Support positive economic, social and environmental links between urban, peri-urban and rural areas by strengthening national and regional development planning

learning opportunities for all" does not have a direct link to LAS, however, it can be efficiently achieved using a modern LAS based on a spatially accurate cadastre that demonstrates the distribution of people based on their age as well as educational institutes and their information across a country.

Our study showed that countries should meet the following requirements for supporting SDGs and targets:

- **Requirement 1** - Provide equal access to ownership and control over land and property;
- **Requirement 2** - Provide secure tenure rights to land with legally recognized documentation (title, deed, etc.);
- **Requirement 3** - Develop an accurate cadastral data set (parcel fabric) as a fundamental layer; and
- **Requirement 4** - Utilize information and communications technology for modernizing LAS.

The next section provides an overview of the Victorian LAS modernization journey to meet the above-mentioned requirements for supporting SDGs.

21.4 Case Study of Victoria, Australia

Victoria is Australia's most densely populated state (highlighted in Figure 21.2) and its second-most populous state overall with population of 6,430,000 [1].

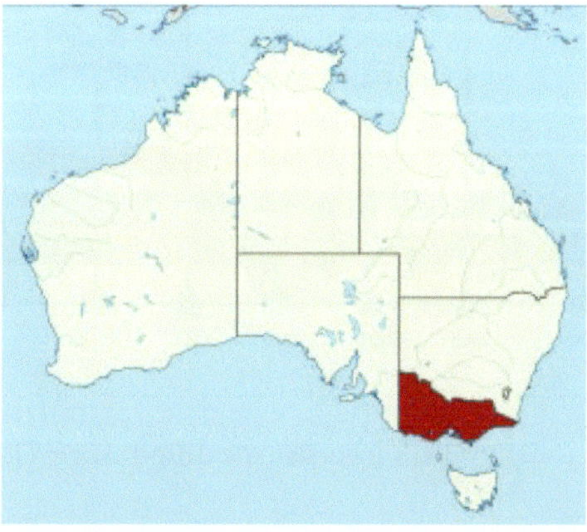

FIGURE 21.2
The location of State of Victoria in Australia

The Victorian Department of Environment, Land, Water and Planning (DELWP), through Land Use Victoria (LUV), is responsible for all land information and administration activities in Victoria including registration of land transactions, property information, surveying, valuation, geographic names, spatial services, government land and the government land monitor.

LUV is also responsible for maintaining the Victorian cadastre (VicMap Property), which is currently a 2D analogue representation of the State's property boundaries, based on property title information, and provides the foundation for Victoria's primary mapping and spatial information systems and services.

To support the SDGs, LUV has been constantly modernizing the LAS using the information and communications technology, as recommended by [9]. Figure 21.3 illustrates the Victorian LAS modernization journey. Before the 1990s, cadastral plans were all lodged in paper. VicMap Property was created in the early 1990s from the digitization of paper-based map records held by Melbourne Water (metropolitan area) and the State government (rural area). VicMap Property comprises more than 3 million land parcels and associated property attributes, such as lot and plan number, and crown description, in the State of Victoria.

Land title information was migrated from paper to the Victorian Online Title System (VOTS) in 2000. VOTS contains a record of all Victorian titles registered under the Torrens System [6]. The system is maintained by LUV and is used to accept, create and register land transaction lodgements, and to update land holdings and registered interests on title as well as create new titles.

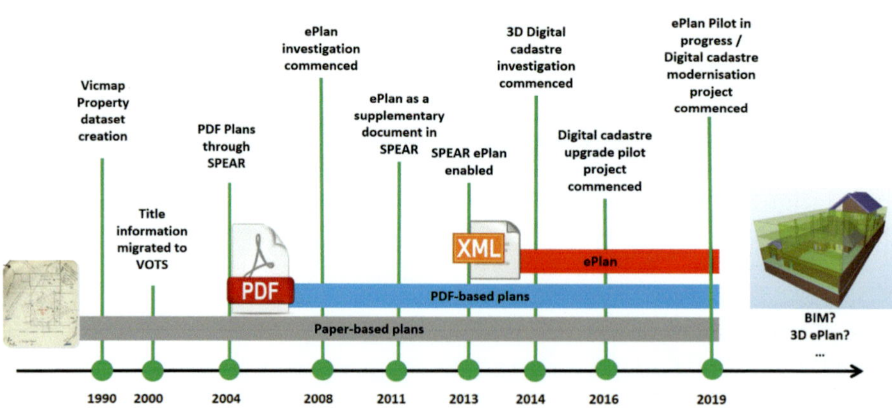

FIGURE 21.3
LAS modernization journey in Victoria (modified after [12])

Prior to the launch of the Surveying and Planning through Electronic Applications and Referrals (SPEAR) pilot in 2004, subdivision applications could only be processed via paper. Diagram (a) in Figure 21.4 illustrates this process. This was generally a lengthy and protracted process that was instigated by the surveyor, on behalf of their client (developer). The process and application milestones are well defined by the Planning and Environment Act 1987 [4] and Subdivision Act 1988 [5], however achieving these milestones, in a paper environment, where there are multiple stakeholders involved in the decision-making process led to delays, errors, and poor transparency between the interested parties.

Although there was a lot of scope to improve efficiencies throughout the life of the application, there was no means of implementing these changes for the benefit of all parties, due to the technology constraints of a pre 'world wide web' era.

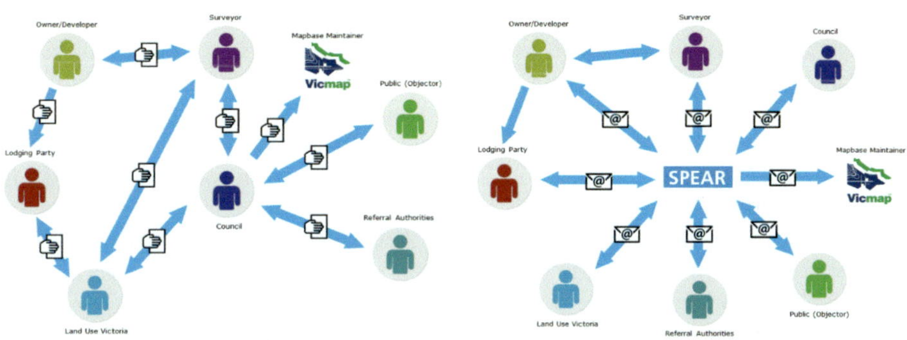

(a) Subdivision process before SPEAR (b) Subdivision process after SPEAR

FIGURE 21.4
Subdivision process before SPEAR (diagram a) vs. after SPEAR (diagram b).

SPEAR revolutionized the way subdivision applications were handled, by introducing online end-to-end processing and tracking of plan applications from their initial submission with local government, right through to registration at LUV. Diagram (b) in Figure 21.4 illustrates the subdivision process after the introduction of SPEAR. A surveyor can use SPEAR to apply for any plan-based dealing under the Subdivision Act 1988, and the planning permit to subdivide under the Planning and Environment Act 1987.

SPEAR introduced invaluable transparency and accountability to the subdivision application process by streamlining the approval process for plans of subdivision, and the associated planning permit to subdivide. The system is now being used by all 79 Victorian local governments, 200 surveying firms, 74 referral authorities, and LUV, which, in total, represents over 4000 users to view the progress of applications.

In addition to SPEAR project, the investigations to the ePlan project commenced in 2008 in Victoria (See Figure 21.3). LUV collaborated with the ICSM ePlan Working Group on developing a national data model to cover all Australian jurisdictions' cadastral and survey requirements [16]. In 2011, SPEAR enabled surveyors interested in ePlan to upload an ePlan LandXML file along with their PDF application. From 2011 to 2013, ePlan was piloted in Victoria by LUV, the surveying industry and software vendors. In May 2013, SPEAR incorporated ePlan services including visualization, validation, data viewer and data download [11, 10]. The ePlan road map defined by LUV has the following visions [12]:

- **Longterm vision**: Implement ePlan for all Victorian cadastral plans and surveys by 2025.
- **Short-term vision**: Provide the infrastructure and services to enable the submission and registration of ePlan for all 2D Victorian cadastral plans by 2020.

Recent advancements in the demand for high precision, and data driven spatial information have led to the need to modernize and digitize Victoria's cadastre. The Digital Cadastre Modernization project is underway and will deliver a fully digital state-wide cadastre over the next 5 years. This will unlock significant new capability and innovation in Victoria for next-generation spatial services.

All 2D plans under the Subdivision Act 1988 are supported in ePlan. However, strata plans (building subdivision plans) which include overlapped ownership rights are not yet supported.

As indicated in Figure 21.3, the investigations around the 3D digital cadastre to support the building subdivisions in ePlan format commenced in 2014. Following the release of the ICSM's strategy on Cadastre 2034, LUV has started to investigate the technical requirements for supporting 3D building subdivisions in ePlan including the potential use of Building Information Modelling (BIM), 3D data visualization, validation and storage. As part of these studies, several prototypes were developed and are under development to evaluate the implementation. As an example, Figure 21.5 presents LUV 3D ePlan Prototype.

Currently, a 3D digital cadastre road map is under development in Victoria following the ePlan long-term vision and goal 4 of ICSM Cadastre 2034 Strategy. The road map aims to show the major milestones and timeframes towards the implementation of a 3D digital cadastre by 2025. In addition, the institutional, technical and legal aspects of a 3D digital cadastre should link together to clarify the connection and relationships of the interests of the property industry, to build a comprehensive framework for implementation [14].

In conclusion, the current status of Victorian LAS in terms of addressing SDGs-related requirements is shown in Table 21.2.

TABLE 21.2
Current status of the Victorian LAS in terms of addressing SDGs-related requirements

Requirement	Victorian LAS current status
Provide equal access to ownership and control over land and property	The current LAS allows both men and women to have equal access to ownership and control over land and property.
Provide secure tenure rights to land with legally recognized documentation (title, deed, etc.)	Victoria's LAS is based on Torrens title system which works on three principles: • The land titles Register accurately and completely reflects the current ownership and interests about a person's land. • Because the land titles Register contains all the information about the person's land, it means that ownership and other interests do not have to be proved by long complicated documents, such as title deeds. • Government guarantee provides for compensation to a person who suffers loss of land or a registered interest.
Develop an accurate cadastral data set (parcel fabric) as a fundamental layer	The digital cadastre modernization project is currently underway. This project aims to build a spatially accurate 2D digital cadastre for Victoria by 2024.
Utilize information and communications technology for modernizing LAS	Both ePlan and 3D digital cadastre projects are currently underway aiming at providing services to enable the submission of digital cadastral data to LUV. These projects leverage the information and communication technologies to develop required services for land administration stakeholders in Victoria.

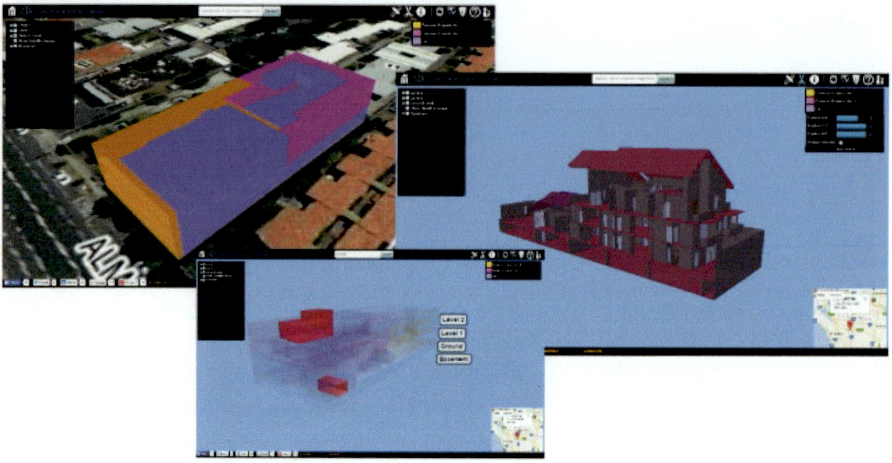

FIGURE 21.5
LUV 3D ePlan prototype (www.spear.land.vic.gov.au/spear/pages/eplan/3d-digital-cadastre/3dprototype/prototype.html) [15]

21.5 Conclusion

This chapter explored the role and requirements of a modern LAS for supporting SDGs. It was discussed that SDGs, depending on their nature, have either a direct or an indirect relationship with LAS. Direct relationship means that a specific goal cannot be achieved at all without a LAS (e.g. Goal 11. Make cities and human settlements inclusive, safe, resilient and sustainable). Whereas, an indirect relationship means that a specific goal might not be efficiently achieved without a LAS (e.g. Goal 4. Ensure inclusive and equitable quality education and promote lifelong learning opportunities for all).

The chapter reviewed the LAS modernization journey in Victoria to support SDGs. The study showed that the current Victorian LAS can meet most of the four requirements identified herein. However, to have access to a fully modern LAS in Victoria, the following research and development projects should be considered: a) fully implement 2D ePlan services for all plan-based dealing types by 2020, b) finalize and endorse 3D digital cadastre road map, and c) implement 3D digital cadastre by 2025.

Bibliography

[1] ABS. Australian demographic statistics, 2018.

[2] Agnieszka Dawidowicz and Ryszard Źróbek. Land administration system for sustainable development–case study of poland. *Real Estate Management and Valuation*, 25(1):112–122, 2017.

[3] FIG. The bathurst declaration on land administration for sustainable development, 1999.

[4] Victorian Government. Planning and environment act 1987, 2019.

[5] Victorian Government. Subdivision act 1988, 2019.

[6] Victorian Government. Torrens titles, 2019.

[7] ICSM. Cadastre 2034, powering land & real property, 2015.

[8] C Lemman. *A Domain Model for Land Administration*. PhD thesis, Delft University of Technology, 2012.

[9] United Nations. Global indicator framework for the sustainable development goals and targets of the 2030 agenda for sustainable development, 2018.

[10] Hamed Olfat, Amul Jani, Davood Shojaei, Ashley Darvill, Mark Briffa, Abbas Rajabifard, and Farshad Badiee. Tackling the challenges of visualising digital cadastral plans: The victorian cadastre experience. *Land Use Policy*, 83:84–94, 2019.

[11] Hamed Olfat, D Shojaei, M Briffa, and A Rajabifard. The current status and ongoing investigations of 2d and 3d digital cadastre (eplan) in victoria, australia. In *Proceedings of the Academic Research Stream at the Annual Conference Locate, Research@ Locate*, 2017.

[12] Hamed Olfat, Davood Shojaei, Mark Briffa, Susannah Maley, and Abbas Rajabifard. Strategic actions for increasing the submission of digital cadastral data by the surveying industry based on lessons learned from victoria, australia. *ISPRS International Journal of Geo-Information*, 7(2):47, 2018.

[13] Abbas Rajabifard, Daniel Steudler, and AIEN Ali. The cadastral template 2.0, from design to implementation. In *FIG CONGRESS*, 2014.

[14] Davood Shojaei, Mohsen Kalantari, Ian D Bishop, Abbas Rajabifard, and Ali Aien. Visualization requirements for 3d cadastral systems. *Computers, Environment and Urban Systems*, 41:39–54, 2013.

[15] Davood Shojaei, Hamed Olfat, Abbas Rajabifard, and Mark Briffa. Design and development of a 3d digital cadastre visualization prototype. *ISPRS International Journal of Geo-Information*, 7(10):384, 2018.

[16] Davood Shojaei, Hamed Olfat, Abbas Rajabifard, Ashley Darvill, and Mark Briffa. Assessment of the australian digital cadastre protocol (eplan) in terms of supporting 3d building subdivisions. *Land Use Policy*, 56:112–124, 2016.

[17] Daniel Steudler. *CADASTRE 2014 and Beyond*. International Federation of Surveyors (FIG), 2014.

[18] Ian Williamson. Using cadastres to support sustainable development. *International Federation of Surveyors (FIG), Article of the Month*, 2008.

[19] Ian Williamson, Stig Enemark, Jude Wallace, and Abbas Rajabifard. *Land administration for sustainable development*. Citeseer, 2010.

Index

Numbers and Symbols

3D reconstruction models, 95
5ps Model, 304

A

Action Framework for Implementation of NUA (AFINUA), 10, 201, 203, 204, 205, 208
Aerial photography, data gathered by, 257–258
Aerial systems, unmanned, 22
AfriGIS, 234, 235, 236
Analytics, 4, 22
Application Programming Interfaces (APIs), 233, 248
ArcticSDP, 238–241
Artifical intelligence (AI), 21, 27, 96, 97, 104
 decision-making by, 133, 137
Association of Flemish Cities and Municipalities (VVSG), 203–204
Australia, case example on geospatial information management, 231–233
Australian Bureau of Statistics (ABS), 231
Australian Urban Research Infrastructure Network (AURIN), 207, 247, 248
Austrian Statistical Geography Standard (ASGS), 248
Autonomous cars, 133

B

Bhoomi project, 55, 56
Big Data, 21, 22, 27, 230
Big geo-data, 96
Brainware, 213 (*see also* education ecosystem)
Building information management system (BIMS), 98
Building Information Modelling (BIM), 332
Bureaucracy, Weberian view of, 167
Business models, corporate, 136

C

Cadastre systems, 45
 disaster resilience, importance to, 84
 Switzerland, 51–53, 58
Case study, Syrian crisis, night-light remote sensing, 275–277
CitiGenius, 264
Citizen science, 316–317, 319
City Performance Initiative (CPI), 10
City Prosperity Initiative (CPI), 203, 204–205, 208
CityGML standards, 263, 264
Cloud computing, 21, 22
Coastal zones
 geospatial information, need for updated, 110
 Hurricane Harvey case example, 119–121
 resilience, 110
 resilient development, 109
 sustainable development, 109
 threats to, 109
Community mapping, 50
Computer literacy, 40 (*see also* digital divide)
Conference on Human Development, United Nations, 16–17

Conference on Sustainable Development, United Nations (Rio+20), 17, 67
Conrad Blucher Institute for Surveying and Science, 119
Consensus-based decision-making, 313
Contextual Analysis Questionnaire, 90
Copernicus Master in Digital Earth, 216
Country Action Plan Template, 90
Crowd2Map, 317
Crowdsourcing methodologies, 50
Cultural-cognitive institutions, 166
Cyclomedia, 263

D

Damage index (DI), 99
Damage index model (DIM), 99
Damage Index of Building (DIoB), 99
Data analytics, 230
Data flow, framework for, 22–23
Data fusion, 230
Data generation, growth in, 22
Data mining, 137
Defense Meteorological Satellite Program/Operational Linescan System (DMSP/OLS) night-light, 267, 268, 271–272
Development agenda, United Nations
  elements of, 3–4
Digital Cadastre Database (DCDB), 326–327
Digital Cadastre Modernization, 331
Digital divide, 7
  closing, 26, 27–28
  Data Ecosystem Gap, 26–27
  Digital Access Gap, 25–26
  Digital Adoption Gap, 26
  Digital Value Gap, 26
  enormity of, between developed and developing countries, 25
  geospatial, 15
    barriers, overcoming, 28
    bridging, 24–25, 28–30
    extension of digital divide, as, 26
    Geospatial Policy Gap, 28
    Geospatial Technology Gap, 27
  Internet connectivity, 135
Digital Earth, Copernicus Master in, 216
Digital economies, 130
  cybercrimes, 132
  digital tracking, 133
  disputes, 133
  encryption, 132
  hacking, 132
  policies and laws relating to, 132–133
  societal implications, 138
Digital elevation models (DEMs), 112
Digital navigation, 239–240
Digital surface model (DSM), 112
Digital terrain models (DTMs), 112
Disaster resilience, 74
  3D reconstruction models, 95
  displacement issues, 82, 85
  earthquakes, 94–95, 96
  enhancing, 82
  geospatial data and systems, physical resilience, 86–87
  geospatial information pertaining to, 85–86
  goals of, 84–85
  Hyogo Framework for Action (*see* Hyogo Framework for Action)
  increasing events, 85, 94, 96
  land administration systems, importance of, 83–84
  land information pertaining to, 85–86
  land records, importance of, 83
  land resilience, 85–87, 89–91
  mismanagement, 95–96
  restoration efforts, 83
  risk management, 85, 86
  risk reduction initiatives, 87
  Sendai Framework (*see* Sendai Framework)
  stakeholders, 85, 91
  types of disasters, 94–95, 95–96
Distruptive technologies, 21
DJI Phantom series, 111
Drones, use of in human search and rescue, 99–102

E

E-commerce, 14
E-government, 14
E-service, 14
Earth Remote Observation System-B (EROS-B), 267
Earth Summit, 17, 47
East Africa Tagging Guidelines, 320
Economic and Social Council (ECOSOC), United Nations
  establishment of UN-GGIM, role in, 13
Economic and Social Council resolution 2016/27, 3
Economic cycle, 290
Economic divide, 288
Economic policies, traditional resources, 130–131
Education ecosystem, 214–215, 216–217
Electronic Navigational Charts (ENC), 240
Environmental regulations, 102–103, 104–105
Eren, Kamil, 263
ESA CoastColour, 319
ESRI, 65
Evidence-based decision making, 23–24, 24–25, 67, 307

F

Federal Emergency Management Agency (FEMA), 99
FIG, 283
Fit-for-purpose formalization, 292
Formalization policies, 291, 292, 293–294
Foundational Guide to NIA-instruments for Geospatial Information Management, 143
Future Cities Pilot, 237
Future We Want, 303
Future We Want, The, 17

G

General Assembly resolution 70/1, 3
General Assembly, United Nations, 17
  2030 Agenda, passage of, 17, 245 (*see also* Transforming our World: The 2030 Agenda for Sustainable Development)
GEO Aqua Watch, 319
GEO Wetlands Initiative, 319
GeoEngine, 97
Geographical Information Systems (GIS)
  disaster response, use in, 94–95
  engineering parameters, use in determining, 99
  mapping, 112
Geography Markup Language (GML), 230
Geospatial Data Infrastructure, 307, 308, 309
Geospatial information
  academic education in, 214 (*see also* brainware)
  awareness regarding, need for, 25, 67
  coastal zones, pertaining to (*see* coastal zones)
  data infrastructure, 299
  data leadership regarding, 14–15
  data sharing, 103
  datasets, 305
  developing countries, issues accessing data and technology, 14–15
  digital divide (*see* digital divide)
  disaster reduction, role in (*see* disaster resilience)
  disruptive technologies, 21
  education ecosystem regarding, need for, 215–216
  global connectivity, 21, 93–94, 298–299
  ICT standards and, 228–229
  importance of, 28
  integration of data and techniques with technology, 96–97
  management (*see* geospatial information management)
  Nigeria-Africa case study, 298, 299
    data fragmentation, 307
    GI4SD mission statement, 305

land administrations, 299–300
SDG goals, 304, 306–309
physical resilience, importance of, 86–87
rapid visual screening techniques, 98–99, 103
regulations regarding, 102–103, 104–105
role of, 305–306
SDGs, importance to, 4–5, 6–7, 14, 17, 25, 81–82, 102, 257–258
SDGs, interconnection, 300–301, 302–303
technological innovations, 21–22, 94, 95, 96–97, 98–99, 99–102
technology implementation, 304
updated, need for, 110
Working Group on National Institutional Arrangements (WG-NIA) recommendations (*see* Working Group on National Institutional Arrangements (WG-NIA))
Geospatial information management
ArcticSDP, 238–241
Australia case example, 231–233
business models, diverse, 158–159
capacity building, 149–150
capacity challenges, 159
change processes, 157–158
collaboration, 166–167
collective decision-making, 148
coordinating functions/entities, 146–147, 166–167
cultural challenges, 159
data, open, 158, 230
digitization, 225–227
Experts, Committee of, 224–225
financial management, 148–149
framework, 30–31, 144, 151
hierarchical arrangements, 145, 148, 169–170
importance, to bridging digital divide, 28
information exchange and sharing, systems for, 147–148
institutional arrangements, 151–152, 158, 166, 170
instruments, 145–146
inter-organizational culture, 149, 158
interoperability, 228
knowledge management, 149
legal framework, 147
maritime sector, 238–241
models for, 157
national institutional arrangements, 141–142, 144–145 (*see also* Working Group on National Institutional Arrangements (WG-NIA))
New Zealand case example, 233–234, 237
partnerships, 148
regulated markets, 147
South Africa case example, 234–237
spatial data infrastructures (SDIs) (*see* spatial data infrastructures (SDIs))
stakeholders, 167, 169
standards, 224, 225–227, 228, 229–230, 231
strategic planning, 148, 158
trends, 157
urban environment, in, 237–238
Geospatial Way to a Better World, 4
Geotech (company), 263–264
Geotechnical Extreme Events Reconnaissance (GEER) project, 121
Global Goals, 65
Global navigation satellite system (GNSS), 111
GNU General Public License (GPL), 315
Google, 263
Government, responsibilities of, 40
Gross domestic product (GDP), 291
Ground sample distance (GSD), 114–115
Group on Earth Observations (GEO), 20

H

Harrison, Benjamin, 239
Here, 263

Highway Tag Africa Typology of Road Network, 320
Hongbo, Wu, 302
Hyogo Framework for Action, 87, 88

I

Improving Resilience and Resilience Impact of National Land and Geospatial Systems, 89
Inertial measurement unit (IMU), 111
Informal building, 283–284, 290–291
Informal property rights, 290–291, 292, 293
Information and communications technology (ICT)
  accessibility, 148
  data management relationship between, 15
  digital divide, as part of, 25, 26
  infrastructure of, 21
  smart citiies, as part of (*see* smart cities)
  standards, 228–229
  systems, improving, 46
Information civilizations, 137
Information systems, 4
Institutionalization, defining, 144
Integrated Geospatial Information Framework (IGIF), United Nations, 40–41, 87, 89
Intellectual property rights, 132
Inter-Agency and Expert Group on SDG Indicators (IAEG-SDGs), 245
Inter-agency Expert Group on Sustainable Development Goal Indicators (IAEG-SDGs), 20, 68
Interlis, 52
International Federation of Surveyors (FIG), 67
International finance corporations (IFCs), 55
International Hydrographic Conference, 239
International Hydrographic Organization (IHO), 239–240
International Marine Conference, 239
International Maritime Organization (IMO), 227
International Society for Photogrammetry, 258
International Space Station, 267–268
International Training Centre for Photogrammetry (the ITC), 259
Internet, 22
Internet literacy, 26
Internet of things (IoT), 4, 22, 27, 104, 182, 183, 189
ISO 19115, 248
ISO 37120, 248
Istanbul Declaration on Human Settlements, 202

J

Jilin1-03B, 267

K

Karnataka, 58

L

Land administration systems (LAS), 168–169, 325–326, 327
Land Resilience Maturity Index Assessment, 90
Land right registries, 45
Land tenure and administration
  corruption in, 56
  development and building permits, 46, 47
  disaster resilience, importance to, 83–84
  institutional systems, 46
  land taxation, 46
  land tenure reform projects, 45–46
  land tenure security, 45, 48, 49, 54, 56, 58, 83–84
  mining rights, 46
  natural resources systems (*see* natural resources systems)
  overview, 45–46
  people-to-land relationships, 82

property rights, protecting, 83
public-private partnerships (PPP), land reform projects (*see* public-private partnerships (PPP), land reform projects)
reform, 7–8
SDGs, importance to, 5–6, 45–46
spatial planning, 45, 73–74
water concession registries, 46
Land titling computerisation project (LTCP), Philippines, 54–55
Land Use Victoria (LUV), 329
Legal and policy frameworks, SDGs
data, government, open access to, 136–137
education and feeding of children, 135
government power, protecting citizens against, 134–135
legal controls, role of, 131–132
order, maintaining, 135–136
overview, 129–130
wealth disparity and opportunity issues, 137–139
Linked Data API, 233
Living labs, 182–183
Luojia 1-01, 267, 270

M

Machine learning, 27, 97, 104, 137
Mapping
aerial photography, data gathered by, 257–258
Europe, in, 261–262
global progress in, 258–260
high-resolution satellite imagery, 263
mobile mapping, 263
new technologies for, 262–263
oblique imagery, 3D, 263–264
photogrammetric, 258–259
stereo satellite, 264
terrestrial surveys, 258
urban areas, large-scale, 260–261
Measurement Analytics (MANTIS) Lab, 119, 120
Methontology, 248
Millennium Development Goals (MDGs), 17, 47, 177, 200, 202–203
achievements of, 297
geospatial data, on, 224
implementation, 300, 308
SDGs, transforming to, 206, 250–251
Missing Maps, 317
Mission Planner, 113
Mobile services, 22
Monopolies, 133–134
Multi-View Stereo (MVS) algorithm, 117

N

National Address Management Framework (NAMF), 231
National Exposure Information System (NEXIS), 231
National Geospatial Information Infrastructure (NGII), 104
National Science Foundation (NSF), 121
National Spatial Data Infrastructures (NSDIs), 23, 32, 49, 206, 302
boundaries, 216
enabling, 302
managing, 213
National Statistical Offices, 23
National Strategic Geospatial Policy Framework
action plans, 41
aspirations of, 40
benefits of, 39
components, 29–30
consolidation of key areas under, 28
deliberate approach of, 30
goals of, 30
guidance offered by, 28–29
implementation, 36, 40–41
mission, 31
principles, 32
stakeholders, 30–31
strategic drivers, 32, 36
vision, 31
Natural resources systems, land tenure security, as part of, 45–46
New Public Management (NPM), 167
New Urban Agenda (NUA), 10

Action Framework, 10, 201, 203, 204, 205, 208
environmental sustainability, 205
equality/social inclusion, 204–205
implementation, 200, 203, 208–209
infrastructure, 204
legislation and regulations, 200
local economies, 200
municipal finance, 200
overview, 199–200
planning and design, 200
productivity, 204
quality of life, 204
SDGs pertaining to, 289
spatial enablement, 201
urban policy, 200, 205

New Zealand, case example on geospatial information management, 233–234, 237

Night-light remote sensing (*see also specific sensing systems*)
applications, 273–275
case study, Syrian crisis, 275–277
overview, 267–268

Normative institutions, 166

O

OGC Geography Markup Language (GML), 230
OGC WaterML, 237
Open data, 158, 230
Open Data Policy, 170
Open Geospatial Consortium (OGC), 207, 248
Open Source Geospatial Foundation (OSGeo), 315, 316, 319
Open source software, 315
Openness, defining, 313–314
OpenStreetMap (OSM), 313, 317, 318, 319, 320
Orthorectification, 112
Orthormosaics, 112, 116
Ownership rights, weak, 287

P

Photogrammetry, 116–117
Predictive analytics, 137
Property rights, protecting, 290 (*see also* land tenure and administration)

Public-private partnerships (PPP), land reform projects, 47
Australia, case example, 54
Canada, case example, 53–54
financial arrangements, 53
funding, 51
India, case example, 55–56, 58
involvement in land administration, 50–51
revenue, 58–59
SDGs, aligning with, 56, 58–59
segmentation, 58
strategies, 59
Switzerland, case example, 51–53, 58

R

Raster cells, 112
Regulative institutions, 166
Remote sensing integration, 98
Rights, restrictions, and responsibilities (LRR), 325
Rio+20 (*see* Conference on Sustainable Development, United Nations (Rio+20))
Rivalry, 130
Robotics, 27
Rohingya Refugee Emergency at a Glance, 65
Rural Access Index (RAI), 320, 322
Rwanda, institutional arrangements in, 170

S

Safety of Life at Sea (SOLAS), 227
Satellite remote sensing, 110
Scale invariant feature transform (SIFT), 117
Schermerhorn, William, 258, 259
Search and rescue operations, 99–102
Sector-ring Histogram of Oriented Gradients (SRHOG), 100–102
Sendai Framework, 3, 14, 87, 88–89
SenseFly eBee, 120

Smart cities
communities, knowledgeable, 183–184
defining, 179–180
governance, 184–185, 186
growth, smart, 186–187
holistic framework, 176
human-driven method (TDM), 180
ICT elements, 179, 181, 183, 185, 189
implementation, 179
indicators, 179
infrastructure, 181, 183, 184
innovation, 185–186
living labs, 182–183
OCG's Smart Cities Domain Working Group, 237
overview, 175–176
quality of life, 189
SDGs, complexity of implementing, 188–189
social sustainability, 179
socio-economic ecosystem, 187–188
strategy of, 181
sustainable development, 180–181, 189
technical and digital support, 181–182
techno-partnership, 187
technology-driven method (TDM), 180
Smith, Adam, 130
South Africa, case example on geospatial information management, 234–237
Spatial data infrastructures (SDIs), 66, 67
advances in, 206–208
coordination challenges, 168–169
governance arrangements, 168
implementation, 170
networked coordination systems, 168
Spatially enabled SDGs (*see also* Sustainable Development Goals (SDGs))
defining, 66–67
land ownership issues, 73–74
overview, 65–66
social impact of, 71, 73
transdisciplinary, 68
Spatially Enabled Society, 67
Stakeholders, 30
Statistical Spatial Framework (SSF), 232, 234
Structure-from-Motion (SfM) photogrammetry, 116–118, 121
Surveying and Planning through Electronic Applications and Referrals (SPEAR) pilot, 330–331
Sustainable development (*see also* Transforming our World: The 2030 Agenda for Sustainable Development)
origins of the concept, 16–17
wicked problem of, 164–165
Sustainable Development Goals (SDGs)
2030 Agenda, as part of, 18–19, 19–21, 23, 176 (*see also* Transforming our World: The 2030 Agenda for Sustainable Development)
adoptions, 47–48
boundaries of, 216
challenges of, 4, 9, 70, 297–298
connectivity of, 7, 66
context of, UN, 3
controversies, 48–49
cross-sector mobilization, 69, 70
defining, 257
development of, 17, 244
disaster prevention, 99 (*see also* disaster resilience)
framework for social investment, sustainable development, 45
geospatial information, relationship between, 257–258 (*see* geospatial information)
goals, 129–130, 244, 298
holistic approach, 65
housing, access to, 68
implementation, 4, 15, 25, 28, 30, 41, 97, 104, 176, 178, 179, 185, 201
indicator frameworks, 246, 251–252, 319

indivisible, 297–298
interactions framework, 303–304
issues, range of, 93
key performance indicators, 217
land tenure and administration (*see* land tenure and administration)
land tenure security (LTS), 45
land, access to, 68
legal frameworks (*see* legal and policy frameworks, SDGs)
local-to-global outlook, 69
measurement of, 15, 68
metrics, 23
monitoring, 69, 315–316
policy frameworks (*see* legal and policy frameworks, SDGs)
resiliency, 65, 301–302
SDG1 (No poverty), 45, 49, 58, 59, 73–74, 288
SDG13 (Disaster risk reduction), 74
SDG16 (Promote peaceful inclusive societies), 177–178
SDG2 (Zero hunger), 45, 59, 109, 288–289
SDG5 (Gender equality), 45, 48–49, 59, 74, 289
SDG6 (Clean water and sanitation), 237
SDG8 (Decent work and economic growth), 178
SDG9 (Industry, Innovation and Infrastructure), 177–178, 181–182, 314
services, access to, 68
smart cities (*see* smart cities)
social results sought by, 46
spatial enablement, 8–9, 69, 201
spatially enabled SDGs (*see* spatially enabled SDGs)
stakeholders, 69
sustainable cities and communities, 109–110
targets, 25, 48, 68–69, 307
three-tier, 177
transnational partnerships, 69
transdisciplinary, 68, 69–70
transnational partnerships, 70–71

Sustainable Development Goals Report, 23

T

Tenure rights, 288–289
Teranet, 53
Tesla, 263
Tomtom, 263
Torrens System, 330
Transforming our World: The 2030 Agenda for Sustainable Development (*see also* Sustainable Development Goals (SDGs))
analysis, 67–68
challenges, 164–165
data acquisition and integration approaches, 87–88
data requirements, 18–19
development policy of, 17–18, 20–21
digital infrastructure, need for, 253
disaster resilience, 87
emergency responses, 102
follow-up, 19, 23
geospatial information, importance of to, 15
global development agenda, 14
global indicator framework, 19–20
goals of, 7, 70, 74, 163
localising, 253
passage of, 17
planning issues, 164–165
progress toward, 318–319
resolution, 3–4
SDGs of (*see* Sustainable Development Goals (SDGs))
sustainable development, evolution of, 15
system outcomes, 32
targets, 18, 23
wicked problem, as, 164–165, 170
Transparency, 130

U

UK Economic and Social Research Council, 207

UN Center for Human Settlements (UNCHS), 202
UN-Habitat, 10
UNECE region, land record experiences with, 283–287
UNHCR, 65
United Nations
    member states, 143, 144, 151, 153, 157, 225, 244
United Nations Conference on Housing and Sustainable Urban Development (Habitat III), 200, 201, 206
United Nations Conferences on Human Settlements, 202
United Nations Economic and Social Council, 143
United Nations General Assembly (*see* General Assembly, United Nations)
United Nations Global Geospatial Information Management (UN-GGIM), 151 (*see also* Working Group on National Institutional Arrangements (WG-NIA))
    disaster loss reduction, 93
    educational help from, 104
    establishment of, 3, 13
    Experts, Committee of, 13, 141, 142, 143
    geospatial information management, 307–308
    harmonisation, 310
    IAEG-SDGs, work with, 20
    issues addressed by, 141–142
    Knowledge Base, 152
    mapping, interest in, 259 (*see also* mapping)
    operating principles, 244
    regions, 151
    spatial enablement of SDGs, identification of, 71
    strategic framework, 244
United Nations Resolution on Housing, Building and Planning, 202
United Nations Statistics Division, 23
United Nations World Geospatial Information Congress (UNWGIC), 4
Unmanned aerial vehicles (UAVs), 100
Unmanned aircraft systems (UAS), 109
    3D modeling, 115–116
    aerial mapping with, 112, 116
    applications, 110
    digital elevation models, use of, 112
    evolving, 122–124
    fixed-wing, 110–111
    flight design, 113–116
    geopositional waypoints, 113
    ground control points (GCPs), 117
    Ground Control Station (GCS), 111–112
    ground sample distance (GSD), 114–115
    Hurricane Harvey case example, 119–121
    image collection, 114, 116
    mission planning, 113
    overview of technology, 110–111
    real-time kinematic (RTK) GNSS, 118
    regulations regarding, 118–119
    RGB values, 117
    rotary craft, 110, 111
    sensors, navigating, 111
    Structure-from-Motion (SfM) photogrammetry (*see* Structure-from-Motion (SfM) photogrammetry)
    system components, 111–112
    visual line of sight (VLOS), 113
Urban Analytics Data Infrastructure (UADI), 10, 243, 246–247, 248–250, 250–251, 253–254
Urban Big Sata Centre (UBDC), 207
Urban Centre for Computation and Data (UrbanCCD), 208
Urban populations, 73
Urban regeneration, 293
Urbanization
    inclusiveness, 7
    inequality, 202
    SDGs, relationship between, 4
    slums, problem of, 202–203
    views on, 202

V

Vancouver Declaration on Human Settlements, 202
Victorian Department of Environment, Land, Water and Planning (DELWP), 329
Visible Infrared Imaging Radiometer Suite (VIIRS) Day/Night Band DNB (DNB), 267, 269–270, 272–273
Volunteered geographic information (VGI), 314

W

Wealth disparity, 137–138
Web searches, 230
WHO Interactive Air Pollution Maps, 319
Working Group on National Institutional Arrangements (WG-NIA), 142–143, 143–144, 147, 148, 150, 151–152, 153, 157 (*see also* geospatial information management)
World Bank, 11, 41, 45, 50, 52, 55, 71, 206, 283, 319
World Commission on Environment and Development, 17
World Council of City Data (WCCD), 244

Y

YouthMappers, 317

For Product Safety Concerns and Information please contact our EU representative GPSR@taylorandfrancis.com Taylor & Francis Verlag GmbH, Kaufingerstraße 24, 80331 München, Germany

Batch number: 10793307

Printed by Printforce, the Netherlands